The Social Construction of Technological Systems

The Social Construction of Technological Systems

New Directions in the Sociology and History of Technology

anniversary edition

edited by Wiebe E. Bijker, Thomas P. Hughes, and Trevor Pinch

foreword by Deborah G. Douglas

The MIT Press
Cambridge, Massachusetts
London, England

MIT Press books may be purchased at special quantity discounts for business or sales promotional use. For information, please email special_sales@mitpress.mit.edu.

This book was set in Stone Sans and Stone Serif by Toppan Best-set Premedia Limited. Printed and bound in the United States of America.

Library of Congress Cataloging-in-Publication Data

The social construction of technological systems : new directions in the sociology and history of technology / edited by Wiebe E. Bijker, Thomas P. Hughes, and Trevor Pinch; foreword by Deborah G. Douglas. — Anniversary ed.
    p. cm.
Includes bibliographical references and index.
ISBN 978-0-262-51760-7 (pbk. : alk. paper)
1. Technology—Sociological aspects—Congresses. I. Bijker, Wiebe E. II. Hughes, Thomas Parke. III. Pinch, T. J. (Trevor J.)
T14.5.S6375 2012
303.48′3—dc23

2011042098

10 9 8 7 6 5 4 3 2

Contents

Foreword

Deborah G. Douglas

The ideas of this book are everywhere. Collectively, the authors of these essays have captured the most ancient and most modern notions of history and the telling of the story of technology. It is, quite simply, a treasure. However, every author, every publisher when contemplating a reprint has a fundamental concern: is this book now a relic or does it still have relevance and the capacity to shape a discipline? By analogy, I found myself thinking about a quarter-century-old Boston road map, which I recently discovered in my car's glove compartment. It was fascinating as a historical document about the city I have called home off-and-on since 1973, but it is no longer a good guide to the current terrain. For sentimental reasons, I was thrilled that this book would remain in print and accessible, but I wondered, how would new readers consider this volume?

This book was astonishing when it was first published. "Combustible," "constructive," "catalytic," and "creative" were the alliterative quartet of adjectives that I wrote down in my seminar notes in the fall of 1987. The book was hot off the press when Professor Arnold Thackray assigned it to all incoming graduate students in the introductory seminar of the History and Sociology of Science department at the University of Pennsylvania. Some of our older classmates had read drafts of one or two of these essays or had heard Professor Tom Hughes describe the 1984 workshop at the University of Twente, but our class was the first to perceive the sociology of technology not as a provisional idea but rather as received knowledge. Within weeks we cracked the binding of our respective copies as we considered—"discussed" and "argued" are more accurate ways of putting

Deborah G. Douglas is the curator of science and technology at the MIT Museum. A specialist in aerospace history, Douglas is the author of *American Women and Flight since 1940* and has curated more than two dozen exhibitions on a wide variety of science and technology topics.

it—the ideas herein. The phrase "shaped and shaped by" quickly became part of our vernacular and identity as scholars.

Not everyone has been quite as enamored by the social construction of technology as my Penn cohorts. It is, after all, a bold thing to lay claim to a new discipline, so it is important to acknowledge that this book proved as irritating to some as it was exhilarating to others. As a young and impressionable scholar, I read the book's introduction as a descriptive (and exciting!) account of scholarly practice. For me, the thirteen essays were not, as some have since implied, dogmatic treatises but rather were like accounts of the various ascents up a mountain. Each offered a new insight, a pathway that enabled one to get to a place with a new and meaningful view of an important historical topic.

As a scholar at the midpoint of my academic career, I now have a much deeper appreciation of the fact that those who commit themselves professionally to the study of technology revel in the field's interdisciplinary qualities and limited licensure for practice. There is great freedom in heterogeneity. So looking at the book today, it is more audacious than I had originally thought. It is one thing to announce that there is such a field of inquiry as the "sociology of technology" and quite another to assert a formal method of inquiry. Since this book's original publication, some of its contributors have become forceful and articulate advocates for the latter, and I see now that the intellectual agenda of what is now called the SCOT Program is unambiguous.

The source of all epistemic power is based in constraint. All models, all methods of analysis, and all belief systems, no matter the duration of the scholar's embrace, demand the acceptance of a particular structural framework to the exclusion of all other alternatives. For there to be *technology*, one has to be willing to offer at least a provisional definition; for there to be a field of scholarly endeavor, there has to be a community willing to identify with a common set of ideas. Whether or not one agrees with the ideas of social construction, the certainty and power with which anyone today can claim to being a scholar of technology owes much to this book.

It is possible to read this new edition merely as a kind of historiographical homage, but I believe it remains a work of opportunity. As contemporary science and technology practitioners embrace interdisciplinarity as a fundamental attribute, the essays in this book have gained a new vitality and reward close study. Personally, the ideas in these essays have increased in relevance and inform every aspect of my current job as curator of science and technology for the MIT Museum from what to collect to how to interpret those artifacts. Coincidentally, at this particular moment, this book has become an artifact, embedded in a larger display about the MIT Press and its most influential titles that is part of our exhibition celebrating MIT's 150th anniversary.

That exhibition is atypical for a science and technology museum because its focus is not on explaining how things work but rather on *why they matter*. Teaching others to frame the study of technology in this manner—why things matter—is both the grand ambition and great achievement of this book. To that end, it is worth observing that this book has shaped a generation of scholars. Now with this reprint, we have the opportunity to reflect on how their scholarship has shaped the ideas of the social construction of large technological systems and technology more broadly. It is still a combustible, constructive, catalytic, and creative collection of ideas. Like the editors and contributors, I look forward with enthusiasm to a renewed and invigorating discourse.

Preface to the Anniversary Edition

Twenty-five years ago a small international workshop on new developments in the Social and Historical Studies of Technology was held at the University of Twente in the Netherlands. That workshop led to the first edition of this book. We think enough time has passed to produce a new edition. In this new edition we have left all the chapters and the original introduction unchanged—all we have added is this new 2011 preface and the preceding foreword by Deborah G. Douglas of MIT.

In this preface we reflect on what happened during the subsequent twenty-five years. Some recollections of the beginnings will help to position the volume in the landscape of the 1980s' history and sociology of technology. We then trace some of the developments that first appeared in this book and helped shape the emerging field of social studies of technology.

Our original 1987 introduction told the story of how this volume came about in some detail, ranging from pink champagne in the Austrian mountains to rental bikes in the Dutch polders. With the benefit of hindsight, it is worth highlighting some additional aspects. One key step was to broaden the initial gathering of sociologists to include historians of technology, too. This suggestion came from Donald MacKenzie and led to Wiebe Bijker and Trevor Pinch contacting Tom Hughes, who eventually agreed to join us as an editor.

We should explain why this preface is being written by only two of the three original editors. Sadly, Tom Hughes is too ill to be able to write with us today; while he is physically healthy, his memory has faded and so he is not able to contribute to these reflections. Tom had agreed with much enthusiasm to take part in this new edition when the idea was first suggested in 2007, but to our great misfortune we waited too long to actually sit down together and start writing. As part of our aim was to recapitulate where the field had gone and what new directions were looming we are acutely aware that this preface can only do justice to the views of two of us. Beneath the veneer of the Southern gentleman, Tom is such a strong character that we dare not presume to speak for him. We have tried below to summarize as best as we can some of the developments in the systems approach he is best known for, but readers should be aware that Tom is a man full of surprises. Indeed one of the biggest surprises of our careers was his agreeing to edit the original book with us.

We both recall that it was with much trepidation that we approached him back in 1985. Tom seemed the best person to collaborate with: he was enthusiastic about sociology, had an encyclopedic knowledge of history of technology—at the drop of a hat he could always come up with the perfect example to illustrate some point—but he was Tom Hughes. To us he was almost a god—an established world leader of the field and a professor at a major American university. And we were virtual unknowns who did not have regular university appointments. Much to our surprise Tom said "yes," and even said "yes" with such enthusiasm that we soon found ourselves being summoned by Tom to join him at various venues around the world to make sure the introduction got written. Some seasoned American academics later commented that they had been surprised that such a prominent scholar as Tom Hughes had been willing to add his name to such a wild project and to collaborate with two young and unknown Europeans. We can only think that it testifies to Tom's broad vision and the true excitement generated during the workshop.

A key question was, of course, the volume's title. The title of the workshop was a bit bland: "New Developments in the Social Studies of Technology." It was Gerard de Vries who suggested the eventual title, "The Social Construction of Technological Systems"—a perfect Dutch compromise between two of three major approaches in the book, social construction of technology and systems. (We will use the acronym SCOTS in the remainder of this new preface to denote the volume and the work and authors related to it.)

When planning the workshop, we did not have publication in mind. Whether this was because of less pressure to publish back then, or because we were just happily enthusiastic, young, and not bothering about careers, we do not know. But toward the end of the workshop the genie was clearly out of the bottle, and the participants agreed that Wiebe and Trevor should investigate the possibilities of an edited volume. Having Tom on board no doubt helped to convince Larry Cohen, the MIT Press editor, of the feasibility and promise of the book proposal.

When asked about his recollections, Cohen responded: "I can't reconstruct my initial reaction. I'd like to think it was positive, but most likely it was 'Oy, another edited volume. What's the kindest way to make it go away?' But I do give myself credit for realizing fairly quickly that there was something special in it. And that it was something that could help me create the sort of STS publishing program I had been struggling to realize."[1] Cohen had been asked by the Press's director, Frank Urbanowski, to sort through projects that fell outside recently chosen core areas and decide about continuing or canceling. One easy cancellation would be, the director suggested, to cut all STS-related projects. But Cohen disagreed, mainly because he "felt that a well-structured STS program could help define the special nature of publishing at a technical institute." These core areas— economics, computer science/AI, cognitive science, and architecture—all had a component that was STS-related in a broad sense; what the MIT Press needed, Cohen decided, was a small program that would bring this work together so that it became something coherent. "What I knew of the sociology and philosophy of science and technology at the time did not excite me," reminisces Cohen. "What we needed, I thought, was a new sort of rubric for publishing about science and engineering. What I saw in SCOTS was the map to that new program."

Larry in his quiet but assertive way got behind this volume and insisted that all the papers be individually refereed, as well as the volume as a whole, and further that we add short introductions to the sections to ensure coherence. The book thus has what we called "intermezzos" (Wiebe's term), and these have proven indispensible when the book is used as a teaching text. After the book got published, Larry made his next move. He worked with us and Tom's former graduate student, Bernie Carlson, to launch the MIT Press series Inside Technology—a series that is still going strong with more than fifty titles published as of 2011, demonstrating that when sociologists and historians collaborate, something truly exciting can happen. Why produce a new edition of this old chestnut now and leave the content virtually unchanged? The major reason is that the book itself has achieved a somewhat iconic status. It is regularly cited and used in research and teaching; it is one of the best-selling edited collections ever published by the MIT Press; it was included in the list of thirty most influential titles ever published by the Press, which are on display at the MIT Museum as part of the MIT's sesquicentennial anniversary celebration. But more to the point, the book is seen as having launched the new field of social studies of technology, a key part of the emerging wider discipline of Science and Technology Studies. The paperback book, now widely used, even has a nickname among graduate students in the United States—it is referred to as the "school bus book" in recognition of its distinctive yellow and black cover. The first editions of the book in hardback were published in an even brighter yellow and black dust cover, so the first students using the book in the United States referred to it as the "New York taxi book"! That a book with a fake (i.e., rigged) Victorian bicycle picture on its front cover should become known in terms of two other modes of transport is itself a sign perhaps that this book was to have an unusual history. Whether the field itself has changed from the exciting but risky adventure of a ride in a New York cab to the safer trip home on the school bus we leave for readers to judge.

It is always difficult to imagine what things were like before something obvious from today's perspective happens. But back in the early 1980s the social studies of technology was not a legitimate field of inquiry. The major journal of the field of STS, *Social Studies of Science,* did not publish work on how technological artifacts were developed and changed. A rare exception was our own joint paper on the Social Construction of Technology published in the journal in 1984, but that paper was published as a "discussion paper" rather than a regular research paper. We recall David Edge as late at the mid-1980s agonizing over whether the journal should publish its first-ever paper concerned solely with the development of specific pieces of technology (MacKenzie and Spinardi 1988). Happily David agreed to do so and the rest, as they say, is history. We should also note here that we as scholars owe David Edge a great debt. It was David who came up with the acronym SCOT while editing our joint paper in his office in Edinburgh. He said to us, "You can't call it "Joint Programme in the Sociology of Science and Technology; why not call it Social Construction of Technology or SCOT for short?" Nice one, David. (Wiebe vividly remembers Trevor's warning to always address letters to David to "Edinburgh, *Scotland,*" rather than "England," as a Dutchman might naively do.)

Common Traits in SCOT, LTS, ANT

The three approaches that together were to constitute the "new sociology of technology"—social construction of technology (SCOT), systems (later known as Large-Scale Technological Systems [LTS]), and actor-network theory (ANT)—stood out clearly only when we started editing the volume: as distinct approaches they had not informed our shaping of the workshop. Indeed we only encountered one of these approaches for the first time at the workshop. Bruno Latour, Michel Callon, and John Law all presented

Figure 1
From left to right: Gerard de Vries, Michel Callon, Bruno Latour, Trevor Pinch, Wiebe
Bijker.
Photo courtesy of Ellen van Oost.

the first ideas in what was later to become known as ANT. Bruno presented
at the workshop but felt that his paper (which eventually became a chapter
of *Science in Action* [Latour 1987]) was not quite ready for publication. One
of the funniest incidents at the workshop is captured in the photograph
in figure 1. Trevor is in the chair and Michel Callon is presenting his paper
(on electric cars—third chapter) in French with Bruno (hastily summoned
and not even with a seat at the table) translating into English. As soon
became apparent to the two Dutch onlookers, Wiebe and Gerard de Vries
(who being Dutch, understood both French and English), Bruno was not
only translating Michel's paper but adding his own gloss on the issues!
Perhaps this was the first taste of what "translation" would come to mean
in actor-network theory.

These three approaches did not completely represent the state of "new"
sociology of technology in the mid-1980s. Feminist studies of technology
had taken off as marked by the edited volume by Joan Rothschild (1983),
and were to become one of the mainstays of the new technology studies.

And even the chapters contained in this volume did not all fit snugly into the three-approach format. Characterizing this volume only in terms of SCOT, LTS, and ANT does not do full justice to the importance of all contributions. Ruth Cowan's chapter, for example, in which she introduces the "consumption junction" can now be read as having been ahead of its time in indicating a way to engage feminist scholarship with consumerism and market issues. For this new preface to the anniversary edition, however, the set of three approaches seems to provide an adequate, if not wholly satisfying, framework.

It is important to understand that in the late 1980s the field had not splintered into the sometimes sparring camps we find today. Sociologists such as Harry Collins and historians such as Edward Constant all comingled at the conference, sensing the excitement in the air of something new emerging. Despite some of the vigorous intellectual debates in later years, such as the infamous "chicken debate" about the distinction between human and nonhuman actors (Callon and Latour 1992; Collins and Yearley 1992), the common traits among the three approaches were then more important than the differences. Certainly the actor-network approach with its suggestion of symmetry between humans and nonhumans raised difficult and perplexing issues for some scholars schooled in the more familiar Edinburgh Strong Program and Bath School approaches. We can recall asking Bruno, Michel, and John whether they really believed the slide projector in the conference room should be treated as an actor equivalent to the excited sociologists and historians gathered around it. John Law's paper on the Portuguese expansion, where he treated trade winds and ships as unproblematic natural entities, ran into the early critical attention of Harry Collins when he famously asked John about whether the presence or lack of the Bermuda Triangle made any difference to the analysis. Collins was pointing out that, from his perspective, epistemological issues still had priority.

Obviously the most important of the common traits in the three approaches was to put *technology* on the agenda of social studies in the first place. As we describe in the introduction to the original volume, inspiration from the history of technology was crucial here. In hindsight it is, as Paul Forman has noted (Forman 2007), remarkable that so many scholars turned to technology at the same time. The wider political context of Britain, which had been a cauldron for the sociology of scientific knowledge (SSK), may have been an influence here. Neither Pinch nor Bijker had permanent jobs, and one reason Pinch had worked on technology at all was because University of Twente had offered him a six-month fellowship—a break from unemployment in the United Kingdom. Times in the United Kingdom in the 1980s were much harder with the newly elected Margaret Thatcher, who was pushing those whom she considered to be ivory tower academics to "apply" their knowledge. The "turn to technology" (Woolgar 1991) thus may have indirectly stemmed from the harsher political realities of the 1980s.

The second common trait to be found is the shared appreciation for Tom's "seamless web" metaphor. Tom loves examples, and one of his favorites was Edison moving between his Menlo Park laboratory and Wall Street to invent technology and raise capital in a seamless way. Whether questioning the fixed boundary between humans and nonhumans, or paying attention to the nontechnical aspects when studying the history of technical systems—all three approaches embraced the methodological principle of paying attention to how the borders between the social and the technical were drawn by actors, rather than assuming that these borders are pre-given and static. This also brings out the common element of a *constructivist* perspective. Rather than taking an essentialist view of technologies and their contexts, we all agreed that describing the activities of actors—whether in the form of relevant social groups (SCOT), systems builders (LTS), or actants (ANT)—was more interesting than a promethean history of technology that emphasized how heroic inventors and engineers stole great ideas about technology from the gods and gave them to mere mortals.

The final common trait, and one that was important for Larry Cohen when he decided on publication, was the integration of empirics and theory—the authors in the volume, if they did nothing else, all had compelling stories to tell. As Cohen relates:

What appealed to me in SCOT was that it provided a set of tools for structuring the telling of complex stories. But the stories were what ultimately mattered, and that is why I always tried to put pressure on our authors to let their theory emerge from the story rather than having the story appear as a pendant to the theory.[2]

We recall that during the workshop John Law's concept of "heterogeneous engineering" appealed to everyone as capturing aspects of the seamless web and constructivism. The empirical research into the social processes whereby actors engaged with technologies was to be done by what we called "thick description," implicitly citing Clifford Geertz (1973). This method underlined the clearly *antideterministic* goal of the volume. One other way in which we often described the common program in the volume was by using the metaphor of "opening the black box": and looking into the black boxes of indeed both technology and society was part of the agenda. Although the term *interpretative flexibility* is part of the SCOT vocabulary (drawing on the empirical program of relativism, EPOR), we would argue that it does characterize the general raison d'être of all three approaches: if technical artifacts would have no interpretative flexibility, their one and only meaning could probably be best read by engineers, so that there would be no room for social analysis at all.

These common traits, we feel, still make sense after twenty-five years. For some audiences the differences between the three approaches just are much less relevant than what can be learned when using a general "constructivist" approach to looking at technology. Bijker, Latour, and Steve Woolgar (who was not at the original workshop but contributed to the volume), for example, collaborated in a European project on innovation in freight container technologies (Hommels, Peters, and Bijker 2007). When explaining our work to the "harbor barons" of Rotterdam, we presented our joint approach as being constructivist and opening the black box, rather than highlighting the differences between the three of us about, for example, actor networks and relevant social groups. (This went down quite well, probably because, in addition to having an engineering background, these top managers had much experience with the key role of nontechnical issues in innovation.) Similarly, when Tom Hughes gathered several social constructivists and students of technical systems together for a series of "large technological systems" conferences in the late 1980s, the broadly shared seamless web approach proved more important than possible differences between an actor-oriented social constructivist perspective and a more structural systems approach (Mayntz and Hughes 1988; Summerton 1994). In that sense, this volume's title, *The Social Construction of Technological Systems,* still seems to make much sense, even though it does not contain an explicit reference to actor networks.

Stressing these commonalities does not imply that we do not recognize important differences. Indeed, we frequently warn over-enthusiastic students who often want to combine elements from the different approaches into one common theoretical framework to be careful. Merging, for example, actor networks with relevant social groups and technological frames while forgetting about the distinctly different backgrounds in semiotics (ANT) and sociology (SCOT) will quite likely produce either an inconsistent or a trivial and vacuous set of concepts. Each approach brings to light different aspects of how people shape and use technology; while SCOT highlights group dynamics, ANT requires that we take into account how nonhuman actants (both machines and natural forces) can shape the trajectory of technological systems.

Criticism of SCOTS

Having pointed out the themes common to all three approaches, it is important to point out that they have also run into some common criticisms. We will outline some of the most prominent ones here without delving into criticisms specific to each approach.

Often from the standpoint of conventional sociology, all these approaches are deemed to be guilty of missing some key element of the sociological enterprise. Because such approaches start with artifacts and their relationships with people whether conceived of in terms of social groups, actor networks, or systems, such approaches are hard to map onto conventional sociological ways of cutting up the cake. For instance, sociologists conventionally want to talk about structure and agency, conceiving of human actors as being located within wider social structures. They often conceive of society in terms of structuring factors such as class, race, and gender. They also want to use methodological techniques such as social network approaches and have rather conventional representations of macro structures somehow shaping micro domains.

Structure and agency become much more complicated when dealing with technologies. For instance, a car or a bicycle will offer humans far more agency in the sense that their capacities to act and where and when they act are changed. More fundamentally, where human agency ends and some sort of nonhuman agency begins is not always clear. For instance, when a complicated sociotechnical system fails, such as when an airplane crashes, accident inquiries often debate whether it is the pilot who caused the crash by failing to manipulate some control correctly (i.e., human agency) or whether the cause of the crash was some improperly designed control (part of the nonhuman technology). Boundaries between humans and nonhumans are often attributed or performed in the course of social life. Similarly, wider social structures cannot be conceived of as merely the social container that somehow constrains social action—social structure is itself material and technological and this again complexifies traditional analyses of social structure and power. Gender, class, and race are similarly not independent of materiality and technology (including space and time);

rather for SCOTS researchers it is more usual to talk about gender, race, and class as outcomes or as being materially performed.

Just as the traditional concepts of sociology are muddied, so are its methods. Social interaction must be studied by including interactions with nonhumans such as machines and pieces of infrastructure. A network analysis that restricts itself to merely social relationships will fail to pick up how technology and materiality mediate the character of such relationships. Similarly a systems approach that defines systems to be merely "social systems" will have no need of the rich analytical vocabulary developed by Tom Hughes to deal with the entanglement of humans and materiality in systems whereby say, the material properties of an element of a light bulb—just one element in a Hughesian system—can become what Hughes calls a "reverse salient" threatening the system as a whole.

The SCOTS volume took a critical stance vis-à-vis conventional sociology's failure to deal with technology and materiality in general. Even if sociologists have given up on materiality and technology, technology and materiality are too important to give up on sociology. Thus when one examines any in-depth sociology one finds that materiality and technology although little noticed are always there. Thus one research program stemming from the SCOTS volume would be to reexamine and re-embed the hidden materiality and technologies present in more conventional sociological accounts. As we show below, the various approaches in the SCOTS volume, however, tend to offer different reactions to conventional sociology. Actor-network theory has the most radical response in offering a sociology of associations that rejects much conventional sociological work. The fundamental distinction between human actors and the rest of the world is transcended with the concept of actant, by which both human and nonhuman entities can have agency. The central question then becomes how network associations between different actants can describe and explain the development of socio-technology. In contrast, SCOT is better able to accommodate to more conventional sociological ideas. Although one can see overlaps and mappings at many points: for example actor (or actant) networks can be mapped onto more conventional social networks, relevant social groups can include groups associated with class, gender, and race; but to be sure, conventional sociology and SCOTS do not always make the best of bedfellows.

Other challenges to SCOTS have come from philosophy of technology and ethics. This critique is best exemplified by Langdon Winner who bemoans the social studies of technology for having opened up the black box of technology only to have left it "empty" (Winner 1991, 1993).

Winner critiques the lack of a general ethical stance on technology within the new social studies of technology and compares it with the highly critical stances taken by social commentators of technology such as Martin Heidegger, Jacques Ellul, Lewis Mumford, and Winner himself. To us, it seems that Winner wants SCOTS to condemn whole classes of technology such as nuclear power on a priori grounds (Pinch 1996). It is not that SCOTS does not deal with politics and ethics—indeed it does, and we'll return to this below—but there is no easy way to produce universalistic claims of either approbation or blame which can be attributed to technologies. The sensitivity to the variety of actors involved (human and nonhuman) and the actual use of technologies tends to prevent blanket commendations and condemnations. The absurdity of such *tout court* proclamations becomes clear if one takes a technology like food. Although one might be opposed to say the wanton use of sulfur dioxide to preserve certain foods such as peas, it would be odd to proclaim being anti-food or anti-peas! SCOTS by its focus on what social groups and actor networks actually say and do with technology make it unlikely that an analysis would start from such general proclamations.

Related to the criticism of Winner is the charge leveled by feminists and others that somehow SCOTS does not provide any form of political action in regard to technology. It merely describes the world but does not attempt to change it. This again is something we will deal with below in more detail. This is an issue that goes back to the strong program in the sociology of scientific knowledge from which SCOTS developed. Because of the principle of symmetry, whereby analysts refuse to impute asymmetrical explanations, it is deemed that somehow this fails to give SCOTS any firm place from which to provide the necessary grounds for political criticism. But this criticism assumes that only one form of criticism is available—by taking sides in a sociotechnical controversy. SCOTS can still be political in all sorts of ways by pointing say to the entanglement of humans and nonhumans, surely itself a principle that could lead to political action.[3]

Changes in Approaches since 1987

Especially when using this volume for teaching, it is worth warning students that SCOT, LTS, and ANT have continued to evolve since this volume first appeared in 1987. In this section we will briefly review some of these changes. Inevitably the emphasis in our discussion will be on changes in SCOT, but we will try to indicate some more general trends. For more authoritative accounts of changes in LTS and ANT we refer to authors who more extensively work with those approaches (Hughes 1998, 2004; Hughes, Hughes, Allen, and Hecht 2001; Latour 2005; Law and Hassard 1999).

One important trend, already hinted at above, is the broadening of the constructivist perspective. Much of recent SCOTS work addresses a variety of construction processes, including the construction of social institutions or the co-construction of technology and society. For instance, Bernie Carlson argued that the rise of General Electric was largely the story of simultaneously creating both new electric power systems as well as new business practices (Carlson 1991). The title of the edited volume from the second Twente workshop, *Shaping Technology/Building Society,* was also a reflection of this expansion of including both institutions and artifacts (Bijker and Law 1992). A broadening of the unit of analysis from artifact to sociotechnical ensemble completed this change (Bijker 1995). This can be interpreted as resulting from a further cross-fertilization between SCOT, LTS, and ANT, like the recent work on co-production in science studies (Jasanoff 2004), though the typical ANT step of making no ontological distinction between human and nonhuman actants is not made in SCOT (Bijker 2010).

Another trend in SCOT is the broadening of the intellectual agenda to more explicitly include normative questions and issues of politics. To be sure, these political issues were not far away in this volume either— engineers doing sociology, the politics of developing a missile guidance system, and regulatory science—but they were made much more explicit in later work. The second Twente workshop took the first step. Then SCOT, LTS, and ANT all developed their own version of this broadened agenda. In the SCOT and closely allied "social shaping of technology" tradition (MacKenzie and Wajcman 1985), Donald MacKenzie and Graham Spinardi (MacKenzie and Spinardi 1995) argued for the possibility to uninvent the atomic bomb; Bijker (1995) developed a concept of power that drew on a social constructivist view of technology; and Anique Hommels (2005) described how within a constructivist view also the hardness and obduracy of technology can be accounted for and thus helped to bring back the societal impact of technology into the research agenda that had been temporarily pushed out by criticizing technological determinism (see also Smith and Marx 1995 and Wyatt 2008). The LTS conferences also drew political scientists and general sociologists like Renate Mayntz, Charles Perrow, and Todd LaPorte into the discussion, although the thrust of the resulting volumes was still primarily historical (Mayntz and Hughes 1988; Summerton 1994); Gabrielle Hecht's (1998) history of French nuclear power was clearly political, and Hughes's *Rescuing Prometheus* (1998) addresses political issues more explicitly than ever.

Current Issues

New Domains

This volume tells stories about such varied technologies as thirteenth-century galleys, eighteenth-century cooking ranges, nineteenth-century bicycles, electricity systems, dyes, and plastics, and twentieth-century electric cars, missiles, turbojets, ultrasound, medical drugs, and expert systems. No wonder that one of the legacies of this volume has been an unrestrained exploration of ever new empirical domains. Without citing studies in detail, the new stories range from nanotechnologies to handloom weaving, from high-tech medical scanners to music synthesizers, from Indian non-chemical pesticides to American farming tractors, from Portuguese dams to Indian water resource management. It is difficult to think of a technology that has not already been studied from a SCOTS perspective in the past twenty-five years; it is impossible, we maintain, to think of a technology that cannot be studied in that way.

Although the original version of this book came out nearly a quarter of a century ago, its message is only just starting to reach some fields and is still resisted in others. The fundamental message—that technology and society are entangled together—continues to be difficult to absorb in those parts of the academy that have a vested interest in studying either one side of the equation of technology and society or the other. Thus in engineering schools, while there are growing numbers of STS and ethics courses, students still tend to split the world up into separate technical and social systems rather than appreciate how engineers create and run sociotechnical systems. This is a pity, as a richer understanding of how engineers shape both the social and technical dimensions would do much to enhance the social and professional standing of engineering. There are, however, an increasing number of important exceptions. Fields such as Information Science and Human Computer Interaction (HCI), which routinely deal with the interfaces between humans and machines, are much more receptive, and these days they do indeed contribute toward social studies of technology (Suchman 1987; Coopmans 2011). In nanotechnologies there is a broad recognition among scientists, engineers, and policymakers of the need to look beyond the laboratory (Bijker, Bal, and Hendriks 2009; McCarthy and Kelty 2010). Leonardi (2010) traces how car engineering, industrial organization, and regulation in automotive safety together co-evolve with practices and theories of crash testing. Studies of robotics not only bridge the gap between the technical and the social but also venture into fundamental questions of action, knowledge, body, and mind (Alač 2009; Collins 2010).

It is interesting that people who act as intermediaries between engineers and the wider social, economic, and legal institutions are the ones who most often understand the value of taking a SCOTS approach. Thus one of our students who left graduate school to work for a Swiss venture capitalist reported that, to his great surprise, the "school bus book" was prominently displayed on the shelf in the venture capitalist's office where he worked.

The empirical zeal to study technology which one finds within STS departments and programs contrasts with the relative paucity of studies of technology to be found within the traditional social sciences. Sociology and political science have yet to take materiality and technology seriously. There are of course exceptions, notably in the fields of economic sociology, and organizational sociology as practiced in business schools. The work of scholars such as Karin Knorr Cetina and Urs Bruegger (2000), Donald MacKenzie (2006), Daniel Beunza and David Stark (2004), and Michel Callon (Muniesa, Millo, and Callon 2007), who have turned to the study of financial markets, makes a strong case for bringing together economic sociology with STS (see also Pinch and Svedberg 2008). Within organizational sociology there has been a recent interest in STS through scholars such as Raghu Garud and Peter Karnoe (Garud and Ahlstrom 1997; Garud and Karnoe 2001), and Wanda Orlikowski and Joanne Yates (Orlikowski 1992; Yates and Orlikowski 1992). Indeed the turn to technology in these fields can be seen as part of a broader turn to take "materiality" more seriously.

Last, it is worth mentioning that new interdisciplinary ventures provide another sort of forum to bring out the issues discussed in this volume. For instance, the newly emerging field of sound studies is heavily indebted to SCOTS (Bijsterveld 2008; Pinch and Bijsterveld 2011; Pinch and Trocco 2002; Thompson 2004). By treating sound as materially and technologically produced, stored, transmitted, and received, many of the issues underlying how technology and society get entangled together reappear in this new domain. Similarly fields such as architecture and archaeology—fields that are dense with material artifacts—at some point have to deal with the inevitable linkages between these artifacts and social formations (see Latour and Yaneva 2008 for an ANT approach toward architecture). Everywhere where we encounter and listen to humans and their machines we need a way to think and talk about both—the tangled relationships are never obvious and never easy to tease out, but the now old and familiar tools offered here still have relevance, especially in an era where the fixation on the technologically new seems to provide a brighter allure than ever.

The Turn to Ontology

There is in current science studies a revival of the notion of ontology. Stemming from so-called posthumanist scholars such as Bruno Latour (1999), Karen Barad (2003), Andy Pickering (1995), John Law (2002), and Annemarie Mol (2002), the suggestion is that approaches that deal with representation—such as social constructivism—are in some ways inadequate or limited because they do not deal with the material stuff of the world in its own right. According to this view, looking at the meanings given to technological artifacts is to focus too much on humans and how humans conceive of or interpret technology. Drawing strongly on ANT, the posthumanists argue that we are failing to grant technology its own independent powers or ability to "act" or "resist" independent of human representations of those powers. Another way to talk about this issue is to focus on "becoming" and to criticize approaches such as SCOT for being human-centered and paying insufficient attention to how new entities emerge in the world or neglecting the "dance of agency" between humans and nonhumans. For scholars such as Barad and Pickering, the emergence of new entities and objects is a continuous process and this again points to a limitation of conventional "humanistic" approaches.

The problem for all such posthuman approaches is of course that there is no ontology without epistemology. As soon as one talks about nonhuman entities one is representing nonhuman entities. Furthermore, when it appears to be otherwise (i.e., that nonhuman actants have agency), it is humans who have achieved this distinction. In view of this it seems unclear why SCOTS needs to delve into ontology at all. Since the contestation of entities in the wider field of STS has included some of the most fundamental entities of matter—e.g., quarks, neutrinos, DNA, cold fusion, parapsychology—it would seem odd for SCOT scholars to start their analysis by building a whole edifice of the stuff of the world. Furthermore, there is no need to do so. The principle we (now speaking as SCOT researchers) need in order to deal with ontology is similar to the principle of "methodological relativism." This stance in relation to epistemological relativism advocates that for the purpose of our studies we bracket epistemological claims—in other words, we do not assume that there is some sort of epistemological criterion that will provide the royal road to scientific truth during a scientific controversy (Collins 1983; Pinch and Svedberg 2008). In our investigations we act as if both sides of the controversy had truth on their side. Similarly, when our respondents make ontological claims and these claims are contested we do not have to endorse one particular set over another. In other words, SCOT recommends ontological agnosticism. This does not mean that SCOT denies there is stuff in the world any more than methodological relativism denies that scientists can reach the "truth" over their scientific claims (Bijker 2010; Pinch 2010).

The Turn to Politics and the Normative

One of us has argued that SCOTS-related research began as an academic detour after the somewhat frustrating experiences in the activist "science and society" movement in northern Europe in the 1970s, and that by the mid-1990s it was time to revisit that original political agenda (Bijker 1995). For some STS researchers that political agenda had never been far off, as the comprehensive review by Woodhouse, Hess, Breyman, and Martin shows (2002). However, browsing the issues of *Social Studies of Science* and *Science, Technology & Human Values* since the 1990s clearly displays an increasing attention to questions of politics and normativity (see also Bijker 2003 and the debate under the label of "third wave": Collins and Evans 2002, 2003; Jasanoff 2003; Rip 2003; Wynne 2003). This even brings a new convergence between the empirical SCOTS studies of technology and the philosophical studies of technology, as some of the recent work on emerging technologies shows (Swierstra, Est, and Boenink 2009). In addition to SCOTS, another important stepping stone for this work is the rediscovery of the earlier work of Don Ihde (1979) and post-phenomenology approaches in the philosophy of technology (Verbeek 2005).

The recent turn to politics and normative questions does not only include academic studies of the relationship between technology and politics but also involves STS researchers as advisers or consultants. Thus Tom Hughes and Wiebe Bijker found each other, admittedly somewhat to their surprise, together in an advisory panel in Japan. They advised the Japanese Central Research Institute of Electric Power Industry (CRIEPI) on "Social Decision Making Process for Energy Technology Introduction." Also the European Commission and various national governments have sought advice from STS researchers (Harremoës et al. 2002; Wynne et al. 2007). And STS researchers combine their research with active intervention and engagement (Zuiderent-Jerak 2010; Zuiderent-Jerak, Strating, Nieboer, and Bal 2009).

The Further Globalization of SCOTS

Science and technology studies are increasingly turning more global, rather than staying limited to the northern and western hemispheres. Journals, book series, graduate programs, and regional societies: all these elements indicate the coming of age of STS in Asia and Latin America. The participation of scholars from Asia and Latin America in conferences of the Society for Social Studies of Science (4S) and the Society for the History of Technology (SHOT) has been increasing over the past two decades. This culminated in the first 4S meeting held outside Europe and the United States: in Tokyo in 2010, where there was much participation from Korea, China, Taiwan, and Japan in addition to the usual suspects from the north and west.

This globalization of STS leads, for example, to addressing long-standing questions in other domains freshly from the typical SCOTS perspective on the relationships between science, technology, and society. One example is cold war history (Hecht 2011), another the question of development aid in Africa (Rottenburg 2009). Possibly even more exciting than applying the received SCOTS perspectives to new geographies and questions is the promise of truly new insights by extending the research into these domains and questions and engaging in collaboration with scholars and practitioners in Asia, Africa, and Latin America. Such insights will, we expect, range from the empirical to the theoretical, and from the epistemological to the political (see, for example, the essay by Visvanathan [1998]).

What Does SCOTS Now Stand For?

When this volume was published twenty-five years ago, it was targeted at the empirical studies of the inside of the black box of technology and on theoretical explorations of that technology's relations to society. We have described how Larry Cohen then helped to position the new series Inside Technology at those crossroads between empirics and theory. One way of answering the question of how SCOTS has developed over the past twenty-five years is to trace the development of the series of which this volume formed a foundation and starting point.

The Inside Technology series has expanded its scope since conception, and these forms of broadening all reflect important developments in the field of SCOTS. The object matter of the books in the series is now broader than the technical artifacts and systems with which we began. Scientific research collaboration (Gorman 2010; Shrum, Genuth, and Chompalov 2007) and colonial Cold War history (Hecht 2011) are now found among the themes of study—and they do so as a natural extension of the SCOTS perspective that highlights the entangled character of technology, society, and science. The second broadening is in the books' style: where the series initially stressed detailed empirical work as a defining characteristic, it now includes more essayistic and philosophical approaches (Callon, Lascoumes, and Barthe 2009; Feenberg 2010; Nowotny 2008). Finally, the series increasingly publishes books that engage with political and normative issues of technology and society, such as the role of scientific advice in democracy (Bijker, Bal, and Hendriks 2009), and generally questions of ethics and technology (Johnson and Wetmore 2009). The Inside Technology demonstrates the growth and expansion of SCOTS at large.

Thus, here we find ourselves twenty-five years after the first publication of this volume. In 1987 the volume signaled an exciting new way of studying technology in society. In 2012 the second edition marks, we hope, an active engagement of SCOTS scholarship with making a better world.

Wiebe Bijker and Trevor Pinch
Maastricht, the Netherlands, and Ithaca, New York
September 2011

Notes

We thank Bernie Carlson and Larry Cohen for their close reading and helpful comments.

1. Email correspondence, 10 February 2011.

2. Ibid.

3. For an example of an exchange over this very point (applied to Collins and Pinch's analysis of the Challenger accident), see Kochan (2006), Feenberg (2006), and Collins and Pinch (2007).

References

Alač, M. 2009. "Moving android: On social robots and body-in-interaction." *Social Studies of Science* 39(4): 491–528.

Barad, K. 2003. "Posthumanist performativity: Toward an understanding of how matter comes to matter. *Signs* 28(3): 801–831.

Beunza, D., and Stark, D. 2004. "Tools of the trade: The socio-technology of arbitrage in a Wall Street trading room." *Industrial and Corporate Change* 13(2): 369–400.

Bijker, W. E. 1995. *Of Bicycles, Bakelites, and Bulbs: Toward a Theory of Sociotechnical Change*. Cambridge, Mass.: MIT Press.

Bijker, W. E. 2003. "The need for public intellectuals: A space for STS." *Science, Technology & Human Values* 28(4): 443–450.

Bijker, W. E. 2010. "How is technology made? That is the question!" *Cambridge Journal of Economics* 34(1): 63–76.

Bijker, W. E., Bal, R., and Hendriks, R. 2009. *The Paradox of Scientific Authority: The Role of Scientific Advice in Democracies*. Cambridge, Mass.: MIT Press.

Bijker, W. E., and Law, J., eds. 1992. *Shaping Technology / Building Society. Studies in Sociotechnical Change*. Cambridge, Mass.: MIT Press.

Bijsterveld, K. 2008. *Mechanical Sound: Technology, Culture, and Public Problems of Noise in the Twentieth Century*. Cambridge, Mass.: MIT Press.

Callon, M., Lascoumes, P., and Barthe, Y. 2009. *Acting in an Uncertain World: An Essay on Technical Democracy*. Cambridge, Mass.: MIT Press.

Callon, M., and Latour, B. 1992. "Don't throw the baby out with the bath school!" A reply to Collins and Yearley, in *Science as Practice and Culture*, A. Pickering, ed. Chicago: University of Chicago Press, 343–368.

Carlson, B. W. 1991. *Innovation as a Social Process: Elihu Thomson and the Rise of General Electric, 1870–1900*. New York: Cambridge University Press.

Collins, H. M. 1983. "An empirical relativist programme in the sociology of scientific knowledge," in *Science Observed*, K. D. Knorr Cetina and M. Mulkay, eds. Beverly Hills: Sage, 85–114.

Collins, H. M. 2010. *Tacit and Explicit Knowledge*. Chicago: University of Chicago Press.

Collins, H. M., and Evans, R. 2002. "The third wave of science studies: Studies of expertise and experience." *Social Studies of Science* 32(2): 235–296.

Collins, H. M., and Evans, R. 2003. "King Canute meets the Beach Boys: Responses to the third wave." *Social Studies of Science* 33(3): 435–452.

Collins, H. M., and Pinch, T. J. 2007. "Who is to blame for the *Challenger* explosion?" *Studies in the History and Philosophy of Science* 38: 254–255.

Collins, H. M., and Yearley, S. 1992. "Epistemological chicken," in *Science as Practice and Culture*, A. Pickering, ed. Chicago: University of Chicago Press, 301–326.

Coopmans, C. 2011. "Face value: New medical imaging software in commercial view." *Social Studies of Science* 41(2): 155–176.

Feenberg, A. 2006. "Symmetry, asymmetry, and the real possibility of radical change: Reply to Kochan." *Studies in History and Philosophy of Science* 37: 721–727.

Feenberg, A. 2010. *Between Reason and Experience: Essays in Technology and Modernity*. Cambridge, Mass.: MIT Press.

Forman, P. 2007. "The primacy of science in modernity, of technology in postmodernity, and of ideology in the history of technology." *History and Technology: An International Journal* 23(1): 1–152.

Garud, R., and Ahlstrom, D. 1997. "Technology assessment: a socio-cognitive perspective." *Journal of Engineering and Technology Management* 14: 25–48.

Garud, R., and Karnoe, P. 2001. *Path Dependence and Creation*. Mahwah, N.J.: Erlbaum.

Geertz, C. 1973. *The Interpretation of Cultures*. New York: Basic Books.

Gorman, M. E. 2010. *Trading Zones and Interactional Expertise: Creating New Kinds of Collaboration*. Cambridge, Mass.: MIT Press.

Harremoës, P., Gee, D., MacGarvin, M., Stirling, A., Wynne, B., and Vaz, S. G., eds. 2002. *The Precautionary Principle in the 20th Century: Late Lessons from Early Warnings*. London: Earthscan Publications and European Environment Agency.

Hecht, G. 1998. *The Radiance of France: Nuclear Power and National Identity after World War II*. Cambridge, Mass.: MIT Press.

Hecht, G. 2011. *Entangled Geographies: Empire and Technopolitics in the Global Cold War*. Cambridge, Mass.: MIT Press.

Hommels, A. 2005. *Unbuilding Cities: Obduracy in Urban Sociotechnical Change*. Cambridge, Mass.: MIT Press.

Hommels, A., Peters, P., and Bijker, W. E. 2007. "Techno therapy or nurtured niches? Technology studies and the evaluation of radical innovations." *Research Policy* 36(7): 1088–1099.

Hughes, T. P. 1998. *Rescuing Prometheus*. New York: Pantheon Books.

Hughes, T. P. 2004. *Human-Built World: How to Think about Technology and Culture*. Chicago: University of Chicago Press.

Hughes, T. P., Hughes, A. C., Allen, M. T., and Hecht, G. 2001. *Technologies of Power: Essays in Honor of Thomas Parke Hughes and Agatha Chipley Hughes*. Cambridge, Mass.: MIT Press.

Ihde, Don. 1979. *Technics and Praxis*. Dordrecht: Reidel.

Jasanoff, S. 2003. "Breaking the waves in science studies: Comment on H. M. Collins and Robert Evans, 'The Third Wave of Science Studies.'" *Social Studies of Science* 33(3): 389–400.

Jasanoff, S., ed. 2004. *States of Knowledge: The Co-production of Science and Social Order*. New York: Routledge.

Johnson, D. G., and Wetmore, J. M. 2009. *Technology and Society: Building Our Sociotechnical Future*. Cambridge, Mass.: MIT Press.

Knorr Cetina, K., and Bruegger, U. 2000. "The market as an object of attachment: Exploring postsocial relations in financial markets." *The Canadian Journal of Sociology* 25(2): 141–168.

Kochan, J. 2006. "Feenberg and STS: Counter-reflections on bridging the gap." *Studies in History and Philosophy of Science* 37: 702–720.

Latour, B. 1987. *Science in Action: How to Follow Scientists and Engineers through Society.* Cambridge, Mass.: Harvard University Press.

Latour, B. 1999. *Pandora's Hope. Essays on the Reality of Science Studies.* Cambridge, Mass.: Harvard University Press.

Latour, B. 2005. *Reassembling the Social: An Introduction to Actor-Network-Theory (ANT).* Oxford: Oxford University Press.

Latour, B., and Yaneva, A. 2008. "Give me a gun and I will make all buildings move: An ANT's view of architecture," in *Explorations in Architecture: Teaching, Design, Research*, R. Geiser, ed. Basel: Birkhäuser, 80–89.

Law, J. 2002. *Aircraft Stories: Decentering the Object in Technoscience.* Durham, N.C.: Duke University Press.

Law, J., and Hassard, J., eds. 1999. *Actor Network Theory and After.* Oxford: Blackwell.

Leonardi, P. M. 2010. "From road to lab to math: The co-evolution of technological, regulatory, and organizational innovations for automotive crash testing." *Social Studies of Science* 40(2): 243–274.

MacKenzie, D. 2006. *An Engine, Not a Camera: How Finance Models Shape Markets.* Cambridge, Mass.: MIT Press.

MacKenzie, D., and Spinardi, G. 1995. "Tacit knowledge and the uninvention of nuclear weapons," in *Knowing Machines: Essays on Technical Change*, D. MacKenzie, ed. Cambridge, Mass.: MIT Press, 215–260.

MacKenzie, D. S., and Spinardi, G. 1988. "The shaping of nuclear weapon system technology: US fleet ballistic missile guidance and navigation: 1: From Polaris to Poseidon." *Social Studies of Science* 18: 419–463.

MacKenzie, D., and Wajcman, J., eds. 1985. *The Social Shaping of Technology: How the Refrigerator Got Its Hum.* Milton Keynes: Open University Press.

Mayntz, R., and Hughes, T. P., eds. 1988. *The Development of Large Technical Systems.* Boulder, Colo.: Westview.

McCarthy, E., and Kelty, C. 2010. "Responsibility and nanotechnology." *Social Studies of Science* 40(3): 405–432.

Mol, A. 2002. *The Body Multiple: Ontology in Medical Practice*. Durham, N.C.: Duke University Press.

Muniesa, F., Millo, Y., and Callon, M. 2007. "An introduction to market devices." *The Sociological Review* 55: 1–12.

Nowotny, H. 2008. *Insatiable Curiosity: Innovation in a Fragile Future*. Cambridge, Mass.: MIT Press.

Orlikowski, W. 1992. "The duality of technology: Rethinking the concept of technology in organizations." *Organization Science* 3: 397–427.

Pickering, A. 1995. *The Mangle of Practice*. Chicago: University of Chicago Press.

Pinch, T. 1996. "Social construction of technology: A review," in *Technological Change: Methods and Themes in the History of Technology*, R. Fox, ed. Amsterdam: Harwood, 17–36.

Pinch, T. 2010. "On making infrastructure visible: Putting the non-humans to rights." *Cambridge Journal of Economics* 34(1): 77–89.

Pinch, T., and Bijsterveld, K., eds. 2011. *The Oxford Handbook of Sound Studies*. Oxford: Oxford University Press.

Pinch, T., and Svedberg, R., eds. 2008. *Living in a Material World: Economic Sociology Meets Science and Technology Studies*. Cambridge, Mass.: MIT Press.

Pinch, T., and Trocco, F. 2002. *Analog Days. The Invention and Impact of the Moog Synthesizer*. Cambridge. Mass.: Harvard University Press.

Rip, A. 2003. "Constructing expertise: In a third wave of science studies?" *Social Studies of Science* 33(3): 419–434.

Rothschild, J., ed. 1983. *Machina Ex Dea: Feminist Perspectives on Technology*. New York: Pergamon.

Rottenburg, R. 2009. *Far-Fetched Facts: A Parable of Development Aid*. Cambridge, Mass.: MIT Press.

Shrum, W., Genuth, J., and Chompalov, I. 2007. *Structures of Scientific Collaboration*. Cambridge, Mass.: MIT Press.

Smith, M. R., and Marx, L., eds. 1995. *Does Technology Drive History? The Dilemma of Technological Determinism*, second edition. Cambridge, Mass.: MIT Press.

Suchman, L. 1987. *Plans and Situated Actions: The Problem of Human Machine Communication*. Cambridge: Cambridge University Press.

Summerton, J., ed. 1994. *Changing Large Technical Systems*. Boulder, Colo.: Westview.

Swierstra, T., van Est, R., and Boenink, M. 2009. "Taking care of the symbolic order: How converging technologies challenge our concepts. *Nanoethics* 3: 269–280.

Thompson, E. 2004. *The Soundscape of Modernity. Architectural Acoustics and the Culture of Listening in America, 1900–1933.* Cambridge, Mass.: MIT Press.

Verbeek, P.-P. 2005. *What Things Do: Philosophical Reflections on Technology, Agency, and Design.* University Park: Pennsylvania State University Press.

Visvanathan, S. 1998. "A celebration of difference: Science and democracy in India." *Science* 280(5360): 42–43.

Winner, L. 1991. "Upon opening the black box and finding it empty: Social constructivism and the philosophy of technology," in *The Technology of Discovery and the Discovery of Technology: Proceedings of The 6th International Conference of The Society for Philosophy and Technology*, C. Pitt and E. Lugo, eds. Blacksburg, Va.: Society for Philosophy and Technology, 503–519.

Winner, L. 1993. "Upon opening the black box and finding it empty: Social constructivism and the philosophy of technology." *Science, Technology, & Human Values* 18(3): 362–378.

Woodhouse, E., Hess, D., Breyman, S., and Martin, B. 2002. "Science studies and activism: Possibilities and problems for reconstructivist agendas." *Social Studies of Science* 32(2): 297–319.

Woolgar, S. 1991. "The turn to technology in social studies of science." *Science, Technology, and Human Values* 16(1): 20–50.

Wyatt, S. 2008. "Technological determinism is dead; long live technological determinism," in *The Handbook of Science and Technology Studies*, third edition, E. J. Hacket, O. Amsterdamska, M. Lynch, and J. Wajcman, eds. Cambridge, Mass.: MIT Press, 165–180.

Wynne, B. 2003. "Seasick on the third wave? Subverting the hegemony of propositionalism: Response to Collins and Evans (2002)." *Social Studies of Science* 33(3): 401–417.

Wynne, B., Felt, U., Callon, M., Gonçalves, M. E., Jasanoff, S., Jepsen, M., et al. 2007. *Taking European Knowledge Society Seriously.* Brussels: Report of the Expert Group on Science and Governance to the Science, Economy and Society Directorate, Directorate-General for Research, European Commission.

Yates, J., and Orlikowski, W. J. 1992. "Genres of organizational communication: A structurational approach to studying communication and media." *The Academy of Management Review* 17(2): 299–326.

Zuiderent-Jerak, T. 2010. "Embodied interventions, interventions on bodies: Experiments in practices of science and technology studies and hemophilia care." *Science, Technology & Human Values* 35(5): 677–710.

Zuiderent-Jerak, T., Strating, M., Nieboer, A., and Bal, R. 2009. "Sociological refigurations of patient safety; ontologies of improvement and 'acting with' quality collaboratives in healthcare." *Social Science & Medicine* 69(12): 1713–1721.

Acknowledgments

Many collections of papers that stem from conferences do not see the light of day until long after everyone has forgotten what all the excitement was about. We have made every effort to prevent this from happening with the current volume. We have almost met our target of getting these papers into print within a period of two years after they were first presented at a workshop held at the University of Twente, the Netherlands, in July 1984. Furthermore, we have been fortunate in that during the intervening period intellectual excitement and interest in the new field of social studies of technology has grown apace. Therefore we hope this volume proves timely.

We are indebted to a number of people and organizations who have helped bring this project to fruition. First and most important, we would like to thank all the contributors for meeting our many unreasonable deadlines and calls for revisions. Our intention has always been to make this collection more than just another conference volume, and we are particularly grateful to all the contributors for their efforts to bring out the links between their papers and others in the collection. The contributors have made the editors' task both pleasurable and easy. We would also like to acknowledge the part played by a number of referees who put in a lot of work reading and commenting on the many papers. A list of all those who kindly acted as referees is given at the end of the preface. The volume as a whole has been read by W. Bernard Carlson. Naturally the responsibility for any mistakes or errors in this volume lies with us and the contributors and not with any of the referees. The workshop from which this volume stems was made a success by its participants; we have also appended a list of all those who attended. As part of our attempt to provide more coherence in this collection, we have compiled a common bibliography. We are especially grateful to Gerdien Linde for her help with this and with other secretarial and typing tasks. The staff at the MIT Press has done everything possible to help us bring out this volume quickly and to ensure that it is a quality production. Finally, we would like to acknowledge financial support from a variety of organizations—the Netherlands Organization for the Advancement of Pure Research (ZWO), the Twente University of Technology, the United Kingdom Economic and Social Research Council, and the British Council.

One month before the manuscript of this volume was delivered to the publisher, we were saddened by the sudden and untimely death of Peter Boskma. Peter Boskma was professor of philosophy of science and technology and head of the science and technology studies program of De Boerderij at the Twente University of Technology. A driving force behind the workshop on which this volume is based, Boskma played a crucial role in realizing the workshop, as well as in making it into a success by his stimulating contributions to the discussions. We dedicate this book to his memory.

Wiebe E. Bijker
Thomas P. Hughes
Trevor Pinch

Referees

James Beniger
Stuart Blume
Edward W. Constant II
David Edge
Ron Johnston
Bruno Latour
Rachel Laudan
John Law
Wesley Shrum
David Travis
Gerard H. de Vries
Brian Wynne
Steven Yearley

Workshop Participants

Henk van den Belt
Boel Berner
Wiebe E. Bijker
Henk J. H. W. Bodewitz

Peter Boskma
Michel Callon
H. Floris Cohen
Harry Collins
Edward W. Constant II
David Edge
Boelie Elzen
Christien M. Enzing
Thomas P. Hughes
Jaap Jelsma
Bruno Latour
John Law
Harry Lintsen
Donald MacKenzie
Trevor Pinch
Arie Rip
Ruth Schwartz Cowan
Wim A. Smit
Ruud Smits
Philip J. Vergragt
Gerard H. de Vries
Charles Weiner
Peter Weingart
Edward Yoxen
Sjerp Zeldenrust

General Introduction

The origin of this book can be traced back to the Burg Landsberg, an old castle crowning a steep hill in Deutschlandsberg, Austria. This is a remote place without worldly temptations, and the participants of the first meeting of the newly formed European Association for the Study of Science and Technology (EASST) could but stick together and work hard.[1] It was here, during a cocktail session in the early evening of September 25, 1982, that two of the three editors of this book met for the first time. Encouraged by their concurring interests in a constructivist approach to the study of technology and by drinking the famous regional pink champagne, Trevor Pinch, a sociologist of science, and Wiebe Bijker, a sociologist of technology, decided to start a joint project. The object would be to bring together Pinch's detailed studies of the development of science with Bijker's studies of technology.

Pinch obtained a scholarship and moved for half a year from the University of Bath to the Twente University of Technology. Eventually this research resulted in a number of papers in which Pinch and Bijker argued that an integrated approach to the social study of science and technology would be feasible and fruitful. Specifically, they proposed a social constructivist approach, thereby extending the relatively new but already well-established sociology of scientific knowledge into the realm of technology. In passing, they discarded most of the existing approaches to technology advocated by historians, philosophers, and economists. This general condemnation, although it had the benefit of defining their stance quite clearly, obviously lacked subtlety. Whatever its cause—youthful enthusiasm or the pink champagne—they would soon be corrected.

The first results of the joint project were presented to an *atelier de recherche* in Paris, March 1983. At this meeting were a number of French and British sociologists of science. The response was somewhat more than Bijker and Pinch had expected. Clearly, among students of scientific knowledge there was an emerging interest in the social study of technology. A snowball started rolling. In some Paris restaurant a group of four agreed to convene a meeting later that year in order to pursue this new approach—a sociology of technology. Two months later a letter was dashed off to inform more colleagues of this meeting. The number of participants by this stage had increased to fifteen. In their reactions some of these sociologists argued for the need to invite historians of technology also. Now being a couple of months older and a little wiser, Bijker and Pinch happily agreed. Tom Hughes was spotted in Europe, and other American historians of technology were contacted during a visit by Bijker to the United States in November 1983. The number of participants increased to twenty.

Now the rolling snowball was growing almost out of control. Offers to participate in the workshop started to come in from unexpected corners. We felt like football trainers who have to decide which players are allowed on the field and which are condemned to the substitutes' bench. The only solution we saw to this problem was to continue in the "pink champagne style": without any formal guidelines and with just our own implicit criteria about the works' being interesting. We had decided to adhere strictly to a maximum of thirty participants, because we thought that a larger number would hamper the emergence of the kind of collective discussion we hoped for. Inevitably this excluded many scholars who could have contributed greatly. However, we did not want to run the risk of the workshop breaking down into small subgroups with parallel lines of discussion. After all, we were organizing a workshop, not a conference.

Drawing on our experiences in the Burg Landsberg, we located the workshop in De Boerderij ("The Farm") on the campus of the Twente University of Technology. The modern architecture of De Boerderij, with its combination of high ceilings and low timbers, into which many a head bumped, constrained and concentrated our intellectual endeavor. Escape from this building would lead you into the Dutch countryside with even fewer worldly temptations than the Austrian mountains could offer. A heavy program, rental bicycles to move between the hotel and De Boerderij, and the food and drink at hand further defined the circumscribed environment. The work ethic prevailed.

As has been noted elsewhere,[2] this pressure cooker could have easily exploded. The differences in historical, sociological, and philosophical approaches and the idiosyncrasies of six different nationalities meant that the workshop could have been a dialogue of the deaf. However, quite the contrary happened. All the participants appeared to have come not only to talk but also to listen. Almost everyone played an active part, either by presenting a paper or by acting as a discussant. The workshop was marked by an open and sympathetic intellectual atmosphere, and discussions of a genuinely interdisciplinary nature occurred. More specifically, a high tolerance was shown for different and sometimes even incompatible concepts and approaches. If a new idea was not found to be immediately convincing, it was not knocked down but discussed critically and then "carried along" to later sessions while we suspended final judgment. New strands of discussion emerged, linking different papers together on a level that no one had envisaged in advance. At the end of the workshop the intellectual excitement was such that it was decided to publish a special journal issue or a book on the basis of the papers and the discussions. This volume is the result.

The intellectual excitement, if we may use these grandiose words once more, has continued long after the workshop. Authors have been stimulated by the discussions and have revised their contributions substantially. In some chapters new empirical results have been incorporated. Of course, the workshop cannot claim all, or even most, of the credit for this. On the contrary, the success of the workshop is symptomatic of a generally emerging interest in a new type of technology study. Other indications of this new interest are the recently published books edited by Rachel Laudan (1984a) and by Donald MacKenzie and Judy Wajcman (1985).

This new type of technology study can be characterized by three trends in the sort of analysis attempted. Authors have been concerned with moving away from the individual inventor (or "genius") as the central explanatory concept, from technological determinism, and from making distinctions among technical, social, economic, and political aspects of technological development. The last point has been aptly summarized by using the metaphor of the "seamless web" of society and technology. The two edited collections mentioned provide an adequate account of the origins of these trends. Moreover, together they provide a fairly comprehensive review of the relevant literature of previous traditions in the history, sociology, and philosophy of technology. Hence we have restricted ourselves in this book to giving only a short sketch of the well-established traditions from which we define our point of departure (Pinch and Bijker, this volume).

"Technology" is a slippery term, and concepts such as "technological change" and "technological development" often carry a heavy interpretative load.[3] It seems unfruitful and indeed unnecessary to devote much effort to working out precise definitions at least at this stage of the research in progress. Three layers of meaning of the word "technology" can be distinguished (MacKenzie and Wajcman 1985). First, there is the level of *physical objects* or *artifacts*, for example, bicycles, lamps, and Bakelite. Second, "technology" may refer to *activities* or *processes*, such as steel making or molding. Third, "technology" can refer to what people *know* as well as what they do; an example is the "know-how" that goes into designing a bicycle or operating an ultrasound device in the obstetrics clinic. In practice the technologies dealt with in this collection cover all three aspects, and often it is not sensible to separate them further. Also, instead of trying to distinguish technology from science (or indeed from any other activity) in general terms, it seems preferable to work from a set of empirical cases that seem intuitively paradigmatic. In this volume a broad range of technologies is examined, including bicycles, missiles, ships, electric vehicles, electric power systems, the cooking stove, pharmaceuticals, ultrasound, dyes, and expert systems.

Having said what we are trying to get away from, where are we heading? It is too early to specify this with any precision without violating the fruitful and stimulating heterogeneity of the emerging field. In the context of the workshop, however, it can safely be said that three approaches played a more or less dominant role and hence have guided the studies in this volume. The first, the social constructivist approach, has been inspired by recent studies in the sociology of scientific knowledge.[4] Key concepts within this approach are "interpretative flexibility," "closure," and "relevant social groups." One of the central tenets of this approach is the claim that technological artifacts are open to sociological analysis, not just in their usage but especially with respect to their design and technical "content." The second approach, stemming largely from the work of the historian of technology Thomas Hughes, treats technology in terms of a "systems" metaphor. This stresses the importance of paying attention to the different but interlocking elements of physical artifacts, institutions, and their environment and thereby offers an integration of technical, social, economic, and political aspects. Moreover, the key concepts of "reverse salient" and "critical problem," which define the parts of a system where at certain stages innovative energy is focused, enable us to link the micro- and macrolevels of analysis. Or, to put it more concretely, they enable us to link, say, Edison's laboratory to the wider society. The third approach, associated with the work of Michel Callon, Bruno Latour, and John Law at L'Ecole des Mines, Paris, attempts to extend this perspective one step further. It does this by breaking down the distinction between human actors and natural phenomena. Both are treated as elements in "actor networks." Also, this approach ostensibly reverses the usual relationship between participant and analyst and casts the engineers as sociologists. In other words, in trying to extend successfully the actor network, the engineers attempt to mold society. The three chapters in part I outline these different general approaches.

A characteristic that all these approaches share is the emphasis on "thick description," that is, looking into what has been seen as the black box of technology (and, for that matter, the black box of society). Such an emphasis may not be new to the history of technology, but it is new to areas of technology studies that have more theoretical aims, whether they are guided by sociology or by economics. This thick description results in a wealth of detailed information about the technical, social, economic, and political aspects of the case under study. In order to make any sense out of this complexity, it is necessary to employ some structuring and simplifying concepts. In part II three chapters deal specifically with a number of such "middle-range" concepts. Setting aside a special section for this topic should not be seen as an attempt on our part to imply that the general concepts outlined in part I, such as "system," "actor network," and "interpretative flexibility," do not have any explanatory power. Nor should it be concluded that chapters in later parts of this volume have no explanatory objectives.

The studies in part III exemplify strategic research sites for working out in detail the approaches put forward in parts I and II. To some extent the concepts introduced in the previous chapters are applied to the case studies in this third part. However, as the heterogeneity of the field guarantees, these studies offer much more than mere applications of the previously presented concepts. In some respects, considering the importance of detailed empirical studies in the emergence of a new approach in technology studies, this section forms the backbone of the volume.

In the fourth and last part of the collection we hope to show that the new approach may also yield results beyond the classical boundaries of technology studies. In the two chapters in this section, the object of study is artificial intelligence and "expert systems." It is claimed that our new understanding of technological development may actually contribute to this rapidly growing new technology. Questions concerning technology transfer and the role of tacit knowledge are dealt with in this context. But there is more to it. In part I of this book the integrated study of society and technology is advocated; parts II and III provide the specific tools and research sites where this can be carried out; throughout, as we have pointed out, it is intuitively clear what is meant by "technological artifact," "technological system," etc. In part IV, however, even this intuition about the identity of "technology" is called into question: The distinctions between human and machine, knowledge and action, engineering and the study of engineering practices are all "blown up." We find that sociologists of technology are actually contributing to the development of technology! In some respect, the circle is closed, and we are back at the beginning of this book. The seamless character of the web of society and technology is reestablished.

Each part of the volume is preceded by an introduction in which we try to bring out some of the common themes in the chapters that follow and their relation to other parts of the book.

Notes

1. Some of the papers presented at this first meeting of the European Association for the Study of Science and Technology have been published in Gotschl and Rip (1984).

2. Three news reports on the workshop have been published: Law (1984a), Pinch (1985), and de Vries (1984).

3. We do not try to (re)establish the useful distinction between "technology" and "technics" or "technique" into English usage. This difference, which does exist in Dutch, German, and French, is analogous to the distinction between "epistemology" and "knowledge."

4. See, for example, Barnes and Edge (1982), Collins (1982), and Collins (1983c).

I Common Themes in Sociological and Historical Studies of Technology

Introduction

System builders are no respecters of knowledge categories or professional boundaries. In his notebooks Thomas Edison so thoroughly mixed matters commonly labeled economic, technical, and scientific that his thoughts composed a seamless web. Charles Stone and Edwin Webster, founders of Stone & Webster, the consulting engineering firm, took as the company logotype the triskelion, to symbolize the thoroughly integrated functions of financing, engineering, and construction performed by their company, an organization responsible for many mammoth twentieth-century engineering projects. Nikolai Lenin, a technological enthusiast, also had a holistic vision when he wrote, "Soviet power + Prussian railroad organization + American technology + the trusts = Socialism" (quoted in Gillen 1977, p. 214). Perhaps Walther Rathenau, the head of Allgemeine Elektrizitäts-Gesellschaft, Germany's largest manufacturing, electrical utility, and banking combination before World War I, epitomized the drive for integration and synthesis, both in his person and in the organizations he created and directed. He envisaged the entire German economy as functioning like a single machine, and he observed that in 1909 "three hundred men, all acquainted with each other [of whom he was one], control the economic destiny of the Continent" (Kessler 1969, p. 121). In *Man Without Qualities* (1930), the novelist Robert Musil characterized his Rathenau-like protagonist, Arnheim, as the embodiment of the "mystery of the whole."

Many engineers, inventors, managers, and intellectuals in the twentieth century, especially in the early decades, created syntheses, or seamless webs. The great technological systems, utility networks, trusts, cartels, and holding companies are evidence of their integrating and controlling aspirations. Essays and books calling for the displacement of the mechanical with the organic also testify to these yearnings. The desire for systems and networks may have resulted in part from the rise of electrical and chemical engineering and the spread of a mode of thinking and organization associated with them. Electrical and chemical relationships, in contrast to mechanical or linear ones, are conceived of in terms of circuits, networks, and systems. Gear trains, cams, and followers are the linear interconnections common to the mechanical.

"Technology/science," "pure/applied," "internal/external," and "technical/ social" are some of the dichotomies that were foreign to the integrating inventors, engineers, and managers of the system- and network-building era. To have asked problem-solving inventors if they were doing science or technology probably would have brought an uncomprehending stare. Even scientists who thought of themselves as pure would not have set up barriers between the internal and the external, if these would have prevented the search for solutions wherever the problem-solving thread might have led. Entrepreneurs and system builders creating regional production complexes incorporated such seemingly foreign actors as legislators and financiers in networks, if they could functionally contribute to the system-building goal. Instead of taking multidivisional organizational layouts as airtight categories, integrating managers such as Stone and Webster saw a seamless web.

Historians and sociologists who want to study technology in this way should choose as their subject matter such inventors, engineers, managers, and scientists or the organizations over which they presided or of which they were an integral part. The dichotomies would promptly evaporate. Historians and sociologists choosing such subjects would do research and writing in which the technical, scientific, economic, political, social, and other categories would overlap and become soft. Some historians of science and technology still take the categories and dichotomies seriously because they write about non-problem-solving, category-filling academics.

In this first part of our book we present three different approaches to deal with this seamless web of technology and society. Also, these chapters explicate the need and possibility of synthesizing ideas and method from the disciplines of sociology and history for studying technology. Together the chapters demarcate a research program for studying the development of technological artifacts and systems—a research program that aims at contributing to a greater understanding of the social processes involved in technological development while respecting the seamless web character of technology and society.

The first chapter of part I begins with a brief critical review of the technology studies literature. Trevor Pinch and Wiebe Bijker outline their social constructivist approach using the case study of the bicycle. Thomas Hughes, in the second chapter, describes his approach to "technological systems" by specifying such terms as "technological style," "reverse salient," and "momentum." Hughes's argument is richly illustrated with a variety of examples. In the third chapter Michel Callon uses his study of electric vehicle development in France to sketch the actor-network approach. He casts engineers as practicing sociologists and concludes that sociology in general would benefit from a sociology of technology that seeks to apply the engineers' own methods.

Of the themes addressed in this part of the volume, the seamless web concept is most pronounced. In this respect, all authors address the science/technology dichotomy. Pinch and Bijker argue that both science and technology are socially constructed cultures and that the boundary between them is a matter for social negotiation and represents no underlying distinction.

Thomas Hughes stresses that the science and technology labels arc imprecise and do not convey the messy complexity of the entities named. He also defines science in part as knowledge about technology and technology as embodied knowledge, so that the distinctions again tend to fade. In addition, Hughes sees some scientists as developing technology, a function usually associated with engineers, and some engineers as doing research in ways usually associated with scientists. Hughes also believes that enthusiastic problem solvers and dedicated system builders are no respecters of disciplinary and knowledge boundaries. From his essay the reader may conclude that Hughes, as historian, would prefer to avoid the middle-level abstractions, such as science and technology, and write of particular actors doing particular things. His frequent use of case histories gives evidence of this.

Michel Callon proposes in his essay that the question of who is a scientist and who is a technologist is negotiable according to circumstances. He, like Hughes, believes that "the fabric has no seams." Callon asks why one should categorize the elements in a system or network "when these elements are permanently interacting, being associated, and being tested by the actors who innovate." Faced by the abstract categories problem—science, technology, economics, politics, etc.—Callon takes a different tack from Hughes, who prefers specific cases or examples in lieu of middle-level categories. Callon uses a higher abstraction, "actors," that subsumes science, technology, and other categories. Actors are the heterogeneous entities that constitute a network.

Callon's actors include electrons, catalysts, accumulators, users, researchers, manufacturers, and ministerial departments defining and enforcing regulations affecting technology. These and many other actors interact

through networks to create a coherent actor world. Callon does not, therefore, distinguish the animate from the inanimate, individuals from organizations. The actor world shapes and supports the technical object, an
electric vehicle in the case Callon presents. Electricité de France (EDF) in
fulfilling its program for developing an electric vehicle (VEL) virtually
writes a script or provides a scenario in which the actors' roles are so
defined and their relationships so bounded that the VEL is conceived by
and becomes a coextension of the actor world. In concepts reminiscent of
Martin Heidegger, the VEL is the physical artifact that the actors are destined to bring forth, enframe, and sustain (Heidegger 1977, p. 19). Callon
believes that there is no outside/inside (that is, social/technology)
dichotomy.

Pinch and Bijker, in contrast, preserve the social environment. The web
is not seamless in this regard, and Pinch and Bijker develop other conceptual themes. The social environment, for instance, shapes the technical
characteristics of the artifact. With their emphasis on social shaping, Pinch
and Bijker deny technological determinism. Borrowing and adapting from
the sociology of knowledge, they argue that the social groups that constitute the social environment play a critical role in defining and solving the
problems that arise during the development of an artifact. Their emphasis
on problem solving during the development of technology is like Hughes's
on reverse salients and critical problems. Pinch and Bijker point out that
social groups give meaning to technology and that problems—Hughes's
reverse salient—are defined within the context of the meaning assigned
by a social group or a combination of social groups. Because social groups
define the problems of technological development, there is flexibility in
the way things are designed, not one best way. This approach is like that
in "the Empirical Programme of Relativism," a sociology of knowledge
program stressing that scientific findings are open to more than one
interpretation.

Pinch and Bijker, drawing again on the "Empirical Programme," also
introduce the concept of closure. Closure occurs in science when a consensus emerges that the "truth" has been winnowed from the various
interpretations; it occurs in technology when a consensus emerges that a
problem arising during the development of technology has been solved.
When the social groups involved in designing and using technology decide
that a problem is solved, they stabilize the technology. The result is closure.
Closure and stabilization, however, are not isolated events; they occur
repeatedly during technological development. To use Hughes's language,
a reverse salient has been corrected, but countless others will emerge as

the technology is invented, developed, expanded, and improved. Returning again to the role of social groups stressed by Pinch and Bijker, various groups will decide differently not only about the definition of the problem but also about the achievement of closure and stabilization.

From the early history of the bicycle, Pinch and Bijker provide examples of closure and stabilization, social shaping, interpretative flexibility, and the influence of social groups. In this case history they present technological development as a nondetermined, multidirectional flux that involves constant negotiation and renegotiation among and between groups shaping the technology. Their model is far from the rigid, categorized, linear one sometimes presented for technological development. They, like Callon and Hughes, find it difficult to use conventional categorizing language to describe continuous change. They need a language analogous to calculus.

Pinch, Bijker, Hughes, and Callon are searching for a language and for concepts to express their new understanding of technological change. In addition to the seamless web, systems, actors, networks, closure, stabilization, and social construction, they explore conservative and radical change, balances and imbalances in evolving technological systems, translation, heterogeneity, and research sites. With regard to conservative and radical change, Pinch, Bijker, and Hughes note that inclusion in a group, organization, or bureaucracy dampens the originality of inventors and innovators (Bijker, this volume). High inclusion brings mission orientation or commitment to incremental improvements in the evolving technological system with which the group, organization, or bureaucracy has identified. The outsiders, Hughes believes, create the radical inventions that must stand initially without substantial organizational support. Radical inventions are often stifled by organizations that consider them a threat to the technology that they nurture. But radical inventions are often the geneses of new systems.

Dynamic imbalance and reverse salients in systems are comparable concepts found in the Callon and the Hughes essays. The use of the concept follows from conceiving of technology as a growing system or network. Because actors or components in a system are functionally related, changes in one or more cause imbalances or reverse salients in the advancing system front until the other components cascade and adjust to achieve an optimal interaction. Because the technological systems are growing or changing, the analysis should be analogous to dynamics (the study of motion and equilibrium) rather than to statics (the study of rest and equilibrium). System components interacting harmoniously—without imbalances or reverse salients—while the system grows can be thought of as being in dynamic equilibrium.

Callon argues that a new actor world and the technology it sustains are not, as has often been said of invention, a new combination of old entities or components. One cannot simply shop in an imaginary technology-component supermarket and then assemble a combination. The actors, whether consumers, fuel cells, or automobile manufacturers (as in the case of Callon's electric vehicle example), must have their attributes defined for them, or translated,[1] so that they can play their assigned roles in the scenario conceived of by the actor-world designer. To use Hughes's language—and Edison's approach—each component in the system has to be designed to interact harmoniously with the characteristics of the others. Callon and Hughes speak as one when they insist that organizations as well as physical artifacts have to be invented for systems and actor worlds. If existing organizations of artifacts are to be used, then they must be translated. Callon, however, stresses how difficult this is once the translation has been made to fix or stabilize it. The heterogeneous actors in the network tend to revert to their former roles or to take on others, and so the network, or system, breaks down. These failures should, Pinch and Bijker believe, be of as much interest to historians and sociologists of technology as the success stories. In saying this they are borrowing from the "Strong Programme" in the sociology of scientific knowledge that calls for sociologists to be impartial to the truth or falsity of beliefs so that they can be explained symmetrically (Bloor 1973).

Pinch and Bijker also borrow from the sociology of knowledge as they recommend that scholars interested in the development of technology choose controversy as one important site for research. The controversy in question is over the truth or falsity of belief or about the success or failure of a technology in solving problems. They show that different groups will define not only the problem differently but also success or failure. They reinforce the wisdom that there is not just one possible way, or one best way, of designing an artifact.

Pinch and Bijker also urge historians and sociologists of technology to borrow from the sociologists of knowledge, who deal with the content of science. They urge historians and sociologists to open the so-called black box in which the workings of technology are housed. Citing M. J. Mulkay, they acknowledge that it is easier to show that the social meaning of television depends on the social context in which it is used than to demonstrate that a working television set is also context dependent. They also realize that demonstrating how the technical characteristics, or meaning, of artifacts, such as electrical generators and transformers, are socially constructed is difficult, but they note that some historians of technology have done case studies of this kind. Callon and Hughes also want the black box pried open. In his essay Hughes provides some instances of the social shaping within the black box.

Note

1. In his contribution to this volume, Callon does not explicitly use the concept "translation," but the idea is implicit in much of his argument. See Callon (1980b, 1981b. 1986) and Latour (1983, 1984).

The Social Construction of Facts and Artifacts: Or How the Sociology of Science and the Sociology of Technology Might Benefit Each Other

Trevor J. Pinch and Wiebe E. Bijker

One of the most striking features of the growth of "science studies" in recent years has been the separation of science from technology. Sociological studies of new knowledge in science abound, as do studies of technological innovation, but thus far there has been little attempt to bring such bodies of work together.[1] It may well be the case that science and technology are essentially different and that different approaches to their study are warranted. However, until the attempt to treat them within the same analytical endeavor has been undertaken, we cannot be sure of this.

It is the contention of this chapter that the study of science and the study of technology should, and indeed can, benefit from each other. In particular we argue that the social constructivist view that is prevalent within the sociology of science and also emerging within the sociology of technology provides a useful starting paint. We set out the constitutive questions that such a unified social constructivist approach must address analytically and empirically.

This chapter falls into three main sections. In the first part we outline various strands of argumentation and review bodies of literature that we consider to be relevant to our goals. We then discuss the two specific approaches from which our integrated viewpoint has developed: the "Empirical Programme of Relativism" (Collins 1981d) and a social constructivist approach to the study of technology (Bijker et al. 1984). In the third part we bring these two approaches together and give some empirical examples. We conclude by summarizing our provisional findings and by indicating the directions in which we believe the program can most usefully be pursued.

Some Relevant Literature

In this section we draw attention to three bodies of literature in science and technology studies. The three areas discussed are the sociology of science, the science-technology relationship, and technology studies. We take each in turn.

Sociology of Science

It is not our intention to review in any depth developments in this field as a whole.[2] We are concerned here with only the recent emergence of the sociology of scientific *knowledge*.[3] Studies in this area take the actual content of scientific ideas, theories, and experiments as the subject of analysis. This contrasts with earlier work in the sociology of science, which was concerned with science as an institution and the study of scientists' norms, career patterns, and reward structures.[4] One major—if not *the* major—development in the field in the last decade has been the extension of the sociology of knowledge into the arena of the "hard sciences." The need for such a "strong programme" has been outlined by Bloor: Its central tenets are that, in investigating the causes of belief, sociologists should be impartial to the truth or falsity of the beliefs, and that such beliefs should be explained symmetrically (Bloor 1973). In other words, differing explanations should not be sought for what is taken to be a scientific "truth" (for example, the existence of x-rays) and a scientific "falsehood" (for example, the existence of n-rays). Within such a program all knowledge and all knowledge claims are to be treated as being socially constructed; that is, explanations for the genesis, acceptance, and rejection of knowledge claims are sought in the domain of the social world rather than in the natural world.[5]

This approach has generated a vigorous program of empirical research, and it is now possible to understand the processes of the construction of scientific knowledge in a variety of locations and contexts. For instance, one group of researchers has concentrated their attention on the study of the laboratory bench.[6] Another has chosen the scientific controversy as the location for their research and have thereby focused on the social construction of scientific knowledge among a wider community of scientists.[7] As well as in hard sciences, such as physics and biology, the approach has been shown to be fruitful in the study of fringe science[8] and in the study of public-science debates, such as lead pollution.[9]

Although there are the usual differences of opinion among researchers as to the best place to locate such research (for instance, the laboratory, the controversy, or the scientific paper) and although there are differences as to the most appropriate methodological strategy to pursue,[10] there is widespread agreement that scientific knowledge can be, and indeed has been, shown to be thoroughly socially constituted. There approaches, which we refer to as "social constructivist," mark an important new development in the sociology of science. The treatment of scientific knowledge as a social construction implies that there is nothing epistemologically special about the nature of scientific knowledge: It is merely one in a whole series of knowledge cultures (including, for instance, the knowledge systems pertaining to "primitive" tribes) (Barnes 1974; Collins and Pinch 1982). Of course, the successes and failures of certain knowledge cultures still need to be explained, but this is to be seen as a sociological task, not an epistemological one.

The sociology of scientific knowledge promises much for other areas of "science studies." For example, it has been argued that the new work has relevance for the history of science (Shapin 1982), philosophy of science (Nickles 1982), and science policy (Healey 1982; Collins 1963b). The social constructivist view not only seems to be gaining ground as an important body of work in its own right but also shows every potential of wider application. It is this body of work that forms one of the pillars of our own approach to the study of science and technology.

Science-Technology Relationship

The literature on the relationship between science and technology, unlike that already referred to, is rather heterogeneous and includes contributions from a variety of disciplinary perspectives. We do not claim to present anything other than a partial review, reflecting our own particular interests.

One theme that has been pursued by philosophers is the attempt to separate technology from science on analytical grounds. In doing so, philosophers tend to posit overidealized distinctions, such as that science is about the discovery of truth whereas technology is about the application of truth. Indeed, the literature on the philosophy of technology is rather disappointing (Johnston 1984). We prefer to suspend judgment on it until philosophers propose more realistic models of both science and technology.

Another line of investigation into the nature of the science-technology relationship has been carried out by innovation researchers. They have attempted to investigate empirically the degree to which technological innovation incorporates, or originates from, basic science. A corollary of this approach has been the work of some scholars who have looked for relationships in the other direction; that is, they have argued that pure science is indebted to developments in technology.[11] The results of the empirical investigations of the dependence of technology on science have been rather frustrating. It has been difficult to specify the interdependence. For example, Project Hindsight, funded by the US Defense Department, found that most technological growth came from mission-oriented projects and engineering R&D, rather than from pure science (Sherwin and Isenson 1966, 1967). These results were to some extent supported by a later British study (Langrish et al. 1972). On the other hand, Project TRACES, funded by the NSF in response to Project Hindsight, found that most technological development stemmed from basic research (Illinois Institute of Technology, 1968). All these studies have been criticized for lack of methodological rigor, and one must be cautious in drawing any firm conclusions from such work (Kreilkamp 1971; Mowery and Rosenberg 1979). Most researchers today seem willing to agree that technological innovation takes place in a wide range of circumstances and historical epochs and that the import that can he attached to basic science therefore probably varies considerably.[12] Certainly the view prevalent in the "bad old days" (Barnes 1982a)—that science discovers and technology applies—will no longer suffice. Simplistic models and generalizations have been abandoned. As Layton remarked in a recent review:

Science and technology have become intermixed. Modern technology involves scientists who 'do' technology and technologists who function as scientists. . . . The old view that basic sciences generate all the knowledge which technologists then apply will simply not help in understanding contemporary technology. (Layton 1977, p. 210)

Researchers concerned with measuring the exact interdependence of science and technology seem to have asked the wrong question because they have assumed that science and technology are well-defined monolithic structures. In short, they have not grasped that science and technology are themselves socially produced in a variety of social circumstances (Mayr 1976). It does seem, however, that there is now a move toward a more sociological conception of the science-technology relationship. For instance, Layton writes:

The divisions between science and technology are not between the abstract functions of knowing and doing. Rather they are social. (Layton 1977, p. 209)

Barnes has recently described this change of thinking:

I start with the major reorientation in our thinking about the science-technology relationship which has occurred in recent years. . . . We recognize science and technology to be on a par with each other. Both sets of practitioners creatively extend and develop their existing culture; but both also take up and exploit some part of the culture of the other. . . . They are in fact enmeshed in a symbiotic relationship. (Barnes 1982a, p. 166)

Although Barnes may be overly optimistic in claiming that a "major reorientation" has occurred, it can be seen that a social constructivist view of science and technology fits well with his conception of the science-technology relationship. Scientists and technologists can be regarded as constructing their respective bodies of knowledge and techniques with each drawing on the resources of the other when and where such resources can profitably be exploited. In other words, both science and technology are socially constructed cultures and bring to bear whatever cultural resources are appropriate for the purposes at hand. In his view the boundary between science and technology is, in particular instances, a matter for social negotiation and represents no underlying distinction. It then makes little sense to treat the science-technology relationship in a general unidirectional way. Although we do not pursue this issue further in this chapter, the social construction of the science-technology relationship is clearly a matter deserving further empirical investigation.

Technology Studies

Our discussion of technology studies work is even more schematic. There is a large amount of writing that falls under the rubric of "technology studies." It is convenient to divide the literature into three parts: innovation studies, history of technology, and sociology of technology. We discuss each in turn.

Most innovation studies have been carried out by economists looking for the conditions for success in innovation. Factors researched include various aspects of the innovating firm (for example, size of R&D effort, management strength, and marketing capability) along with macroeconomic factors pertaining to the economy as a whole.[13] This literature is in some ways reminiscent of the early days in the sociology of science, when scientific knowledge was treated like a "black box" (Whitley 1972) and, for the purpose of such studies, scientists might as well have produced meat pies. Similarly, in the economic analysis of technological innovation everything is included that might be expected to influence innovation, except any discussion of the technology itself. As Layton notes:

What is needed is an understanding of technology from inside, both as a body of knowledge and as a social system. Instead, technology is often treated as a "black box" whose contents and behaviour may be assumed to be common knowledge. (Layton 1977, p. 198)

Only recently have economists started to look into this black box.[14]

The failure to take into account the content of technological innovations results in the widespread use of simple linear models to describe the process of innovation. The number of developmental steps assumed in these models seems to be rather arbitrary (for an example of a six-stage process see figure 1).[15] Although such studies have undoubtedly contributed much to our understanding of the conditions for economic success in technological innovation, because they ignore the technological content they cannot be used as the basis for a social constructivist view of technology.[16]

This criticism cannot be leveled at the history of technology, where there are many finely crafted studies of the development of particular technologies. However, for the purposes of a sociology of technology, this work presents two kinds of problems. The first is that descriptive historiography is endemic in this field. Few scholars (but there are some notable exceptions) seem concerned with generalizing beyond historical instances, and it is difficult to discern any overall patterns on which to build a theory of technology (Staudenmaier 1983, 1985). This is not to say that such studies might not be useful building blocks for a social constructivist view of technology—merely that these historians have not yet demonstrated that they are doing sociology of knowledge in a different guise.[17]

The second problem concerns the asymmetric focus of the analysis. For example, it has been claimed that in twenty-five volumes of *Technology and Culture* only nine articles were devoted to the study of failed technological innovations (Staudenmaier 1985). This contributes to the implicit adoption of a linear structure of technological development, which suggests that

the whole history of technological development had followed an orderly or rational path, as though today's world was the precise goal toward which all decisions, made since the beginning of history, were consciously directed. (Ferguson 1974b, p. 19)

This preference for successful innovations seems to lead scholars to assume that the success of an artifact is an explanation of its subsequent development. Historians of technology often seem content to rely on the manifest success of the artifact as evidence that there is no further explanatory work to be done. For example, many histories of synthetic plastics start by describing the "technically sweet" characteristics of Bakelite; these features are then used implicitly to position Bakelite at the starting point of the glorious development of the field:

God said: "let Baekeland be" and all was plastics! (Kaufman 1963, p. 61)

However, a more detailed study of the developments of plastic and varnish chemistry, following the publication of the Bakelite process in 1909 (Backeland 1909c, d), shows that Bakelite was at first hardly recognized as the

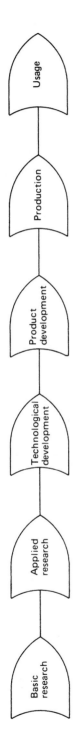

Figure 1

A six-stage model of the innovation process.

marvelous synthetic resin that it later proved to be.[18] And this situation did not change much for some ten years. During the First World War the market prospects for synthetic plastics actually grew worse. However, the dumping of war supplies of phenol (used in the manufacture of Bakelite) in 1918 changed all this (Haynes 1954, pp. 137–138) and made it possible to keep the price sufficiently low to compete with (semi-) natural resins, such as celluloid.[19] One can speculate over whether Bakelite would have acquired its prominence if it had not profited from that phenol dumping. In any case it is clear that a historical account founded on the retrospective success of the artifact leaves much untold.

Given our intention of building a sociology of technology that treats technological knowledge in the same symmetric, impartial manner that scientific facts are treated within the sociology of scientific knowledge, it would seem that much of the historical material does not go far enough. The success of an artifact is precisely what needs to be explained. For a sociological theory of technology it should be the *explanandum*, not the *explanans*.

Our account would not be complete, however, without mentioning some recent developments, especially in the American history of technology. These show the emergence of a growing number of theoretical themes on which research is focused (Staudenmaier 1985; Hughes 1979b). For example, the systems approach to technology,[20] consideration of the effect of labor relations on technological development,[21] and detailed studies of some not-so-successful inventions[22] seem to herald departures from the "old" history of technology. Such work promises to be valuable for a sociological analysis of technology, and we return to some of it later.

The final body of work we wish to discuss is what might be described as "sociology of technology."[23] There have been some limited attempts in recent years to launch such a sociology, using ideas developed in the history and sociology of science—studies by, for example, Johnston (1972) and Dosi (1982), who advocate the description of technological knowledge in terms of Kuhnian paradigms.[24] Such approaches certainly appear to be more promising than standard descriptive historiography, but it is not clear whether or not these authors share our understanding of technological artifacts as social constructs. For example, neither Johnston nor Dosi considers explicitly the need for a symmetric sociological explanation that treats successful and failed artifacts in an equivalent way. Indeed, by locating their discussion at the level of technological paradigms, we are not sure how the artifacts themselves are to be approached. As neither author has yet produced an empirical study using Kuhnian ideas, it is difficult to evaluate how the Kuhnian terms may be utilized.[25] Certainly this has been a pressing problem in the sociology of science, where it has not always been possible to give Kuhn's terms a clear empirical reference.

The possibilities of a more radical social constructivist view of technology have been touched on by Mulkay (1979a). He argues that the success and efficacy of technology could pose a special problem for the social constructivist view of *scientific knowledge*. The argument Mulkay wishes to counter is that the practical effectiveness of technology somehow demonstrates the privileged epistemology of science and thereby exempts it from sociological explanation. Mulkay opposes this view, rightly in our opinion, by pointing out the problem of the "science discovers, technology applies" notion implicit in such claims. In a second argument against this position, Mulkay notes (following Mario Bunge (1966)) that it is possible for false or partly false theory to be used as the basis for successful practical application: The success of the technology would not then have anything to say about the "truth" of the scientific knowledge on which it was based. We find this second point not entirely satisfactory. We would rather stress that the truth or falsity of scientific knowledge is irrelevant to sociological analysis of belief: To retreat to the argument that science may be wrong but good technology can still be based on it is missing this point. Furthermore, the success of technology is still left unexplained within such an argument. The only effective way to deal with these difficulties is to adopt a perspective that attempts to show that technology, as well as science, can be understood as a social construct.

Mulkay seems to be reluctant to take this step because, as he points out, "there are very few studies . . . which consider how the technical meaning of hard technology is socially constructed" (Mulkay 1979a, p. 77). This situation, however, is starting to change: A number of such studies have recently emerged. For example, Michel Callon, in a pioneering study, has shown the effectiveness of focusing on technological controversies. He draws on an extensive case study of the electric vehicle in France (1960–75) to demonstrate that almost everything is negotiable: what is certain and what is not: who is a scientist and who is a technologist; what is technological and what is social; and who can participate in the controversy (Callon 1980a, b, 1981b, and this volume). David Noble's study of the introduction of numerically controlled machine tools can also be regarded as an important contribution to a social constructivist view of technology (Noble 1984). Noble's explanatory goals come from a rather different (Marxist) tradition,[26] and his study has much to recommend it: He considers the development of both a successful and a failed technology and gives a symmetric account of both developments. Another intriguing study in this tradition is Lazonick's account (1979) of the introduction of the self-acting mule: He shows that aspects of this technical development can be understood in terms of the relations of production rather than any inner logic of technological development. The work undertaken by Bijker, Bönig, and Van Oost is another attempt to show how the socially constructed character of the content of some technological artifacts might be approached empirically: Six case studies were carried out, using historical sources.[27]

In summary, then, we can say that the predominant traditions in technology studies—innovation studies and the history of technology—do not yet provide much encouragement for our program. There are exceptions, however, and some recent studies in the sociology of technology present promising starts on which a unified approach could he built. We now give a more extensive account of how these ideas may he synthesized.

EPOR and SCOT

In this part we outline in more detail the concepts and methods that we wish to employ. We start by describing the "Empirical Programme of Relativism" as it was developed in the sociology of scientific knowledge. We then go on to discuss in more detail the approach taken by Bijker and his collaborators in the sociology of technology.

The Empirical Programme of Relativism (EPOR)

The EPOR is an approach that has produced several studies demonstrating the social construction of scientific knowledge in the "hard" sciences. This tradition of research has emerged from recent sociology of scientific knowledge. Its main characteristics, which distinguish it from other approaches in the same area, are the focus on the empirical study of contemporary scientific developments and the study, in particular, of scientific controversies.[28]

Three stages in the explanatory aims of the EPOR can be identified. In the *first stage* the interpretative flexibility of scientific findings is displayed; in other words, it is shown that scientific findings are open to more than one interpretation. This shifts the focus for the explanation of scientific developments from the natural world to the social world. Although this interpretative flexibility can be recovered in certain circumstances, it remains the case that such flexibility soon disappears in science; that is, a scientific consensus as to what the "truth" is in any particular instance usually emerges. Social mechanisms that limit interpretative flexibility and thus allow scientific controversies to be terminated are described in the *second stage*. A *third stage*, which has not yet been carried through in any study of contemporary science, is to relate such "closure mechanisms" to the wider social-cultural milieu. If all three stages were to be addressed in a single study, as Collins writes, "the impact of society on knowledge 'produced' at the laboratory bench would then have been followed through in the hardest possible case" (Collins 1981d, p. 7).

The EPOR represents a continuing effort by sociologists to understand the content of the natural sciences in terms of social construction. Various parts of the program are better researched than others. The third stage of the program has not yet even been addressed, but there are many excellent studies exploring the first stage. Most current research is aimed at elucidating the closure mechanisms whereby consensus emerges (the second stage). Many studies within the EPOR have been most fruitfully located in the area of scientific controversy. Controversies offer a methodological advantage in the comparative ease with which they reveal the interpretative flexibility of scientific results. Interviews conducted with scientists engaged in a controversy usually reveal strong and differing opinions over scientific findings. As such flexibility soon vanishes from science, it is difficult to recover from the textual sources with which historians usually work. Collins has highlighted the importance of the "controversy group" in science by his use of the term "core set" (Collins 1981b). These are the scientists most intimately involved in a controversial research topic. Because the core set is defined in relation to knowledge production in science (the core set constructs scientific knowledge), some of the empirical problems encountered in the identification of groups in science by purely sociometric means can be overcome. And studying the core set has another methodological advantage, in that the resulting consensus can be monitored. In other words, the group of scientists who experiment and theorize at the research frontiers and who become embroiled in scientific controversy will also reflect the growing consensus as to the outcome of that controversy. The same group of core set scientists can then be studied in both the first and second stages of the EPOR. For the purposes of the third stage, the motion of a core set may be too limited.

The Social Construction of Technology (SCOT)

Before outlining some of the concepts found to be fruitful by Bijker and his collaborators in their studies in the sociology of technology, we should point out an imbalance between the two approaches (EPOR and SCOT) we are considering. The EPOR is part of a flourishing tradition in the sociology of scientific knowledge: It is a well-established program supported by much empirical research. In contrast, the sociology of technology is an embryonic field with no well-established traditions of research, and the approach we draw on specifically (SCOT) is only in its early empirical stages, although clearly gaining momentum.[29]

In SCOT the developmental process of a technological artifact is described as an alternation of variation and selection.[30] This results in a "multidirectional" model, in contrast with the linear models used explicitly in many innovation studies and implicitly in much history of technology. Such a multidirectional view is essential to any social constructivist account of technology. Of course, with historical hindsight, it is possible to collapse the multidirectional model on to a simpler linear model; but this misses the thrust of our argument that the "successful" stages in the development are not the only possible ones.

Let us consider the development of the bicycle.[31] Applied to the level of artifacts in this development, this multidirectional view results in the description summarized in figure 2. Here we see the artifact "Ordinary" (or, as it was nicknamed after becoming less ordinary, the "Penny Farthing"; figure 3) and a range of possible variations. It is important to recognize that, in the view of the actors of those days, these variants were at the same time quite different from each other and equally were serious rivals. It is only by retrospective distortion that a quasi-linear development emerges, as depicted in figure 4. In this representation the so-called safety ordinaries (Xtraordinary (1878), Facile (1879), and Club Safety (1885)) figure only as amusing aberrations that need not be taken seriously (figures 5, 6, and 7). Such a retrospective description can be challenged by looking at the actual situation in the 1880s. Some of the "safety ordinaries" were produced commercially, whereas Lawson's Bicyclette, which seems to play an important role in the linear model, proved to be a commercial failure (Woodforde 1970).

However, if a multidirectional model is adopted, it is possible to ask why some of the variants "die," whereas others "survive." To illuminate this "selection" part of the developmental processes, let us consider the problems and solutions presented by each artifact at particular moments. The rationale for this move is the same as that for focusing on scientific controversies within EPOR. In this way, one can expect to bring out more clearly the interpretative flexibility of technological artifacts.

In deciding which problems are relevant, the social groups concerned with the artifact and the meanings that those groups give to the artifact

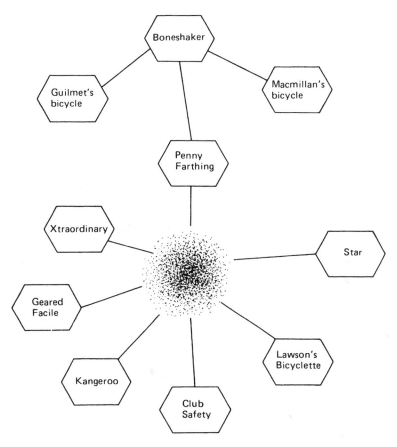

Figure 2
A multidirectional view of the developmental process of the Penny Farthing bicycle.
The shaded area is filled in and magnified in figure 11. The hexagons symbolize
artifacts.

play a crucial role: A problem is defined as such only when there is a social
group for which it constitutes a "problem."

The use of the concept of a relevant social group is quite straightforward.
The phrase is used to denote institutions and organizations (such as the
military or some specific industrial company), as well as organized or
unorganized groups of individuals. The key requirement is that all members
of a certain social group share the same set of meanings, attached to a
specific artifact.[32] In deciding which social groups are relevant, we must
first ask whether the artifact has any meaning at all for the members
of the social group under investigation. Obviously, the social group of

Figure 3
A typical Penny Farthing, the Bayliss-Thomson Ordinary (1878). Photograph courtesy of the Trustees of the Science Museum, London.

"consumers" or "users" of the artifact fulfills this requirement. But also less obvious social groups may need to be included. In the case of the bicycle, one needs to mention the "anticyclists." Their actions ranged from derisive cheers to more destructive methods. For example, Reverend L. Meadows White described such resistance to the bicycle in his book, *A Photographic Tour on Wheels:*

... but when to words are added deeds, and stones are thrown, sticks thrust into the wheels, or caps hurled into the machinery, the picture has a different aspect. All the above in certain districts are of common occurrence, and have all happened to me, especially when passing through a village just after school is closed. (Meadows, cited in Woodforde 1970, pp. 49–50)

Clearly, for the anticyclists the artifact "bicycle" had taken on meaning!

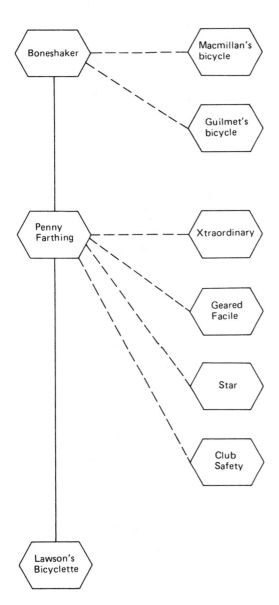

Figure 4
The traditional quasi-linear view of the developmental process of the Penny Farthing
bicycle. Solid lines indicate successful development, and dashed lines indicate failed
development.

Figure 5
The American Star bicycle (1885). Photograph courtesy of the Trustees of the Science Museum, London.

Figure 6
Facile bicycle (1874). Photograph courtesy of the Trustees of the Science Museum, London.

Figure 7
A form of the Kangaroo bicycle (1878). Photograph courtesy of the Trustees of the
Science Museum, London.

Another question we need to address is whether a provisionally defined
social group is homogeneous with respect to the meanings given to the
artifact—or is it more effective to describe the developmental process by
dividing a rather heterogeneous group into several different social groups?
Thus within the group of cycle-users we discern a separate social group of
women cyclists. During the days of the high-wheeled Ordinary women
were not supposed to mount a bicycle. For instance, in a magazine advice
column (1885) it is proclaimed, in reply to a letter from a young lady:

The mere fact of riding a bicycle is not in itself sinful, and if it is the only means
of reaching the church on a Sunday, it may be excusable. (cited in Woodforde 1970,
p. 122)

Tricycles were the permitted machines for women. But engineers and producers anticipated the importance of women as potential bicyclists. In a
review of the annual Stanley Exhibition of Cycles in 1890, the author
observes:

From the number of safeties adapted for the use of ladies, it seems as if bicycling
was becoming popular with the weaker sex, and we are not surprised at it, considering the saving of power derived from the use of a machine having only one slack.
(Stanley Exhibition of Cycles, 1890, pp. 107–108)

Thus some parts of the bicycle's development can be better explained by including a separate social group of feminine cycle-users. This need not, of course, be so in other cases: For instance, we would not expect it to be useful to consider a separate social group of women users of, say, fluorescent lamps.

Once the relevant social groups have been identified, they are described in more detail. This is also where aspects such as power or economic strength enter the description, when relevant. Although the only defining property is some homogeneous meaning given to a certain artifact, the intention is not just to retreat to worn-out, general statements about "consumers" and "producers." We need to have a detailed description of the relevant social groups in order to define better the function of the artifact with respect to each group. Without this, one could not hope to be able to give any explanation of the developmental process. For example, the social group of cyclists riding the high-wheeled Ordinary consisted of "young men of means and nerve: they might be professional men, clerks, schoolmasters or dons" (Woodforde 1970, p. 47). For this social group the function of the bicycle was primarily for sport. The following comment in the *Daily Telegraph* (September 7, 1877) emphasizes sport, rather than transport:

Bicycling is a healthy and manly pursuit with much to recommend it, and, unlike other foolish crazes, it has not died out. (cited in Woodforde 1970, p. 122)

Let us now return to the exposition of the model. Having identified the relevant social groups for a certain artifact (figure 8), we are especially interested in the problems each group has with respect to that artifact (figure 9). Around each problem, several variants of solution can be identified (figure 10). In the case of the bicycle, some relevant problems and solutions are shown in figure 11, in which the shaded area of figure 2 has been filled. This way of describing the developmental process brings out clearly all kinds of conflicts: conflicting technical requirements by different social groups (for example, the speed requirement and the safety requirement); conflicting solutions to the same problem (for example, the safety low-wheelers and the safety ordinaries); and moral conflicts (for example, women wearing skirts or trousers on high-wheelers; figure 12). Within this scheme, various solutions to these conflicts and problems are possible—not only technological ones but also judicial or even moral ones (for example, changing attitudes toward women wearing trousers).

Following the developmental process in this way, we see growing and diminishing degrees of stabilization of the different artifacts.[33] In

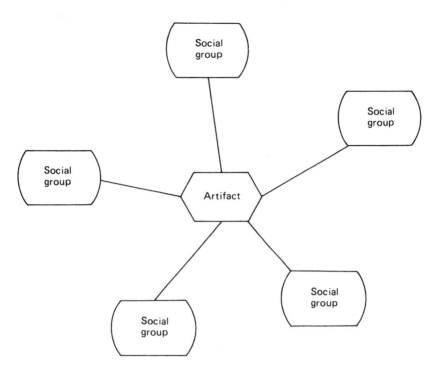

Figure 8
The relationship between an artifact and the relevant social groups.

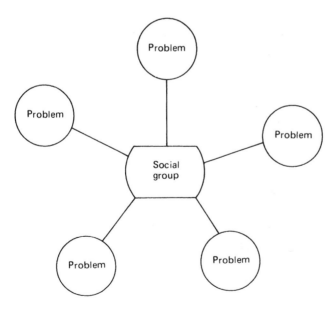

Figure 9
The relationship between one social group and the perceived problems.

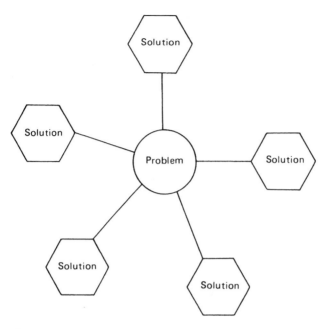

Figure 10
The relationship between one problem and its possible solutions.

principle, the degree of stabilization is different in different social groups. By using the concept of stabilization, we see that the "invention" of the safety bicycle was not an isolated event (1884), but a nineteen-year process (1879–98). For example, at the beginning of this period the relevant groups did not see the "safety bicycle" but a wide range of bi- and tricycles—and, among those, a rather ugly crocodilelike bicycle with a relatively low front wheel and rear chain drive (Lawson's Bicyclette; figure 13). By the end of the period, the phrase "safety bicycle" denoted a low-wheeled bicycle with rear chain drive, diamond frame, and air tires. As a result of the stabilization of the artifact after 1898, one did not need to specify these details: They were taken for granted as the essential "ingredients" of the safety bicycle.

We want to stress that our model is not used as a mold into which the empirical data have to be forced, *coûle que coûle*. The model has been developed from a series of case studies and not from purely philosophical or theoretical analysis. Its function is primarily heuristic—to bring out all the aspects relevant to our purposes. This is not to say that there are no explanatory and theoretical aims, analogous to the different stages of the EPOR (Bijker 1984 and this volume). And indeed, as we have shown, this

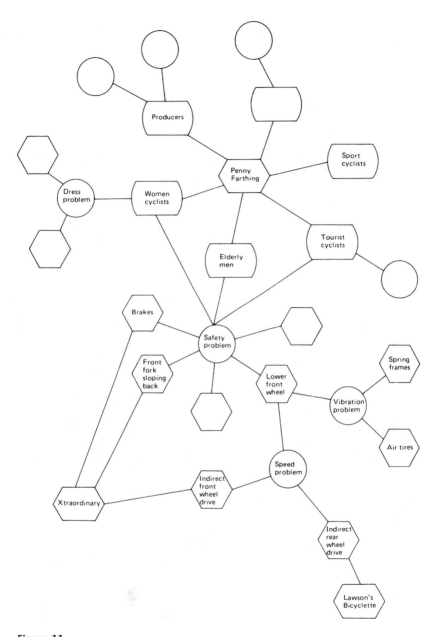

Figure 11
Some relevant social groups, problems, and solutions in the developmental process of the Penny Farthing bicycle. Because of lack of space, not all artifacts, relevant social groups, problems, and solutions are shown.

Figure 12
A solution to the women's dressing problem with respect to the high-wheeled Ordinary. This solution obviously has technical and athletic aspects. Probably, the athletic aspects prevented the solution from stabilizing. The set-up character of the photograph suggests a rather limited practical use. Photograph courtesy of the Trustees of the Science Museum, London.

Figure 13
Lawson's Bicyclette (1879). Photograph courtesy of the Trustees of the Science Museum, London.

model already does more than merely describe technological development: It highlights its multidirectional character. Also, as will be indicated, it brings out the interpretative flexibility of technological artifacts and the role that different closure mechanisms may play in the stabilization of artifacts.

The Social Construction of Facts and Artifacts

Having described the two approaches to the study of science and technology we wish to draw on, we now discuss in more detail the parallels between them. As a way of putting some flesh on our discussion we give, where appropriate, empirical illustrations drawn from our own research.

Interpretative Flexibility

The first stage of the EPOR involves, the demonstration of the interpretative flexibility of scientific findings. In other words, it must be shown that different interpretations of nature are available to scientists and hence that nature alone does not provide a determinant outcome to scientific debate.[34]

In SCOT, the equivalent of the first stage of the EPOR would seem to be the demonstration that technological artifacts are culturally constructed and interpreted; in other words, the interpretative flexibility of a technological artifact must be shown. By this we mean not only that there is flexibility in how people think of or interpret artifacts but also that there is flexibility in how artifacts are *designed*. There is not just one possible way or one best way of designing an artifact. In principle, this could be demonstrated in the same way as for the science case, that is, by interviews with technologists who are engaged in a contemporary technological controversy. For example, we can imagine that, if interviews had been carried out in 1890 with the cycle engineers, we would have been able to show the interpretative flexibility of the artifact "air tyre." For some, this artifact was a solution to the vibration problem of small-wheeled vehicles:

[The air tire was] devised with a view to afford increased facilities for the passage of wheeled vehicles—chiefly of the lighter class such for instance as velocipedes, invalid chairs, ambulances—over roadways and paths, especially when these latter are of rough or uneven character. (Dunlop 1888, p. 1)

For others, the air tire was a way of going faster (this is outlined in more detail later). For yet another group of engineers, it was an ugly looking way of making the low-wheeler even less safe (because of side-slipping) than it already was. For instance, the following comment, describing the Stanley Exhibition of Cycles, is revealing:

The most conspicuous innovation in the cycle construction is the use of pneumatic tires. These tires are hollow, about 2 in. diameter, and are inflated by the use of a small air pump. They are said to afford most luxurious riding, the roughest macadam and cobbles being reduced to the smoothest asphalte. Not having had the opportunity of testing these tires, we are unable to speak of them from practical experience; but looking at them from a theoretical point of view, we opine that considerable difficulty will be experienced in keeping the tires thoroughly inflated. Air under pressure is a troublesome thing to deal with. From the reports of those who have used these tires, it seems that they are prone to slip on muddy roads. If this is so, we fear their use on rear-driving safeties—which are all more or less addicted to side-slipping—is out of the question, as any improvement in this line should be to prevent side slip and not to increase it. Apart from these defects, the appearance of the tires destroys the symmetry and graceful appearance of a cycle, and this alone is, we think, sufficient to prevent their coming into general use. (Stanley Exhibition of Cycles, 1890, p. 107)

Figure 14
Whippet spring frame (1885). Photograph courtesy of the Trustees of the Science Museum, London.

And indeed, other artifacts were seen as providing a solution for the vibration problem, as the following comment reveals:

With the introduction of the rear-driving safety bicycle has arisen a demand for anti-vibration devices, as the small wheels of these machines are conducive to considerable vibration, even on the best roads. Nearly every exhibitor of this type of machine has some appliance to suppress vibration. (Stanley Exhibition of Cycles, 1889, pp. 157–158)

Most solutions used various spring constructions in the frame, the saddle, and the steering-bar (figure 14). In 1896, even after the safety bicycle (and the air tire with it) achieved a high degree of stabilization, "spring frames" were still being marketed.

It is important to realize that this demonstration of interpretative flexibility by interviews and historical sources is only one of a set of possible methods. At least in the study of technology, another method is applicable and has actually been used. It can be shown that different social groups have radically different interpretations of one technological artifact. We call these differences "radical" because the *content* of the artifact seems to be involved. It is something more than what Mulkay rightly claims to be rather easy—"to show that the social meaning of television varies with and depends upon the social context in which it is employed." As Mulkay notes: "It is much more difficult to show what is to count as a 'working television set' is similarly context-dependent in any significant respect" (Mulkay 1979a, p. 80).

We think that our account—in which the different interpretations by social groups of the content of artifacts lead by means of different chains of problems and solutions to different further developments—involves the content of the artifact itself. Our earlier example of the development of the safety bicycle is of this kind. Another example is variations within the high-wheeler. The high-wheeler's meaning as a virile, high-speed bicycle led to the development of larger front wheels—for with a fixed angular velocity one way of getting a higher translational velocity over the ground was by enlarging the radius. One of the last bicycles resulting from this strand of development was the Rudge Ordinary of 1892, which had a 56-inch wheel and air tire. But groups of women and of elderly men gave quite another meaning to the high-wheeler. For them, its most important characteristic was its lack of safety:

Owing to the disparity in wheel diameters and the small weight of the backbone and trailing wheel, also to the rider's position practically over the centre of the wheel, if the large front wheel hit a brick or large stone on the road, and the rider was unprepared, the sudden check to the wheel usually threw him over the handle-bar. For this reason the machine was regarded as dangerous, and however enthusiastic one may have been about the ordinary—and I was an enthusiastic rider of it once—there is no denying that it was only possible for comparatively young and athletic men. (Grew 1921, p. 8)

This meaning gave rise to lowering the front wheel, moving back the saddle, and giving the front fork a less upright position. Via another chain of problems and solutions (see figure 7), this resulted in artifacts such as Lawson's Bicyclette (1879) and the Xtraordinary (1878; figure 15). Thus there was not *one* high-wheeler; there was the *macho* machine, leading to new designs of bicycles with even higher front wheels, and there was the *unsafe* machine, leading to new designs of bicycle with lower front wheels, saddles moved backward, or reversed order of small and high wheel. Thus the interpretative flexibility of the artifact Penny Farthing is materialized in quite different design lines.

Figure 15
Singer Xtraordinary bicycle (1878). Photograph courtesy of the Trustees of the
Science Museum, London.

Closure and Stabilization

The second stage of the EPOR concerns the mapping of mechanisms for
the closure of debate—or, in SCOT, for the stabilization of an artifact. We
now illustrate what we mean by a closure mechanism by giving examples
of two types that seem to have played a role in cases with which we are
familiar. We refer to the particular mechanisms on which we focus as rhetorical closure and closure by redefinition of problem.

Rhetorical Closure

Closure in technology involves the stabilization of an artifact and the
"disappearance" of problems. To close a technological "controversy,"
one need not *solve* the problems in the common sense of that word.
The key point is whether the relevant social groups *see* the problem
as being solved. In technology, advertising can play an important role
in shaping the meaning that a social group gives to an artifact.[35] Thus,
for instance, an attempt was made to "close" the "safety controversy"

Figure 16
Geared Facile bicycle (1888). Photograph courtesy of the Trustees of the Science Museum, London.

around the high-wheeler by simply claiming that the artifact was perfectly safe. An advertisement for the "Facile" (*sic*!) Bicycle (figure 16) reads:

Bicyclists! Why risk your limbs and lives on high Machines when for road work a 40 inch or 42 inch "Facile" gives all the advantages of the other, together with almost absolute safety. (*Illustrated London News,* 1880; cited in Woodforde 1970, p. 60)

This claim of "almost absolute safety" was a rhetorical move, considering the height of the bicycle and the forward position of the rider, which were well known to engineers at the time to present problems of safety.

Closure by Redefinition of the Problem
We have already mentioned the controversy around the air tire. For most of the engineers it was a theoretical and practical monstrosity. For the general public, in the beginning it meant an aesthetically awful accessory:

Messenger boys guffawed at the sausage tyre, factory ladies squirmed with merriment, while even sober citizens were sadly moved to mirth at a comicality obviously designed solely to lighten the gloom of their daily routine. (Woodforde 1970, p. 89)

For Dunlop and the other protagonists of the air tire, originally the air tire meant a solution to the vibration problem. However, the group of sporting cyclists riding their high-wheelers did not accept that as a problem at all. Vibration presented a problem only to the (potential) users of the low-wheeled bicycle. Three important social groups were therefore opposed to the air tire. But then the air tire was mounted on a racing bicycle. When, for the first time, the tire was used at the racing track, its entry was hailed with derisive laughter. This was, however, quickly silenced by the high speed achieved, and there was only astonishment left when it outpaced all rivals (Croon 1939). Soon handicappers had to give racing cyclists on high-wheelers a considerable start if riders on air-tire low-wheelers were entered. After a short period no racer of any pretensions troubled to compete on anything else (Grew 1921).

What had happened? With respect to two important groups, the sporting cyclists and the general public, closure had been reached, but not by convincing those two groups of the feasibility of the air tire in its meaning as an antivibration device. One can say, we think, that the meaning of the air tire was translated[36] to constitute a solution to quite another problem: the problem of how to go as fast as possible. And thus, by redefining the key problem with respect to which the artifact should have the meaning of a solution, closure was reached for two of the relevant social groups. How the third group, the engineers, came to accept the air tire is another story and need not be told here. Of course, there is nothing "natural" or logically necessary about this form of closure. It could be argued that speed is not the most important characteristic of the bicycle or that existing cycle races were not appropriate tests of a cycle's "real" speed (after all, the idealized world of the race track may not match everyday road conditions, any more than the Formula-1 racing car bears on the performance requirements of the average family sedan). Still, bicycle races have played an important role in the development of the bicycle, and because racing can be viewed as a specific form of testing, this observation is much in line with Constant's recent plea to pay more attention to testing procedures in studying technology (Constant 1983).

The Wider Context

Finally, we come to the third stage of our research program. The task here in the area of technology would seem to be the same as for science—to relate the content of a technological artifact to the wider sociopolitical milieu. This aspect has not yet been demonstrated for the science case,[37] at least not in contemporaneous sociological studies.[38] However, the SCOT method of describing technological artifacts by focusing on the meanings given to them by relevant social groups seems to suggest a way forward. Obviously, the sociocultural and political situation of a social group shapes its norms and values, which in turn influence the meaning given to an artifact. Because we have shown how different meanings can constitute different lines of development, SCOT's descriptive model seems to offer an operationalization of the relationship between the wider milieu and the actual content of technology. To follow this line of analysis, see Bijker (this volume).

Conclusion

In this chapter we have been concerned with outlining an integrated social constructivist approach to the empirical study of science and technology. We reviewed several relevant bodies of literature and strands of argument. We indicated that the social constructivist approach is a flourishing tradition within the sociology of science and that it shows every promise of wider application. We reviewed the literature on the science-technology relationship and showed that here, too, the social constructivist approach is starting to bear fruit. And we reviewed some of the main traditions in technology studies. We argued that innovation studies and much of the history of technology are unsuitable for our sociological purposes. We discussed some recent work in the sociology of technology and noted encouraging signs that a new wave of social constructivist case studies is beginning to emerge.

We then outlined in more detail the two approaches—one in the sociology of scientific knowledge (EPOR) and one in the field of sociology of technology (SCOT)—on which we base our integrated perspective. Finally, we indicated the similarity of the explanatory goals of the two approaches and illustrated these goals with some examples drawn from technology. In particular, we have seen that the concepts of interpretative flexibility and closure mechanism and the notion of social group can be given empirical reference in the social study of technology.

As we have noted throughout this chapter, the sociology of technology is still underdeveloped, in comparison with the sociology of scientific knowledge. It would be a shame if the advances made in the latter field could not be used to throw light on the study of technology. On the other hand, in our studies of technology it appeared to be fruitful to include several social groups in the analysis, and there are some indications that this method may also bear fruit in studies of science. Thus our integrated approach to the social study of science and technology indicates how the sociology of science and sociology of technology might benefit each other.

But there is another reason, and perhaps an even more important one, to argue for such an integrated approach. And this brings us to a question that some readers might have expected to be dealt with in the first paragraph of this chapter, namely, the question of how to distinguish science from technology. We think that it is rather unfruitful to make such an a priori distinction. Instead, it seems worthwhile to start with commonsense notions of science and technology and to study them in an integrated way, as we have proposed. Whatever interesting differences may exist will gain contrast within such a program. This would constitute another concrete result of the integrated study of the social construction of facts and artifacts.

Notes

This chapter is a shortened and updated version of Pinch and Bijker (1984).

We are grateful to Henk van den Belt, Ernst Homburg, Donald MacKenzie, and Steve Woolgar for comments on an earlier draft of this chapter. We would like to thank the Stiftung Volkswagen, Federal Republic of Germany, the Twente University of Technology, The Netherlands, and UK SSRC (under grant G/00123/0072/1) for financial support.

1. The science technology divorce seems to have resulted not so much from the lack of overall analytical goals within "science studies" but more from the contingent demands of carrying out empirical work in these areas. To give an example, the new sociology of scientific knowledge, which attempts to take into account the actual content of scientific knowledge, can best be carried out by researchers who have some training in the science they study, or at least by those who are familiar with an extensive body of technical literature (indeed, many researchers are ex-natural scientists). Having gained such expertise, the researchers tend to stay within the domain where that expertise can best be deployed. Similarly, R&D studies and innovation studies, in which the analysis centers on the firm and the marketplace, have tended to demand the specialized competence of economists. Such disparate bodies of work do not easily lead to a more integrated conception of science and technology. One notable exception is Ravetz (1971). This is one of the few works of recent science studies in which both science and technology and their differences are explored within a common framework.

2. A comprehensive review can be found in Mulkay and Milič (1980).

3. For a recent review of the sociology of scientific knowledge, see Collins (1983c).

4. For a discussion of the earlier work (largely associated with Robert Merton and his students), see Whitley (1972).

5. For more discussion, see Barnes (1974), Mulkay (1979b), Collins (1983c), and Barnes and Edge (1982). The origins of this approach can be found in Fleck (1935).

6. See, for example, Latour and Woolgar (1979), Knorr-Cetina (1981), Lynch (1985a), and Woolgar (1982).

7. See, for example, Collins (1975), Wynne (1976), Pinch (1977, 1986), Pickering (1984), and the studies by Pickering, Harvey, Collins, Travis, and Pinch in Collins (1981a).

8. Collins and Pinch (1979, 1982).

9. Robbins and Johnston (1976). For a similar analysis of public science controversies, see Gillespie et al. (1979) and McCrea and Markle (1984).

10. Some of the most recent debates can be found in Knorr-Cetina and Mulkay (1983).

11. The *locus classicus* is the study by Hessen (1931).

12. See, for example, de Solla Price (1969), Jevons (1976), and Mayr (1976).

13. See, for example, Schumpeter (1928, 1942), Schmookler (1966, 1972), Freeman (1974, 1977), and Scholz (1976).

14. See, for example, Rosenberg (1982), Nelson and Winter (1977, 1982), and Dosi (1982, 1984). A study that preceded these is Rosenberg and Vincenti (1978).

15. Adapted from Uhlmann (1978), p. 45.

16. For another critique of these linear models, see Kline (1985).

17. Shapin writes that "a proper perspective of the uses of science might reveal that sociology of knowledge and history of technology have more in common than is usually thought" (1980, p. 132). Although we are sympathetic to Shapin's argument, we think the time is now ripe for asking more searching questions of historical studies.

18. Manuals describing resinous materials do mention Bakelite but not with the amount of attention that, retrospectively, we would think to be justified. Professor Max Bottler, for example, devotes only one page to Bakelite in his 228-page book on resins and the resin industry (Bottler 1924). Even when Bottler concentrates in another book on the *synthetic* resinous materials, Bakelite does not receive an indisputable "first place." Only half of the book is devoted to phenol/formaldehyde

condensation products, and roughly half of that part is devoted to Bakelite (Bottler 1919). See also Matthis (1920).

19. For an account of other aspects of Bakelite's success, see Bijker (this volume).

20. See, for example, Constant (1980), Hughes (1983), and Hanieski (1973).

21. See, for example, Noble (1979), Smith (1977), and Lazonick (1979).

22. See, for example, Vincenti (1986).

23. There is an American tradition in the sociology of technology. See, for example, Gilfillan (1935), Ogburn (1945), Ogburn and Meyers Nimkoff (1955), and Westrum (1983). A fairly comprehensive view of the present state of the art in German sociology of technology can be obtained from Jokisch (1982). Several studies in the sociology of technology that attempt to break with the traditional approach can be found in Krohn et al. (1978).

24. Dosi uses the concept of technological trajectory, developed by Nelson and Winter (1977); see also Van den Belt and Rip (this volume). Other approaches to technology based on Kuhn's idea of the community structure of science are mentioned by Bijker (this volume). See also Constant (this volume) and the collection edited by Laudan (1984a).

25. One is reminded of the first blush of Kuhnian studies in the sociology of science. It was hoped that Kuhn's "paradigm" concept might be straightforwardly employed by sociologists in their studies of science. Indeed there were a number of studies in which attempts were made to identify phases in science, such as preparadigmatic, normal, and revolutionary. It soon became apparent, however, that Kuhn's terms were loosely formulated, could be subject to a variety of interpretations, and did not lend themselves to operationalization in any straightforward manner. See, for example, the inconclusive discussion over whether a Kuhnian analysis applies to psychology in Palermo (1973). A notable exception is Barnes's contribution to the discussion of Kuhn's work (Barnes 1982b).

26. For a valuable review of Marxist work in this area, see MacKenzie (1984).

27. For a provisional report of this study, see Bijker et al. (1984). The five artifacts that are studied are Bakelite, fluorescent lighting, the safety bicycle, the Sulzer loom, and the transistor. See also Bijker (this volume).

28. Work that might be classified as falling within the EPOR has been carried out primarily by Collins, Pinch, and Travis at the Science Studies Centre, University of Bath, and by Harvey and Pickering at the Science Studies Unit, University of Edinburgh. See, for example, the references in note 7.

29. See, for example, Bijker and Pinch (1983) and Bijker (1984 and this volume). Studies by Van den Belt (1985), Schot (1985, 1986), Jelsma and Smit (1986), and Elzen (1985, 1986) are also based on SCOT.

30. Constant (1980) used a similar evolutionary approach. Both Constant's model and our model seem to arise out of the work in evolutionary epistemology; see, for example, Toulmin (1972) and Campbell (1974). Elster (1983) gives a review of evolutionary models of technical change. See also Van den Belt and Rip (this volume).

31. It may be useful to state explicitly that we consider bicycles to be as fully fledged a technology as, for example, automobiles or aircraft. It may be helpful for readers from outside notorious cycle countries such as The Netherlands, France, and Great Britain to point out that both the automobile and the aircraft industries are, in a way, descendants from the bicycle industry. Many names occur in the histories of both the bicycle and the autocar: Triumph, Rover, Humber, and Raleigh, to mention but a few (Caunter 1955, 1957). The Wright brothers both sold and manufactured bicycles before they started to build their flying machines mostly made out of bicycle parts (Gibbs-Smith 1960).

32. There is no cookbook recipe for how to identify a social group. Quantitative instruments using citation data may be of some help in certain cases. More research is needed to develop operationalizations of the notion of "relevant social group" for a variety of historical and sociological research sites. See also Law (this volume) on the demarcation of networks and Bijker (this volume).

33. Previously, two concepts have been used that can be understood as two distinctive concepts within the broader idea of stabilization (Bijker et al. 1984). *Reification* was used to denote social existence—existence in the consciousness of the members of a certain social group. *Economic stabilization* was used to indicate the economic existence of an artifact—its having a market. Both concepts are used in a continuous and relative way, thus requiring phrases such as "the *degree* of reification of the high-wheeler is *higher* in the group of young men of means and nerve than in the group of elderly men."

34. The use of the concepts of interpretative flexibility and rhetorical closure in science cases is illustrated by Pinch and Bijker (1984).

35. Advertisements seem to constitute a large and potentially fruitful data source for empirical social studies of technology. The considerations that professional advertising designers give to differences among various "consumer groups" obviously fit our use of different relevant groups. See, for example, Schwartz Cowan (1983) and Bijker (this volume).

36. The concept of translation is fruitfully used in an extended way by Callon (1980b, 1981b, 1986), Callon and Law (1982), and Latour (1983, 1984).

37. A model of such a "stage 3" explanation is offered by Collins (1983a).

38. Historical studies that address the third stage may be a useful guide here. See, for example, MacKenzie (1978), Shapin (1979, 1984), and Shapin and Schaffer (1985).

The Evolution of Large Technological Systems

Thomas P. Hughes

Definition of Technological Systems

Technological systems contain messy, complex, problem-solving components. They are both socially constructed and society shaping.[1]

Among the components in technological systems are physical artifacts, such as the turbogenerators, transformers, and transmission lines in electric light and power systems.[2] Technological systems also include organizations, such as manufacturing firms, utility companies, and investment banks, and they incorporate components usually labeled scientific, such as books, articles, and university teaching and research programs. Legislative artifacts, such as regulatory laws, can also be part of technological systems. Because they are socially constructed and adapted in order to function in systems, natural resources, such as coal mines, also qualify as system artifacts.[3]

An artifact—either physical or nonphysical—functioning as a component in a system interacts with other artifacts, all of which contribute directly or through other components to the common system goal. If a component is removed from a system or if its characteristics change, the other artifacts in the system will alter characteristics accordingly. In an electric light and power system, for instance, a change in resistance, or load, in the system will bring compensatory changes in transmission, distribution, and generation components. If there is repeated evidence that the investment policies of an investment bank are coordinated with the sales activities of an electrical manufacturer, then there is likely to be a systematic interaction between them; the change in policy in one will bring changes in the policy of the other. For instance, investment banks may systematically fund the purchase of the electric power plants of a particular manufacturer with which they share owners and interlocking boards of directors.[4] If courses in an engineering school shift emphasis from the study of direct current (dc) to alternating current (ac) at about the same time as the physical artifacts in power systems are changing from dc to ac, then a systematic relationship also seems likely. The professors teaching the courses may be regular consultants of utilities and electrical manufacturing firms; the alumni of the engineering schools may have become engineers and managers in the firms; and managers and engineers from the firm may sit on the governing boards of the engineering schools.

Because they are invented and developed by system builders and their associates, the components of technological systems are socially constructed artifacts. Persons who build electric light and power systems invent and develop not only generators and transmission lines but also such organizational forms as electrical manufacturing and utility holding companies. Some broadly experienced and gifted system builders can invent hardware as well as organizations, but usually different persons take these responsibilities as a system evolves. One of the primary characteristics of a system builder is the ability to construct or to force unity from diversity, centralization in the face of pluralism, and coherence from chaos. This construction often involves the destruction of alternative systems. System builders in their constructive activity are like "heterogeneous engineers" (Law, this volume).

Because components of a technological system interact, their characteristics derive from the system. For example, the management structure of an electric light and power utility, as suggested by its organizational chart, depends on the character of the functioning hardware, or artifacts, in the system. In turn, management in a technological system often chooses technical components that support the structure, or organizational form, of management.[5] More specifically, the management structure reflects the particular economic mix of power plants in the system, and the layout of the power plant mix is analogous to the management structure. The structure of a firm's technical system also interacts with its business strategy.[6] These analogous structures and strategies make up the technological system and contribute to its style.

Because organizational components, conventionally labeled social, are system-builder creations, or artifacts, in a technological system, the convention of designating social factors as the environment, or context, of a technological system should be avoided. Such implications occur when scholars refer to the social context of technology or to the social background of technological change. A technological system usually has an environment consisting of intractable factors not under the control of the system managers, but these are not all organizational. If a factor in the environment—say, a supply of energy—should come under the control of the system, it is then an interacting part of it. Over time, technological systems manage increasingly to incorporate environment into the system, thereby eliminating sources of uncertainty, such as a once free market. Perhaps the ideal situation for system control is a closed system that does not feel the environment. In a closed system, or in a system without environment, managers could resort to bureaucracy, routinization, and deskilling to eliminate uncertainty—and freedom. Prediction by extrapolation, a characteristic of system managers, then becomes less fanciful.

Two kinds of environment relate to open technological systems: ones on which they are dependent and ones dependent on them. In neither case is there interaction between the system and the environment; there is simply a one-way influence. Because they are not under system control, environmental factors affecting the system should not be mistaken for components of the system. Because they do not interact with the system, environmental factors dependent on the system should not be seen as part of it either. The supply of fossil fuel is often an environmental factor on which an electric light and power system is dependent. A utility company fully owned by an electrical manufacturer is part of a dependent environment if it has no influence over the policies of the manufacturer but must accept its products. On the other hand, ownership is no sure indicator of dependence, for the manufacturer could design its products in conjunction with the utility.[7] In this case the owned utility is an interacting component in the system.

Technological systems solve problems or fulfill goals using whatever means are available and appropriate; the problems have to do mostly with reordering the physical world in ways considered useful or desirable, at least by those designing or employing a technological system. A problem to be solved, however, may postdate the emergence of the system as a solution. For instance, electrical utilities through advertising and other marketing tactics stimulated the need for home appliances that would use electricity during hours when demand was low. This partial definition of technology as problem-solving systems does not exclude problem solving in art, architecture, medicine, or even play, but the definition can be focused and clarified by further qualification: It is problem solving usually concerned with the reordering of the material world to make it more productive of goods and services. Martin Heidegger defines technology as an ordering of the world to make it available as a "standing reserve" poised for problem solving and, therefore, as the means to an end. This challenging of man to order the world and in so doing to reveal its essence is called enframing (Heidegger 1977, p. 19).

Technological systems are bounded by the limits of control exercised by artifactual and human operators. In the case of an electric light and power system, a load-dispatching center with its communication and control artifacts and human load dispatchers is the principal control center for power plants and for transmission and distribution lines in the system. The load-dispatching center is, however, part of a hierarchical control system involving the management structure of the utility. That structure may itself be subject to the control of a holding company that incorporates other utilities, banks, manufacturers, and even regulatory agencies. An electric utility may be interconnected with other utilities to form a regional, centrally controlled electric light and power system. Regional power systems sometimes integrate physically and organizationally with coal-mining enterprises and even with manufacturing enterprises that use the power and light. This was common in the Ruhr region in the years between World War I and World War II. Systems nestle hierarchically like a Russian Easter egg into a pattern of systems and subsystems.

Inventors, industrial scientists, engineers, managers, financiers, and workers are components of but not artifacts in the system. Not created by the system builders, individuals and groups in systems have degrees of freedom not possessed by artifacts. Modern system builders, however, have tended to bureaucratize, deskill, and routinize in order to minimize the voluntary role of workers and administrative personnel in a system. Early in this century, Frederick W. Taylor's scientific-management program organized labor as if it were an inanimate component in production systems. More recently, some system builders have designed systems that provide labor with an opportunity to define the labor component of a system. The voluntary action does not come to labor as it functions in the system but as it designs its functions. A crucial function of people in technological systems, besides their obvious role in inventing, designing, and developing systems, is to complete the feedback loop between system performance and system goal and in so doing to correct errors in system performance. The degree of freedom exercised by people in a system, in contrast to routine performance, depends on the maturity and size, or the autonomy, of a technological system, as will be shown. Old systems like old people tend to become less adaptable, but systems do not simply grow frail and fade away. Large systems with high momentum tend to exert a soft determinism on other systems, groups, and individuals in society.

Inventors, organizers, and managers of technological systems mostly prefer hierarchy, so the systems over time tend toward a hierarchical structure. Thus the definer and describer of a system should delimit the level of analysis, or subsystem, of interest (Constant, this volume). For instance, interacting physical artifacts can be designated a system, or physical artifacts plus interacting organizations can be so designated. The turbogenerators in an electric power system can be seen as systems with components such as turbines and generators. These artifacts can, in turn, be analyzed as systems with components. Therefore the analyzers of systems should make clear, or at least be clear in their own minds, that the system of interest may be a subsystem as well as one encompassing its own subsystems. In a large technological system there are countless opportunities for isolating subsystems and calling them systems for purposes of comprehensibility and analysis. In so doing, however, one rends the fabric of reality and may offer only a partial, or even distorted, analysis of system behavior.

The definer or describer of a hierarchical system's choice of the level of analysis from physical artifact to world system can be noticeably political. For instance, an electric light and power system can be so defined that externalities or social costs are excluded from the analysis. Textbooks for engineering students often limit technological systems to technical components, thereby leaving the student with the mistaken impression that problems of system growth and management are neatly circumscribed and preclude factors often pejoratively labeled "politics." On the other hand, neoclassical economists dealing with production systems often treat technical factors as exogenous. Some social scientists raise the level of analysis and abstraction so high that it does not matter what the technical content of a system might be.

A technological system has inputs and outputs. Often these can be subsumed under a general heading. For instance, an electric light and power system has heat or mechanical energy as its primary input and electrical energy as its output. Within the system the subsystems are linked by internal inputs and outputs, or what engineers call interfaces. An electrical-manufacturing concern in the system may take electrical energy from the utility in the system and supply generating equipment to the utility. The manufacturing concern may also take income from the profits of the utility and from sale of equipment to the utility and then reinvest in the utility. Both may exchange information about equipment performance for purposes of design and operation. An investment bank may take profits from its investments in a manufacturing company and a utility and then also invest in these enterprises. Financial and technical information about light and power systems is also interchanged. In the examples given, one assumes interlocking boards of directors and management and control.

Pattern of Evolution

Large, modern technological systems seem to evolve in accordance with a loosely defined pattern. The histories of a number of systems, especially the history of electric light and power between 1870 and 1940, display the pattern described in this chapter. The sample is not large enough, however, to allow essentially quantitative statements, such as "most" or "the majority," to be made. Relevant examples from the history of modern technological systems, many from electric light and power, support or illustrate my arguments. I also use a number of interrelated concepts to describe the pattern of evolution. The concept of reverse salient, for instance, can be appreciated only if it is related to the concept of system used in this chapter. The concept of technological style should be related to the concept of technology transfer. The term "pattern" is preferable to "model" because a pattern is a metaphor suggesting looseness and a tendency to become unraveled.

The pattern suggested pertains to systems that evolve and expand, as so many systems originating in the late nineteenth century did. With the increased complexity of systems, the number of components and the problems of control increased. Intense problems of control have been called crises of control (Beniger 1984). Large-scale computers became a partial answer. An explanation of the tendency of systems to expand is offered here. The study of systems contracting, as countless have through history, would by comparison and contrast help explain growth. Historians of systems need among their number not only Charles Darwins but also Edward Gibbons.

The history of evolving, or expanding, systems can be presented in the phases in which the activity named predominates: invention, development, innovation, transfer, and growth, competition, and consolidation. As systems mature, they acquire style and momentum. In this chapter style is discussed in conjunction with transfer, and momentum is discussed after the section on growth, competition, and consolidation. The phases in the history of a technological system are not simply sequential; they overlap and backtrack. After invention, development, and innovation, there is more invention. Transfer may not necessarily come immediately after innovation but can occur at other times in the history of a system as well. Once again, it should be stressed that invention, development, innovation, transfer, and growth, competition, and consolidation can and do occur throughout the history of a system but not necessarily in that order. The thesis here is that a pattern is discernible because of one or several of these activities predominating during the sequence of phases suggested.

The phases can be further ordered according to the kind of system builder who is most active as a maker of critical decisions.[8] During invention and development inventor-entrepreneurs solve critical problems; during innovation, competition, and growth manager-entrepreneurs make crucial decisions; and during consolidation and rationalization financier-entrepreneurs and consulting engineers, especially those with political influence, often solve the critical problems associated with growth and momentum. Depending on the degree of adaptation to new circumstances needed, either inventor-entrepreneurs or manager-entrepreneurs may prevail during transfer. Because their tasks demand the attributes of a generalist dedicated to change rather than the attributes of a specialist, the term "entrepreneur" is used to describe system builders. Edison provides a prime example of an inventor-entrepreneur. Besides inventing systematically, he solved managerial and financial problems to bring his invention into use. His heart, however, at least as a young inventor, lay with invention. Elmer Sperry, a more professional and dedicated inventor than Edison but also an entrepreneur, saw management and finance as the necessary but boring means to bring his beloved inventions into use (Hughes 1971, pp. 41, 52–53).

Invention

Holding companies, power plants, and light bulbs—all are inventions. Inventors, managers, and financiers are a few of the inventors of system components. Inventions occur during the inventive phase of a system and during other phases. Inventions can be conservative or radical. Those occurring during the invention phase are radical because they inaugurate a new system; conservative inventions predominate during the phase of competition and system growth, for they improve or expand existing systems. Because radical inventions do not contribute to the growth of existing technological systems, which are presided over by, systematically linked to, and financially supported by larger entities, organizations rarely nurture a radical invention. It should be stressed that the term "radical" is not used here in a commonplace way to suggest momentous social effects. Radical inventions do not necessarily have more social effects than conservative ones, but, as defined here, they are inventions that do not become components in existing systems.

Independent professional inventors conceived of a disproportionate number of the radical inventions during the late nineteenth and early twentieth centuries (Jewkes et al. 1969, pp. 79–103). Many of their inventions inaugurated major technological systems that only later came under the nurturing care of large organizations; they then stabilized and acquired momentum. Outstanding examples of independent inventors and their radical inventions that sowed the seeds of large systems that were presided over by new organizations are Bell and the telephone, Edison and the electric light and power system, Charles Parsons and Karl Gustaf Patrik de Laval and the steam turbine, the Wright brothers and the airplane, Marconi and the wireless, H. Anschütz-Kaempfe and Elmer Sperry and the gyrocompass guidance and control system, Ferdinand von Zeppelin and the dirigible, and Frank Whittle and the jet engine.[9] Even though tradition assigns the inventions listed to these independent inventors, it should be stressed that other inventors, many of them independents, also contributed substantially to the inauguration of the new systems. For instance, the German Friedrich Haselwander, the American C. S. Bradley, and the Swede Jonas Wenström took out patents on polyphase systems at about the same time as Tesla; and Joseph Swan, the British inventor, should share credit with Edison for the invention of a durable incandescent filament lamp, if not for the incandescent lamp system.

Even though radical inventions inaugurate new systems, they are often improvements over earlier, similar inventions that failed to develop into innovations. Historians have a rich research site among the remains of these failed inventions. Elmer Sperry, who contributed to the establishment of several major technological systems, insisted that all his inventions, including the radical ones, were improvements on the earlier work of others (Sperry 1930, p. 63). The intense patent searches done by independents reinforce his point.

The terms "independent" and "professional" give needed complexity to the concept of inventor. Free from the constraints of organizations, such as industrial or government research laboratories, independent inventors can roam widely to choose problems to which they hope to find solutions in the form of inventions. Independent inventors often have their own research facilities or laboratories, but these are not harnessed to existing systems, as is usually the case with government and industrial research laboratories. Not all independent inventors are "professional"; professional inventors support their inventive activities over an extended period by a series of commercially successful inventions. They are not salaried employees, although they might take consulting fees. Many independents who were not professionals, such as Alexander Graham Bell, gained immense income from several major inventions and then chose to live, or enjoy, life other than as inventors. Elmer Sperry, Elihu Thomson, Edward Weston, Thomas Edison, and Nikola Tesla are outstanding examples of inventors who persisted as professionals for an extended period during the late nineteenth and early twentieth centuries.

The independents who flourished in the late nineteenth and early twentieth centuries tended to concentrate on radical inventions for reasons both obvious and obscure. As noted, they were not constrained in their problem choices by mission-oriented organizations with high inertia. They prudently avoided choosing problems that would also be chosen by teams of researchers and developers working in company engineering departments or industrial research laboratories. Psychologically they had an outsider's mentality; they also sought the thrill of a major technological transformation. They often achieved dramatic breakthroughs, not incremental improvements. Elmer Sperry, the independent inventor, said: "If I spend a life-time on a dynamo I can probably make my little contribution toward increasing the efficiency of that machine six or seven percent. Now then, there are a whole lot of arts that need electricity, about four or five hundred per cent, let me tackle one of those" (Sperry 1930, p. 63). To achieve these breakthroughs, the independents had the insight to distance themselves from large organizations. They rightly sensed that the large organization vested in existing technology rarely nurtured inventions that by their nature contributed nothing to the momentum of the organization and even challenged the status quo in the technological world of which the organization was a leading member. Radical inventions often deskill workers, engineers, and managers, wipe out financial investments, and generally stimulate anxiety in large organizations. Large organizations sometimes reject the inventive proposals of the radicals as technically crude and economically risky, but in so doing they are simply acknowledging the character of the new and radical.

In the 1920s several of the world's major oil companies rejected the proposals made by the French inventor Eugene Joules Houdry for a radically different way of refining gasoline with catalytic agents. The engineering staffs of the established companies justified their rejections by citing the lack of refined engineering detail and the engineering problems not solved in the process as then developed by Houdry. Apparently they did not take into account that this was indeed a characteristic common among radical inventions in the development phase. After development in the 1930s by Sun Oil Company, an innovative, relatively small independent US refiner, the Houdry process brought substantially increased yields of the gasoline fraction from a given amount of crude oil and became the envy of, and model for, the petroleum industry (Enos 1962, pp. 137, 140–141).

Independent inventors such as Houdry have more freedom but consequently more difficulty in identifying problems than inventors and scientists working in large-company engineering departments or industrial research laboratories. On several notable occasions academics stimulated the problem choices of independent inventors who flourished in the late nineteenth and early twentieth centuries. Charles Hall heard his professor of science say that the world awaited the inventor who could find a practical means of smelting aluminum; a professor at the Polytechnic in Graz, Austria, stimulated Nikola Tesla to embark on the search that culminated in his polyphase electrical system (Hughes 1983, p. 113); Professor Carl von Linde of the Munich Polytechnic defined a problem for his student Rudolf Diesel that eventually resulted in Diesel's engine (Diesel 1953, p. 97); and physics professor William A. Anthony of Cornell University outlined several problems for young Elmer Sperry that climaxed in his first major patents.[10] Perhaps the academics' imaginations ranged freely because they, like independent inventors, were not tied to industry but at the same time were broadly acquainted with technical and scientific literature.

Inventors do publish, despite widespread opinion to the contrary. They publish patents, and they often publish descriptions of their patented inventions in technical journals. The technical articles, sometimes authored by the inventors, sometimes by cooperating technical journalists, brought not only recognition but also publicity of commercial value. Whether patent or article, the publication informed the inventive community about the location of inventive activity. This alerted the community about problems that needed attention, for rarely was a patent or invention the ultimate solution to a problem, and experienced inventors realized that a basic problem could be solved in a variety of patentable ways, including their own. So, by keeping abreast of patents and publications, inventors could identify problem areas. This helps explain why patents tend over a period of several years to cluster around problem sites.

Professional inventors have other reasons for their problem choices. In avoiding problems on which engineering departments and industrial research laboratories were working, independents narrowed their problem choice. The challenge of sweet problems that have foiled numerous others often stimulates the independents' problem choices. They believe their special gifts will bring success where others have failed. Not strongly motivated by a defined need, they exhibit an elementary joy in problem solving as an end in itself. Alexander Graham Bell, a professor of elocution and an authority on deafness. seeing the analogy between acoustic and electrical phenomena, pursued the goal of a speaking telegraph despite the advice of friends and advisers who urged him to continue to concentrate on the problem of multiplexing wire telegraphy, a conservative telegraph-industry-defined problem. Another independent, Elisha Gray, who was also working on multiplexing and who also saw the possibility of a speaking telegraph, made the conservative decision and concentrated on multiplexing (Hounshell 1975).

The independent professionals had not only freedom of problem choice but also the less desirable freedom from the burden of organizational financial support. Their response has been ingenious. At the turn of the century they often traded intellectual property for money. In an era before a patent became essentially a license to litigate and before the large companies amassed the resources to involve an independent in litigation to the point of financial exhaustion, independent professionals transformed their ideas into property in the form of patents. Having done this, they sold their intellectual property to persons with other forms of property, especially money. Sometimes the inventor and the financier would each deposit so many patents and so much cash and divide the stock of a new company founded to exploit the patent. In democratic America the ability of a self-made inventor to match wits with the presumedly ill-gotten gains of financiers was believed wonderfully meritocratic.

As the armaments race, especially the naval one, increased in intensity before World War I, inventors turned to the government for development funds. These came as contracts to supply airplanes, wireless, gunfire control, and other high technology artifacts of the day. Governments contracted for a few models that were in essence experimental designs. With income from these contracts the inventors invested in further development. In order to contract with the armed services, many of the inventors allied with financiers to form small companies. The possibility existed that the company would flourish, and then the inventor would be harnessed to a burden of his own making; but many of the companies collapsed, leaving the inventor to savor independence again. The independents also raised funds by setting up as consultants or by organizing small research and development companies that would develop their own and others' inventions. Perhaps the ideal of funding and freedom came when the inventor had licensed sufficient patents over the years to bring a steadily mounting income that could be reinvested in invention. The investment was often in workshop, laboratory facilities, and staff, for contrary to myth independent inventors were not necessarily "lone" inventors.

An aspect of radical invention less understood than problem choice and funding lies at the heart of the matter: the times of inspiration or "Eureka!" moments. There exists a helpful body of literature on the psychology of invention and discovery, but it lacks richly supported and explored case histories of invention.[11] The inventors themselves have rarely verbalized their moments of inspiration. Some promising but unexplored leads to follow exist, however. Frequently, inventors speak of their inventions in terms of metaphor or analogy. An analogy is an invention that carries its creator from the known to the unknown. Inventors often develop a particular mechanism or process that they then formulate as an abstract concept, probably visual, that subsequently becomes a generalized solution. So prepared, the inventor becomes a solution looking for a problem. These clues, however, only tantalize. Historians and sociologists of technology should join psychologists in exploring the act of creation.[12]

Development

Radical inventions, if successfully developed, culminate in technological systems. One inventor may be responsible for most or all of the inventions that become the immediate cause of a technological system; the same inventor may preside over the development of the inventions until they result in an innovation, or a new technological system in use. If one inventor proves responsible for most of the radical inventions and the development of these, then he or she fully deserves the designation inventor-entrepreneur.

Development is the phase in which the social construction of technology becomes clear. During the transformation of the invention into an innovation, inventor-entrepreneurs and their associates embody in their invention economic, political, and social characteristics that it needs for survival in the use world. The invention changes from a relatively simple idea that can function in an environment no more complex than can be constituted in the mind of the inventors to a system that can function in an environment permeated by various factors and forces. In order to do this, the inventor-entrepreneur constructs experimental, or test, environments that become successively more complex and more like the use world that the system will encounter on innovation. Elmer Sperry, for instance, having written, or having had written for him, the equations of his concept of a gyro ship stabilizer gave the concept material form in a model of a rolling ship consisting of a simple pendulum and a laboratory gyroscope. In the next step he redesigned the invention, making it more complex, and experimented with it in an environment incorporating more ship and sea variables than the simple pendulum could provide. In time the model reached a level of complexity that in Sperry's opinion allowed it to accommodate to use-world variables. He tested the ship stabilizer on a destroyer provided by the US Navy. The testing of inventions as mathematical formulas and as models stripped down to scientific abstractions permits small investments and small failures before the costly venture of full-scale trial and ultimate use is attempted.

There are countless examples of independent inventor-entrepreneurs providing their inventions with the economic, political, and other characteristics needed for survival. Edison's awareness of the price of gaslight deeply influenced his design of a competitive electric light system. In the early 1880s in England, Lucien Gaulard and John Gibbs invented a transformer with physical characteristics that allowed the transformer's output voltage to be varied as required by the Electric Lighting Act of 1882 (Hughes 1983, pp. 34–38, 89–90). The Wright brothers carefully took into account the psychology and physiology of the pilots who would have to maintain the stability of their flyer. According to David Noble, digital machine tool systems have built into them the interests of the managerial class (Noble 1979).

Because new problems arise as the system is endowed with various characteristics, radical inventor-entrepreneurs continue to invent during the development period. Because problems arise out of the systematic relationship of the system components being invented, the choice of problems during the development process becomes easier. If, for instance, during development the inventor varies the characteristics of one component, then the other interrelated components' characteristics usually have to be varied accordingly. This harmonizing of component characteristics during development often results in patentable inventions. An entire family of patents sometimes accompanies the development of a complex system.

A large organization inventing and developing a system may assign subprojects and problems to different types of professionals. When the Westinghouse Corporation developed Tesla's polyphase electric power transmission system, it used him as a consultant, but ultimately a talented group of Westinghouse engineers brought the system into use (Passer 1953, pp. 276–282). Physicists, especially academic ones, have sometimes proven more adept at invention than engineers, who often display a preference and a capability for development. Until World War II academic physicists were relatively free of organizational constraints, and during World War II this frame of mind survived, even in such large projects as the Radiation Laboratory in Cambridge, Massachusetts, the Manhattan Project laboratory in Chicago under Arthur Compton, and the Los Alamos laboratory under Robert Oppenheimer. Since the end of the nineteenth century, engineers have been associated with large industrial corporations, or, in the case of academic engineers, they have tended to look to the industrial sector for definition of research problems (Noble 1977, pp. 33–49).

The relationships between engineers and scientists and between technology and science have long held the attention of historians, especially historians of science. From the systems point of view the distinctions tend to fade. There are countless cases of persons formally trained in science and willing to have their methods labeled scientific immersing themselves fully in invention and development of technology.[13] Engineers and inventors formally trained in courses of study called science have not hesitated to use the knowledge and methods acquired. Persons committed emotionally and intellectually to problem solving associated with system creation and development rarely take note of disciplinary boundaries, unless bureaucracy has taken command.

Innovation

Innovation clearly reveals technologically complex systems. The inventor-entrepreneur, along with the associated engineers, industrial scientists, and other inventors who help to bring the product into use, often combines the invented and developed physical components into a complex system consisting of manufacturing, sales, and service facilities. On the other hand, rather than establishing a new company, the inventor-entrepreneur sometimes provides specifications enabling established firms to manufacture the product or provide the service. Many of the independent professionals of the late nineteenth and early twentieth centuries, however, founded their own manufacturing, sales, and service facilities because, in the case of radical inventions, established manufacturers were often reluctant to provide the new machines, processes, and organizations needed for manufacture. Independent inventor-entrepreneurs chose to engage in manufacture because they wanted to introduce a manufacturing process systematically related to the invention. They often invented and developed the coordinated manufacturing process as well as the product. If, on the other hand, the invention was a conservative one, in essence, an improvement in an ongoing system, the manufacturer presiding over this system would often be interested in manufacturing the invention.

George Eastman, for instance, concentrated on the invention and development of machinery for the photography devices invented by him and his partner William Hall Walker. Eastman, while developing a dry-plate system, obtained a patent in 1880 for a machine that continuously coated glass plates with gelatin emulsion. With Walker, Eastman then turned to the invention of a photographic film and a roll holder system to replace the one using glass plates. Later, Eastman concentrated on the design of production machinery while Walker directed his attention to the invention and development of cameras. In the fall of 1884 the two had developed, along with the holder mechanism and the film, the production machinery. Eastman also dedicated his inventive talents to production machinery in the development of the Kodak system of amateur photography (Jenkins 1975).

Edison also provides a classic example of the inventor-entrepreneur presiding over the introduction of a complex system of production and utilization. Edison had the assistance of other inventors, managers, and financiers who were associated with him, but he more than any other individual presided over the intricate enterprise. The organizational chart of 1882 of Edison-founded companies outlines the complex technological system. Among the Edison companies were The Edison Electric Light Company, formed to finance Edison's invention, patenting, and development of the electric-lighting system and the licensing of it; The Edison Electric Illuminating Company of New York, the first of the Edison urban lighting utilities; The Edison Machine Works, founded to manufacture the dynamos covered by Edison's patents; The (Edison) Electric Tube Company, established by Edison to manufacture the underground conductors for his system; and the Edison Lamp Works (Jones 1940, p. 41). When Edison embarked on the invention of an incandescent lighting system, he could hardly have anticipated the complexity of the ultimate Edison enterprise.

System builders, such as Eastmen and Edison, strive to increase the size of the system under their control and to reduce the size of the environment that is not. In the case of the Edison system at the time of the innovation, the utilities, the principal users of the equipment patented by The Edison Electric Light Company and manufactured by the mix of Edison companies, were being incorporated into the system. The same group of investors who owned the patent-holding company owned The Edison

Electric Illuminating Company of New York, the first of the Edison urban utilities. The owners of the Edison companies accepted stock from other utilities in exchange for equipment, thereby building up an Edison empire of urban utilities variously owned and controlled. Similar policies were followed later by the large manufacturers in Germany. The manufacturers' absorption of supply and demand organizations tended to eliminate the outside/inside dichotomy of systems, a dichotomy avoided by Michel Callon in his analysis of actor networks (Callon, this volume).

Once innovation occurs, inventor-entrepreneurs tend to fade from the focal point of activity. Some may remain with a successful company formed on the basis of their patents, but usually they do not become the manager-entrepreneurs of the enterprise. Elihu Thomson (1853–1937), a prolific and important American inventor who acquired 696 patents over five decades, became head of research for the Thomson-Houston Company, an electrical manufacturer founded on the basis of his patents. Afterward he served as principal researcher and inventor for the General Electric Company, formed in 1892 by a merger of Thomson-Houston and The Edison General Electric Company. Thomson's point of view remained that of an inventor, and the contrasts between it and the views of the manager-entrepreneurs taking over the General Electric Company became clear. Diplomatic negotiations on the part of managers such as Charles A. Coffin, early head of GE, reconciled the laboratory with the front office (Carlson 1983). The manager-entrepreneur, after innovation, gradually displaced the inventor as the responder to the principal reverse salients and the solver of critical problems associated with them.

Technology Transfer

The transfer of technology can occur at any time during the history of a technological system. Transfer immediately after innovation probably most clearly reveals interesting aspects of transfer, for the technological system is not laden with the additional complexities that accrue with age and momentum. Because a system usually has embodied in it characteristics suiting it for survival in a particular time and place, manifold difficulties often arise in transfer at another time or to a different environment. Because a system usually needs adaptation to the characteristics of a different time or place, the concepts of transfer and adaptation are linked. Besides adaptation, historians analyzing transfer have stressed the modes of transfer.[14]

Aspects of adaptation can be shown by episodes drawn from the early history of the transformer. As noted, Lucien Gaulard and John Gibbs introduced a transformer with characteristics that suited it to British electric fighting legislation. They organized several test and permanent installations of their transformer in the early 1880s. In 1884 Otto Titus Blàthy and Charles Zipernowski, two experienced engineers from Ganz and Company, the preeminent Hungarian electrical manufacturer, saw the transformer on exhibit in Turin, Italy. They redesigned it for a Ganz system and for Hungarian conditions, under which electrical legislation did not require the complex characteristics embodied in the Gaulard and Gibbs device. The resulting transformer has been designated the world's first practical and commercial transformer (Halacsy and Von Fuchs 1961, p. 121). But such a designation is misleading because the transformer was practical for Hungary, not for the world. In the United States the Westinghouse Company also learned of the Gaulard-Gibbs transformer, acquired the rights to the patent, and had it adapted to American conditions. Westinghouse employed William Stanley, an independent inventor, to develop a transformer system of transmission on the basis of the Gaulard-Gibbs device. Subsequently, the engineering staff at Westinghouse gave the system an American style by presuming a large market and adapting the transformer and the processes for manufacturing it for mass production (Hughes 1983, pp. 98–105).

The case of the Gaulard-Gibbs transformer reveals legislation and market as critical factors in transfer and adaptation, but there are other factors involved, including geographical and social ones (Lindqvist 1984, pp. 291–307). The Gaulard and Gibbs case involves a physical object being transferred and adapted; when a technological system is transferred, organizational components are as well. There are numerous cases of the transfer, successful and unsuccessful, of companies as well as of product so whether the agent of transfer is an inventor, an engineer, a manager, or some other professional depends on the components being transferred and the phase of development of the technological system.

Technological Style
Exploration of the theme of technology transfer leads easily to the question of style, for adaptation is a response to different environments and adaptation to environment culminates in style. Architectural and art historians have long used the concept of style. When Heinrich Wölfflin in 1915 wrote about the problem of the development of style in art, he did not hesitate to attribute style in art and architecture to individual and national character. The concept of style can, on the other hand, be developed without reference to national and racial character, or to *Zeitgeist*. Historians of art

and architecture now use the concept of style warily, for "style is like a rainbow. . . . We can see it only briefly while we pause between the sun and the rain, and it vanishes when we go to the place where we thought we saw it" (Kubler 1962, p. 129).

Historians and sociologists of technology can, however, use the notion of style to advantage, for, unlike historians of art, they are not burdened by long-established and rigid concepts of style, such as those of the High Renaissance and the Baroque that can obfuscate perceptive differentiation. Historians and sociologists can use style to suggest that system builders, like artists and architects, have creative latitude. Furthermore, the concept of style accords with that of social construction of technology. There is no one best way to paint the Virgin; nor is there one best way to build a dynamo. Inexperienced engineers and laymen err in assuming that there is an ideal dynamo toward which the design community Whiggishly gropes. Technology should be appropriate for time and place; this does not necessarily mean that it be small and beautiful.[15]

Factors shaping style are numerous and diverse. After the traumatic Bolshevik Revolution of 1917 and during the shaky beginnings of the new state, the Soviets needed the largest and the fastest technology, not for economic reasons but in order to gain prestige for the regime (Bailes 1976). After comparing the gyrocompass he invented with German ones, Elmer Sperry decided that his was more practical because the Germans pursued abstract standards of performance, not functional requirements. His observation was a comment on style. Charles Merz, the British consulting engineer who designed regional power systems throughout the world, said in 1909 that "the problem of power supply in any district is . . . completely governed by local conditions" (Merz 1908, p. 4).

The concept of style applied to technology counters the false notion that technology is simply applied science and economics, a doctrine taught only a decade or so ago in engineering schools. Ohm's and Joule's laws and factor inputs and unit costs are not sufficient explanation for the shape of technology. The concepts of both the social shaping of technology and technological style help the historian and the sociologist, and perhaps the practitioner, to avoid reductionist analyses of technology.

The concept of style also facilitates the writing of comparative history. The historian can search for an explanation for the different characteristics of a particular technology, such as electric power, in different regions. The problem becomes especially interesting in this century when international pools of technology are available to the designers of regional technology because of the international circulation of patents, internationally circulated technical and scientific literature, international trade in technical goods and services, the migration of experts, technology transfer agreements, and other modes of exchange of knowledge and artifacts. Having noted the existence of an international pool of technology and having acknowledged that engineering science allows laws to be stated and equations to be written that describe an ideal, or highly abstract, electrical system made up of electromotive forces, resistances, capacitors, and inductances that are internationally valid and timeless, we come upon the fascinating problem: Why do electric light and power systems differ in characteristics from time to time, from region to region, and even from nation to nation?

There are countless examples in this century of variations in technological style. A 1920 map of electricity supply in London, Paris, Berlin, and Chicago reveals remarkable variation from city to city in the size, number, and location of the power plants (Hughes 1983, p. 16). The striking variation is not the amount of light and power generated (the output in quantitative terms) but the way in which it is generated, transmitted, and distributed. (Focusing on the quantitative, the economic historian often misses variations in style.) Berlin possessed about a half dozen large power plants, whereas London had more than fifty small ones. The London style of numerous small plants and the Berlin style of several large ones persisted for decades. London, it must be stressed, was not technically backward. In the London and Berlin regulatory legislation that expressed fundamental political values rests the principal explanation for the contrasting styles. The Londoners were protecting the traditional power of local government by giving municipal boroughs authority to regulate electric light and power and the Berliners were enhancing centralized authority by delegating regulatory power to the City of Berlin (Hughes 1983, pp. 175–200, 227–261).

Natural geography, another factor, also shapes technological style. Because regions as traditionally defined are essentially geographical and because geography so deeply influences technology, the concept of regional technological style can be more easily identified than national style. When regulatory legislation applies on a national level, however, regional styles tend to merge into national ones. Before 1926 and the National Grid in Great Britain, for example, there were distinctive regional styles of power systems—London in contrast to the northeast coast; but the grid brought a more national style as legislation prevailed over other style-inducing factors.

Regional and national historical experiences also shape technological style. During World War I a copper shortage in Germany caused power plant designers to install larger and fewer generators to save copper. This learning experience, or acquired design style, persisted after the war, even though the critical shortage had passed. After World War I the Treaty of Versailles deprived Germany of hard-coal-producing areas and demanded the export of hard coal as reparations, so the electric power system builders turned increasingly to soft coal, a characteristic that also persisted after the techniques were learned. Only history can satisfactorily explain the regional style of Ruhr and Cologne area power plants with their post–World War I dependence on lignite and large generating units (Hughes 1983, pp. 413–414).

Technological style is a concept applicable to technologies other than electric light and power and useful to professionals other than historians. Louis Hunter pointed out fascinating contrasts between Hudson River and Mississippi River steamboats (Hunter 1949). Eda Kranakis has written about the French "academic style" of engineering (Kranakis 1982, pp. 8–9), and Edwin Layton has contrasted the US and the French approaches to water-turbine design in the nineteenth century (Layton 1978). In the 1950s the American public became familiar with contrasting American and European styles of automobiles and even with Soviet and US space vehicles of contrasting designs.[16] Recently, Mary Kaldor identified a Baroque style of military technology—in the twentieth century (Kaldor 1981). Aware of the richness and complexity of the concept of style and the possibility of using it to counter reductionist approaches to engineering design, Hans Dieter Hellige has urged the introduction of style into the education of engineers (Hellige 1984, pp. 281–283).

Growth, Competition, and Consolidation

Historians of technology describe the growth of large systems but rarely explore in depth the causes of growth. Explanations using such concepts as economies of scale and such motives as the drive for personal power and organizational aggrandizement can mask contradictions. If by economies of scale one means the savings in material and heat energy that come from using larger containers, such as tanks, boilers, and furnaces, then the economy can be lost if the larger container is not used to capacity. If by economy of scale one simply refers to the number of units produced or serviced, then plant or organization capacity and the spread of the output over time are not taken into account and economy is not adequately measured. For instance, a power plant scaled up to generate twice as many kilowatt-hours per month would increase its unit cost if the increased load were concentrated during a few peak load hours a day. If a larger organization is assumed to bring greater influence and control for the managers, then the distinct possibility that individual initiative will be lost in bureaucratic routine is ignored. Long ago, Leo Tolstoy argued in *War and Peace* that the overwhelming momentum of the huge French army and the image of the all-powerful and victorious Emperor gave Napoleon during the invasion of Russia less freedom of action than the common foot soldier. Small firms and armies are not as likely to smother initiative.

Some designers of technological systems have taken these contradictions into account. Designers of electric power plants decide whether to build a large plant or to construct a number of smaller ones over an extended time. The latter choice often matches growing capacity to increasing load. Utility managers and operators also manage the load to avoid extreme peaks and valleys in output that signify unused capacity. In the past managers of small electric utilities often fought the absorption of their systems by larger ones because they anticipated that in the larger organization bureaucracy would reduce their exercise of authority. The small, technically advanced, and profitable power plants and utilities that flourished in London from about 1900 to the implementation of the National Grid after 1926 give evidence that large-scale output and organizational size are not necessary for profitability and personal power (Hughes, 1983, pp. 259–360). Most of the top managers of the small utilities that have been absorbed into larger ones were destined to play subordinate roles in the bureaucratic recesses of middle management.

Yet in modern industrial nations technological systems tend to expand, as shown by electric, telephone, radio, weapon, automobile production, and other systems. A major explanation for this growth, and one rarely stressed by technological, economic, or business historians, is the drive for high diversity and load factors and a good economic mix. This is especially true in twentieth century systems in which accountants pay close attention to, and managers are informed about, interest on capital investment. The load factor, a concept now applied to many systems, originated in the electrical utility industry in the late nineteenth century. The load factor is the ratio of average output to the maximum output during a specified period. Best defined by a graph, or curve, the load factor traces the output of a generator, power plant, or utility system over a twenty-four-hour period. The curve usually displays a valley in the early morning, before the waking hour, and a peak in the early evening, when business and industry use power, homeowners turn on lights, and commuters increase their use of electrified conveyance. Showing graphically the maximum capacity of the generator, plant, or utility (which must be greater than the highest peak) and tracing the load curve with its peaks and valleys starkly reveal the utilization of capacity. Because many technological systems now using the concept are capital intensive, the load curve that indicates the load factor, or the utilization of investment and the related unit cost, is a much relied on indicator of return on investment.

The load factor does not necessarily drive growth. A small technological system can have a high load factor, for example, if the load, or market, for output is diversified. The load of an electric power system becomes desirably diverse if the individual consumers make their peak demands at different times, some in the late evening, some in the early morning, and so on. When this is not the case, the managers of a technological system try to expand the system in order to acquire a more desirable load or diversity. The load can also be managed by differential pricing to raise valleys and lower peaks. In general, extension over a larger geographical area with different industrial, residential, and transportation loads provides increased diversity and the opportunity to manage the load to improve the load factor. During the twentieth century expansion for diversity and management for a high load factor have been prime causes for growth in the electric utility industry. The load factor is, probably, the major explanation for the growth of capital-intensive technological systems in capitalistic, interest-calculating societies.[17]

The managers of electric power systems also seek an improved economic mix. This results, for instance, in the interconnection of a power plant located in the plains near coal mines with another in distant high mountains. The Rheinisch-Westfälisches Elektrizitätswerk, a utility in the Ruhr Valley of Germany, expanded in the 1920s hundreds of miles until the system reached the Alps in the south. Then, after the spring thaws, it drew low-cost hydroelectric power from the Alps and at other times from the less economical coal-fired plants of the Ruhr. The outputs of the regional plants could also be mixed, the less efficient carrying the peak loads on the system and the more economical carrying a steady base load. The intellectual attraction—the elegant puzzle-solving aspect—that the load factor, economic mix, and load management had for the engineer-managers of rapidly expanding electric power systems becomes understandable. For those more concerned with the traditional drive for power and profit, elegant problem solving was coupled with increased profits, market domination, and organization aggrandizement.

As the systems grew, other kinds of problems developed, some of which can be labeled "reverse salients." Conservative inventions solved these problems, whereas radical ones brought the birth of systems. A salient is a protrusion in a geometric figure, a line of battle, or an expanding weather front. As technological systems expand, reverse salients develop. Reverse salients are components in the system that have fallen behind or are out of phase with the others. Because it suggests uneven and complex change, this metaphor is more appropriate for systems than the rigid visual concept of a bottleneck. Reverse salients are comparable to other concepts used in describing those components in an expanding system in need of attention, such as drag, limits to potential, emergent friction, and systemic efficiency. In an electrical system engineers may change the characteristics of a generator to improve its efficiency. Then another component in the system, such as a motor, may need to have its characteristics—resistance, voltage, or amperage—altered so that it will function optimally with the generator. Until that is done, the motor remains a reverse salient. In a manufacturing system one productive unit may have had its output increased, resulting in all the other components of the system having to be modified to contribute efficiently to overall system output. Until the lagging components can be altered, often by invention, they are reverse salients. During the British Industrial Revolution, observers noted such imbalances in the textile industry between weaving and spinning, and inventors responded to the reverse salients by inventions that increased output in the laggard components and in the overall system. In a mature, complex technological system the need for organization may often be a reverse salient. In the 1920s manager-entrepreneurs saw the need for an organizational form that could preside over the construction, management, and financing of horizontally and vertically integrated utilities. The invention of an appropriate holding-company form corrected the reverse salient.

Entrepreneurs and organizations presiding over expanding systems monitor the appearance of reverse salients, sometimes identifying them by cost-accounting techniques. Having identified the reverse salients, the organization assigns its engineering staff or research laboratory to attend to the situation, if it is essentially one involving machines, devices, processes, and the theory and organized knowledge describing and explaining them. The staff or laboratory has the communities of technological practitioners possessing the traditions of relevant practice (Constant, this volume). Communities of inventors congregate at reverse salient sites, for a number of companies in an industry may experience the reverse salient at about the same time. The inventors, whether engineers or industrial scientists, then define the reverse salient as a set of critical problems, which when solved will correct it. Reverse salients emerge, often unexpectedly; the defining and solving of critical problems is a voluntary action. If the reverse salient is organizational or financial in nature, then the individuals or communities of practitioners who attack the problem may be professional managers or financiers who come forth with their inventive solutions. In each stage in the growth of the system the reverse salients elicit the emergence of a sequence of appropriate types of problem solver inventors, engineers, managers, financiers, and persons with experience in legislative and legal matters (Hughes 1983, pp. 14–17).

Industrial research laboratories, which proliferated in the first quarter of this century, proved especially effective in conservative invention. The laboratories routinized invention. The chemist Carl Duisberg, a director of Bayer before World War I, aptly characterized the inventions of industrial research laboratories (*Etablissements-erfindungen*) as having "Von Gedankenblitz keine Spur" (no trace of a flash of genius) (Van den Belt and Rip, this volume). Unfortunately for the understanding of technological change, the public relations departments and self-promoting industrial scientists persuaded the public, managers, and owners that industrial laboratories had taken over invention from independent investors because the independents were less effective. Considerable evidence shows, to the contrary, that radical inventions in disproportionate number still come from the independents.[18] A mission-oriented laboratory tied to an industrial corporation or government agency with vested interest in a growing system nurtures it with conservative improvements or with inventions that are responses to reverse salients.

The early problem choices of the pioneer industrial laboratories suggest this rigid commitment to conservative inventions and relative disinterest in radical ones. After the Bell Telephone System in 1907 consolidated its research activities in the Western Electric Company and in American Telephone & Telegraph, its staff of scientists and engineers concentrated on reverse salients that arose out of the decision to build a transcontinental telephone line. Attenuation, or energy loss, proved a major reverse salient. The invention of the loading coil reduced attenuation. By 1911 the introduction of improved repeaters for transmission lines became a major problem for the research and development staff.[19] Reverse salients in electric light and power systems attacked by engineers and scientists at the General Electric Research Laboratory at about the time of its founding in 1900 included improved filaments and vacuum for incandescent lamps and improvements in mercury vapor lamps. Even Irving Langmuir, a distinguished GE scientist who was given exceptional freedom in his choice of research problems, did not neglect highly practical problems encountered by the General Electric Company as it expanded its product lines. Willis

R. Whitney, laboratory director, pursued the policy of "responsiveness to business needs" (Wise 1980, p. 429).

When a reverse salient cannot he corrected within the context of an existing system, the problem becomes a radical one, the solution of which may bring a new and competing system. Edward Constant has provided an example of the emergence of a new system out of an established one in which a "presumptive anomaly" was identified. Constant states that presumptive anomalies occur when assumptions derived from science indicate that "under some future conditions the conventional system will fail (or function badly) or that a radically different system will do a much better job" (Constant 1980, p. 15). A presumptive anomaly resembles a presumed reverse salient, but Constant rightly stresses the role of science in identifying it. A notable presumptive anomaly emerged in the late 1920s when insights from aerodynamics indicated that the conventional piston engine–propeller system would not function at the near-sonic speeds foreseen for airplanes. The inventors Frank Whittle, Hans von Ohain, Herbert Wagner, and Helmut Schelp responded with the turbojet engine, the first three working as independents when they conceived of the new engine (Constant 1980, pp. 194–207, 242).

Edison and others presiding over the growth of the dc electric lighting system in the early 1880s failed to solve a reverse salient and saw other inventors and engineers respond to it with radical inventions that inaugurated the ac system. A "battle of the systems" then ensued between the two, culminating in the 1890s, not with victor and vanquished, but with the invention of devices making possible the interconnection of the two systems. These motor-generator sets, transformers, and rotary converters interconnected heterogeneous[20] loads, such as incandescent lamps, arc lamps, induction motors for industry, dc motors for streetcars, or trams, into a universal system[21] supplied by a few standardized polyphase generators and linked by high-voltage transmission and low-voltage distribution lines. The design and installation of universal power systems in the 1890s is comparable to the introduction by AT&T a decade or so later of a universal telephone network and is similar to the recent design by computer manufacturers of large interconnections for diverse systems. These physical linkages were accompanied by the organizational linkages of utilities and manufacturers who had nurtured the competing systems. The Thomson-Houston Company, with its ac system, merged in 1893 with the Edison General Electric Company with its dc system.[22] Consolidation of electric light and power systems occurred throughout the industrial world until the interwar period, when two large manufacturers in the

United States (General Electric and Westinghouse) and two in Germany (Allgemeine Elektrizitäts-Gesellschaft and Siemens) dominated electrical manufacturing. Similarly, large regional utilities prevailed in electrical supply. At about the same time industry-wide standardization of technical hardware created, for instance, standard voltages, frequencies, and appliance characteristics. Similar mergers and standardization took place in the telephone and automobile-production systems during the early twentieth century.

Momentum

Technological systems, even after prolonged growth and consolidation, do not become autonomous; they acquire momentum. They have a mass of technical and organizational components; they possess direction, or goals; and they display a rate of growth suggesting velocity. A high level of momentum often causes observers to assume that a technological system has become autonomous.[23] Mature systems have a quality that is analogous, therefore, to inertia of motion. The large mass of a technological system arises especially from the organizations and people committed by various interests to the system. Manufacturing corporations, public and private utilities, industrial and government research laboratories, investment and banking houses, sections of technical and scientific societies, departments in educational institutions, and regulatory bodies add greatly to the momentum of modern electric light and power systems. Inventors, engineers, scientists, managers, owners, investors, financiers, civil servants, and politicians often have vested interests in the growth and durability of a system. Communities of practitioners, especially engineers maintaining a tradition of technological practice, sometimes avoid deskilling by furthering a system in which they have a stake (Constant, this volume). Actor networks, as defined by Michel Callon, add to system momentum (Callon, this volume). Concepts related to momentum include vested interests, fixed assets, and sunk costs.

The durability of artifacts and of knowledge in a system suggests the notion of trajectory,[24] a physical metaphor similar to momentum. Modern capital-intensive systems possess a multitude of durable physical artifacts. Laying off workers in labor-intensive systems reduces momentum, but capital-intensive systems cannot lay off capital and interest payments on machinery and processes. Durable physical artifacts project into the future the socially constructed characteristics acquired in the past when they were designed. This is analogous to the persistence of acquired characteristics in a changing environment.[25]

The momentum of capital-intensive, unamortized artifacts partially explains the survival of direct current after the "battle of the systems," despite the victory of the competing alternating current. The survival of high-temperature, high-pressure, catalytic-hydrogenation artifacts at the German chemical firm of Badische Anilin- and Soda-Fabrik (BASF) from about 1910 to 1940 offers another example of momentum and trajectory (Hughes 1969). In the BASF case a core group of engineers and scientists knowledgeable about the hydrogenation process through the design of nitrogen-fixation equipment during World War I subsequently deployed their knowledge and the equipment in the production of methanol during the Weimar period and of synthetic gasoline during the National Socialist decade.

From 1910 to 1930 system builders contributed greatly to the momentum of electric light and power systems in the industrialized West. Combining complex experiences and competence, especially in engineering, finance, management, and politics, Hugo Stinnes, the Ruhr stagnate, Emile and Walther Rathenau, the successive heads of Germany General Electric (AEG), and Oskar von Miller, who helped create the Bayernwerk, the Bavarian regional utility, built large German systems. Walter von Rathenau, who was especially fascinated by the aesthetics of system building, said approvingly in 1909 that "three hundred men, all acquainted with each other [of whom he was one], control the economic destiny of the Continent" (Kessler 1969, p. 121). In 1907 his AEG system was "undoubtedly the largest European combination of industrial units under a centralized control and with a centralized organization." In Great Britain consulting engineer Charles Merz presided over the growth of the country's largest electric supply network, the Northeastern Electric Supply Company. In the United States Samuel Insull of Middle West Utilities Company, S. Z. Mitchell of Electric Bond and Share, a utility holding company associated with General Electric, and Charles Stone and Edwin Webster of Stone & Webster ranked among the leading system designers.

Stone and Webster's became an exemplary system. Just graduated from the Massachusetts Institute of Technology in 1880, they founded a small consulting engineering company to advise purchasers of electric generators, motors, and other equipment. Knowing that the two young men were expert in power plant design and utility operation, J. P. Morgan, the investment banker, asked them to advise him about the disposition of a large number of nearly defunct utilities in which he had financial interest. From the study of them, Stone and Webster identified prime and widespread reverse salients throughout the utility industry and became expert in rectifying them. Realizing that money spent prudently on utilities whose ills had been correctly diagnosed often brought dramatic improvement and profits, Stone and Webster in about 1910 were holistically offering to finance, construct, and manage utilities. As a result, a Stone and Webster system of financially, technically, and managerially interrelated utilities, some even physically interconnected by transmission lines, operated in various parts of the United States. In the 1920s Stone and Webster formed a holding company to establish closer financial and managerial ties within the system (Hughes 1983, pp. 386–391). Similar utility holding companies spread throughout the Western world. Some involved the coal-mining companies supplying fuel for the power plants in the system; others included electrical manufacturers making equipment for the utilities. Others established linkages through long-term contractual relations, inter-locking boards of directors, and stock purchases with manufacturing firms and transportation companies that were heavy consumers of electricity. In Germany local government sometimes shared the ownership of the utilities with private investors. Brought into the system, thereby, local government became both regulator and owner.

Such mammoth, high-momentum systems were not limited to the electrical utility field. The system of automobile production created by Henry Ford and his associates provides a classic example of a high-momentum system. Coordinated to ensure smooth flow from raw material to finished automobile ready for sale, interconnected production lines, processing plants, raw material producers, transportation and materials-handling networks, research and development facilities, and distributors and dealers made up the Ford system. Interconnection of production and distribution into systems with high flow or throughput also took place in the chemical industry early in this century.[26]

The high-momentum systems of the interwar years gave the appearance of autonomous technology. Because an inner dynamic seemed to drive their course of development, they pleased managers who wished to reduce uncertainty and engineers who needed to plan and design increased system capacity. After 1900, for instance, the increasing consumption of electricity could be confidently predicted at 6 percent annually. Such systems appeared to be closed ones, not subject to influence from external factors or from the environment. These systems dwarfed the factors of the environment not yet absorbed by them. Subject to the power brokering, the advertising, and the money influence of the system, those who controlled forces in the environment took on the values and objectives of the system.

Appearances of autonomy have proved deceptive. During and immediately after World War I, for instance, the line of development and the characteristics of power systems in England changed appreciably. Before the war the British systems were abnormally small compared to those in the United States and industrial Germany. Utility operators elsewhere called the British system backward. In fact, the British style accorded nicely with prevailing British political values and the regulatory legislation that expressed them. Traditionally, the British placed a high value on the power of local government, especially in London, and electrical utilities were bound within the confines of the small political jurisdictions.[27] World War I in particular and the increasingly apparent loss of industrial preeminence in general brought into question the political and economies values long prevalent in Great Britain. During the war Parliament overrode local government sensibilities and forced interconnection of small electrical systems to achieve higher load factors and to husband scarce resources. With victory the wartime measures could have been abandoned, but influential persons questioned whether the efficiency achieved during the war was not a prerequisite for industrial recovery in peacetime. As a result, in 1926 technological change in electric power systems was given a higher priority than tradition in local government. Parliament enacted legislation that created the first national interconnection, or grid. The political forces that were brought to bear more than matched the internal dynamic of the system.

After World War II, utility managers, especially in the United States, wrongly assumed that nuclear power reactors could easily be incorporated in the pattern of system development. Instead, nuclear power brought reverse salients not easily corrected. Since World War II changes such as the supply of oil, the rise of environmental protection groups, and the decreasing effectiveness of efficiency-raising technical devices for generating equipment have all challenged the electrical utility managers' assumptions of momentum and trajectory.

These instances, in which the momentum of systems was broken, remind historians and sociologists to use such concepts and patterns of envolving systems as heuristic aids and system managers to employ them cautiously as predictive models. Momentum, however, remains a more useful concept than autonomy. Momentum does not contradict the doctrine of social construction of technology, and it does not support the erroneous belief in technological determinism. The metaphor encompasses both structural factors and contingent events.

Conclusion

This chapter has dealt with the patterns of growing or evolving systems. Countless other technological systems in history have arrived at a stage of stasis and then entered a period of decline.[28] In the nineteenth century, for instance, the canal and gas light systems moved into stasis and then decline. Historians and sociologists of technology should also search for patterns and concepts applicable to these aspects of the history of technological systems.

Notes

The Wissenschaftszentrum (Berlin) and the Wissenschaftskolleg zu Berlin provided support for the preparation of this chapter, which is part of a long-term study of technological change.

1. The concept of technological system used in this essay is less elegant but more useful to the historian who copes with messy complexity than the system concepts used by engineers and many social scientists. Several works on systems, as defined by engineers, scientists, and social scientists, are Ropohl (1979), von Bertalanffy (1968), and Parsons (1968). For further references to the extensive literature on systems, the reader should refer to the Ropohl and the Bertalanffy bibliographies. Among historians, Bertrand Gille has used the systems approach explicitly and has applied it to the history of technology. See, for instance, his *Histoire des techniques* (1978).

2. In this chapter "technical" refers to the physical components (artifacts) in a technological system.

3. A coal mine is analogous to the wind in John Law's Portuguese network, for the winds are adapted by sails for use in the system. See Law (this volume).

4. Most of the examples of systems in this essay are taken from my *Networks of Power* (1983). For the relation between investment organizations and electrical manufactures, for instance, see pp. 180–181 and 387–403 of that book.

5. I am grateful to Charles Perrow of Yale University for cautioning me against acceptance of the contingency theory of organization, which holds that an organization simply reflects the pattern of hardware, or artifacts, in a system. Perrow has contributed to the clarification of other points in this essay.

6. In contrast to Alfred D. Chandler, Jr. (1966, pp. 15–19), who locates technological (technical) changes as part of a context, including population and income, within which an organization develops strategy and structure, I have treated technical changes as part of a technological system including organizations. Borrowing from architectural terminology, one can say not only that in a technological system

organizational form follows technical function but also that technical function follows organizational form.

7. The manufacturer, Allgemeine Elektrizitäts-Gesellschaft and the utility Berliner Electrizitäts-Werke were linked by ownership and cooperated systematically in design and operation of apparatus (Hughes 1983, pp. 175–200).

8. For an extended set of case histories supporting the phase and system builder sequence suggested, see my *Networks of Power* (1983).

9. Existing telephone and telegraph companies played a minor role in the early history of the wireless; existing compass makers did not take up the gyrocompass; and existing aircraft manufacturers provided little support for early turbojet inventive activities.

10. Anthony told Sperry that there was a need for an automatically regulated constant-current generator (Hughes 1971, p. 16).

11. See, for instance, Arieti (1976) and the appended bibliography.

12. Arthur Koestler provides imaginative insights in *The Act of Creation* (1964). Arieti (1976) is also stimulating.

13. See, for example, Hoddeson (1981), Wise (1980), and Hughes (1976b). For an analysis of positions taken in the journal *Technology and Culture*, see Staudenmaier (1985, pp. 83–120).

14. An issue of *Technikgeschichte* (1983, vol. 50, no. 3) with articles by Ulrich Troitzsch, Wolfhard Weber, Rainer Fremdling, Lars U. Scholl, Ulrich Wengenroth, Wolfgang Mock, and Han-Joachim Braun, who have written often on transfer, is given over to *Technologietransfer im 19. und 20. Jahrhundert.*

15. Compare the concept of technological frame proposed by Bijker (this volume).

16. I am indebted to Edward Constant for information on style in automobiles and to Alex Roland for information on contrasting styles of Soviet and US space technology.

17. For a further discussion of load and diversity factors, see Hughes (1983, pp. 216–222). Alfred Chandler labels a similar but less graphic concept applied to manufacturing and chemical industries as "throughput" (1977, p. 241).

18. Jewkes et al. (1969) persuasively argue the case for the independents in the past and present.

19. For more on invention (conservative) and the expanding telephone system, see Hoddeson (1981).

20. See Law (this volume) on heterogeneous entities and engineers.

21. I am indebted to Robert Belfield for the concept of universal system, which he encountered in the Charles F. Scott papers at Syracuse University.

22. On the "battle of the systems," see Hughes (1983, pp. 106–135). See also Bijker (this volume).

23. Langdon Winner (1977) has analyzed the question of whether or not technology is autonomous. For a sensible discussion of the questions of autonomy and technological determinism, see the introduction to MacKenzie and Wajcman (1985, pp. 4–15).

24. For a discussion of trajectory, see Van den Belt and Rip (this volume).

25. Edward Constant has explored and explained communities of practitioners. See, for instance, his chapter in this volume.

26. A recent study of the Ford and other systems of production is provided by Hounshell (1984). Chandler (1977) analyzes and describes the integration of production and distribution facilities in several industries, including the chemical industry.

27. For an extended account of the electric utility situation in Great Britain before and after World War I, see Hughes (1983, pp. 227–261, 319–323, 350–362).

28. I am indebted to Richard Hirsh of Virginia Polytechnic Institute and State University for calling my attention to stasis in the post–World War II electrical utilities. Hirsh explores the concept in his unpublished manuscript, "Myths, Managers, and Megawatts: Technological Stasis and Transformation in Electric Power Industry."

Society in the Making: The Study of Technology as a Tool for Sociological Analysis

Michel Callon

Social scientists, whether they are historians, sociologists, or economists, have long attempted to explain the scope, effects, and conditions of the development of technology. They consider technology a specific object that presents a whole range of problems that these experts have tried to solve using a series of different methods available to the social sciences.[1] But at no point have they judged that the study of technology itself can be transformed into a sociological tool of analysis. The thesis to be developed here proposes that this sort of reversal of perspective is both possible and desirable. Not only would it enlarge the methodological range of the social sciences but it would also facilitate the understanding of technological development. To bring this reversal about, I show that engineers who elaborate a new technology as well as all those who participate at one time or another in its design, development, and diffusion constantly construct hypotheses and forms of argument that pull these participants into the field of sociological analysis. Whether they want to or not, they are transformed into sociologists, or what I call engineer-sociologists.

Seeing the process of technological innovation and the role played by engineers in this way defies certain accepted ideas. By taking this perspective I am not simply repeating the already countless criticisms of the notion of innovation as a linear process. This notion describes technological development as a succession of steps from the birth of an idea (invention) to its commercialization (innovation) by way of its development. Everyone now recognizes that the to and fro's of coupling processes that continuously occur between technology and the market are extremely important.[2] Nor in this chapter do I challenge the notion that claims that the role and importance of financial backing or organizational structure varies considerably between periods of elaboration and development of an innovation.[3] What I am questioning here is the claim that it is possible to distinguish during the process of innovation phases or activities that are distinctly technical or scientific from others that are guided by an economic or commercial logic. For example, it is often believed that at the beginning of the process of innovation the problems to be solved are basically technical and that economic, social, political, or indeed cultural considerations come into play only at a later stage.[4] However, more and more studies are showing that this distinction is never as clear-cut. This is particularly true in the case of radical innovations: Right from the start, technical, scientific, social, economic, or political considerations have been inextricably bound up into an organic whole.[5] Such heterogeneity and complexity, which everyone agrees is present at the end of the process, are not progressively introduced along the way. They are present from the beginning. Sociological, technoscientific, and economic analyses are permanently interwoven in a seamless web (Hughes 1983). Using the case study of an innovation, I show how it is possible to use this characteristic in order to transform the study of technology into a tool for sociological analysis; this leads to a new interpretation of the dynamics of technology.

Engineer-Sociologists

To illustrate the capacity of engineers to act as sociologists (or historians or economists), I describe certain aspects of the development of what was intended to be a major innovation: the introduction of an electric car (VEL) in France.[6]

This project was first presented by a group of engineers working for EDF (Electricité de France)[7] in the early 1970s. They outlined the project in a series of technical publications and in applications for funding to government agencies.[8] It is by no means easy to create a new market of this sort in a society organized entirely around the traditional motorcar. The project conjectured not only that the technoscientific problems could be overcome but also that French social structures would change radically.

EDF's engineers presented a plan for the VEL that determined not only the precise characteristics of the vehicle it wished to promote but also the social universe in which the vehicle would function. We will see that in addition to their technical know-how the engineers of EDF used skills more commonly found in social scientists. They resembled their illustrious predecessors from the Renaissance who had deftly played on several registers at the same time (Gille 1978). Like Edison almost a hundred years ago, they continuously mixed technical and social sciences.[9]

First, the EDF defined a certain history by depicting a society of urban post-industrial consumers who were grappling with new social movements. The motorcar occupied a position that was highly exposed, for it formed part of a world that was under attack. Thus it served as a point of departure for the construction of far-reaching and radical demands that would lead to a future that could be discerned only with difficulty. The internal combustion engine is the offspring of an industrial civilization that is behind us. The Carnot cycle and its deplorable by-products were stigmatized in order to demonstrate the necessity for other forms of energy conversion. On the one hand, the motor vehicle was considered responsible for the air pollution and noise that plagues our cities. On the other hand, it was irretrievably linked to a consumer society in which the private car constituted a primordial element of status. However, electric propulsion would render the car commonplace by decreasing its performance and reducing it to a simple, useful object. The electric car could lead to a new era in public transport in the hands of new social groups that were struggling to improve conditions in the city by means of science and technology. The goal would be to put science and technology at the service of the user and to do away with social categories that attempted to distinguish themselves by their styles of consumption. The EDF based this vision on an evaluation of the trajectories of development open to different types of electrochemical batteries.[10] First, public transport could be equipped with improved lead accumulators. Then accumulators and fuel cells could open up the larger market of private transport by enabling the VEL to reach speeds of up to 90 km/h, on the condition that safe catalysts cheaper than platinum could be developed (cheaper but poisonous catalysts had already been found).

By predicting the disappearance of the internal combustion engine as a result of the rise of electrochemical generators and by ignoring traditional consumers so as to better satisfy users who had new demands, the EDP not only defined a social and technological history but also identified the manufacturers that would be responsible for the construction of the new VEL. The CGE (Compagnie Générale d'Electricité) would be asked to develop the electric motor, the second generation of batteries, and to perfect the lead accumulators that would be used in the first generation of the VEL. Renault would mobilize its expertise in the production of traditional automobiles in order to assemble the chassis and make the car bodies. The government would also be enlisted: Such and such a ministry would subsidize those municipalities interested in electric traction. The list went on: Companies that ran urban transport systems were to be put together with research centers, scientists, etc. The EDF defined the roles, and attempted to enroll other entities into them. It also bound the functions of these roles together by building a world in which everything had its own place.

Up to this point the entities are those familiar to the sociologist. There are consumers, social movements, and ministries. But it would be wrong to limit the inventory. There are also accumulators, fuel cells, electrodes, electrons, catalysts, and electrolytes, for, if the electrons did not play their part or if the catalysts became contaminated, the result would be no less disastrous than if the users rejected the new vehicle, the new regulations were not enforced, or Renault stubbornly decided to develop the R5. In the world defined and built by the EDF, at least three new and essential entities must be added: zinc/air accumulators, lead accumulators, and fuel cells with their associated elements (catalysts, electrons, etc.).

The EDF engineers determined not only the repertoire of entities that they enlisted and the histories in which they would take part but also their relative size. For EDF's engineers Renault would no longer be a powerful company seeking to be once more the largest European car manufacturer. Indeed, it would never regain that status. Rather, it was reduced to the level of a modest entity that intervened in the assembly of the VEL. The same is true of the old status groups that would give way to new social movements and their new demands.

The ingredients of the VEL are the electrons that jump effortlessly between electrodes; the consumers who reject the symbol of the motorcar and who are ready to invest in public transport; the Ministry of the Quality of Life, which imposes regulations about the level of acceptable noise pollution; Renault, which accepts that it will be turned into a manufacturer of car bodies; lead accumulators, whose performance has been improved; and post-industrial society, which is on its way. None of these ingredients can be placed in a hierarchy or distinguished according to its nature. The activist in favor of public transport is just as important as a lead accumulator, which can be recharged several hundred times.

This case shows that the engineers left no stone unturned. They went from electrochemistry to political science without transition. The analysis of French society that they proposed was both remarkably incisive and fully elaborated. Five years after the "great cultural revolution" of May 1968[11] and one year before the first oil crisis, they outlined the course of an evolutionary movement that would propel French society from the industrial to the post-industrial age. This change was to occur through pressure from new social movements and with the expected help of electrons.[12]

The sociologist who studies the VEL project cannot but be struck by the similarity between the "sociological" arguments developed by the engineers at EDF and the analyses proposed at the same time by one of the most respected French sociologists, A. Touraine. This similarity, which I come back to shortly, obviously suggests a question: Could not social sciences in some way or another make use of the astonishing faculty engineers possess for conceiving and testing sociological analyses at the same time as they develop their technical devices? It is to answer this question, which supposes that it is possible to compare the sociology of engineers with professional sociology, that I now present the analyses proposed by Touraine and the controversies to which they gave rise.

Sociology and the Problem of Consumption

Where was French society really going in 1973? And, in particular, what destiny lay in store for the traditional motorcar? The engineers at EDF asked themselves these sorts of questions, and they responded to them by conceiving of the VEL project. They were not alone in asking these questions. Sociologists too were trying to answer them, and the analyses they elaborated display great diversity. Several schools confronted each other. For my purposes I need to retain only the opposition between Touraine (1973, 1979) and Bourdieu (1979; Bourdieu and Darbel 1966; Bourdieu and Passeron 1970). These two gave radically different interpretations of the dynamics of consumption.

Touraine is part of a sociological tradition that emphasizes the role of class conflict in making society function and in producing its history. Unlike Marxists, he believes that the central conflict of Western society is no longer the struggle between the working class and the bourgeoisie. Technological development has brought new factors into play. On one side now there are large concerns (big corporations, research and development agencies) that orient scientific research as well as define and control the application of technology. On the other side we find the consumer, whose needs and aspirations are manipulated by the technocrats who run the large concerns. This conflict explains the birth of social movements that challenge (either through categorical demands or through calls for a move "back to basics") the power of the technocracy or its orientations for social and economic development. These movements are relatively widespread and ephemeral. Sociologists must learn to decipher their demands and technocrats must take them into consideration if they wish to safeguard the legitimacy of their choices and decisions. This new type of class conflict defines what Touraine calls post-industrial society.

Bourdieu's vision of society can be arrayed point for point against Touraine's. For Bourdieu, society is not organized around a primordial confrontation between ruling classes and classes that are ruled, fighting for control of technological development. The confrontation is fragmented between various specialist spheres (the field of politics, the field of science, the field of consumption, etc.) that maintain mutual relationships of exchange and subordination. Each field is the site of strategic confrontations between social agents who fight to occupy positions of power. But these different fields, which in their multiplicity embrace the diversity of social practice and express increasing differentiation of societies, are caught in a group logic that lends cohesion to society. This unification is organized around a dominant cultural model, that of the upper classes, in relation to which the other social classes define and orient themselves. Whatever particular field is considered, these classes are in constant competition in order to delineate their differences and to vie for positions on a scale of status. This competition is nowhere more apparent and nowhere more lively than in the field of consumption. The reader will recognize here the essential elements of the theory of social stratification, in which distinction, differentiation, and mobility play an essential role.

Beyond the classical opposition they display between a sociology of social class and a sociology of stratification, Touraine and Bourdieu share the feature that they place the question of consumption at the center of their analyses. Touraine does so in order to show that consumption is largely manipulated by industry and the great technological agencies, Bourdieu to establish its irreducible autonomy. Touraine sees in the definition of demand or need the site of the emergence of new class conflicts, whereas Bourdieu affirms that goods and services, whatever their intrinsic characteristics, are ineluctably reinscribed by consumers into the logic of social distinction.

Although they attribute to consumption the same strategic value, these two analytic schemas lead to two radically different interpretations of its evolution. The automobile and its future provide particularly salient illustrations of this evolution.

If one has a stake in the coming of a post-industrial society, the traditional motorcar is doomed to lose ground because it is an integral part of a social system that is disappearing; it stands as both the symbol and cornerstone of that system. Social movements that diminish the importance of and criticize the use of the automobile anticipate and express the necessity of this evolution. In the Tourainian schema the technocrats/decision makers design products to meet these demands in order to use them for support: This double game, whereby popular protest is used by technocrats to serve their own ends, is the driving force of history. The appearance of a new technology, such as the VEL, is thus much more likely because it introduces a rupture in industrial society and is supported at the same time by social movements and the technocracy.

In Bourdieu's perspective the future of the automobile is inscribed in a different logic. The total banalization of an object of consumption, which plays a central role in struggles for distinction, seems highly improbable. Social movements that protest against the symbol Automobile are without doubt quite right to see in it one of the cornerstones of our societies; but instead of believing in their capacity to create a new era, we should learn the lesson it teaches them against their will. The automobile is at the nerve center of society, so socially embedded that it can be modified only with great care. It must undergo evolution, but this is not purely and simply a case of making it disappear so that it can be replaced with a radically new technology; the only realistic strategy is to transform it gradually through progressive introduction of technical improvements enabling it to respond to new user demands. The best answer that can be given to social movements is to introduce yet more differentiation, not make a tabula rasa of the past.

Who Is Right?

What, in 1970, was the future of the automobile in French society? This question was at the center of the VEL project as it was developed by the engineers at EDF. Furthermore, it is a question that should not have been ignored by sociologists, because, as I have just shown, consumption and its evolution occupied a central place in the theoretical apparatus they had elaborated.

In fact, sociologists were little concerned with the EDF adventure and abstained from establishing some link between their theories and this astonishing story that was unfolding before their eyes. A story so much the more astonishing because, as we will see, the engineers at EDF were to become rapidly engaged in a controversy in which their Tourainian sociology would set itself against the sociology à la Bourdieu employed by the engineers at Renault. The controversy was, however, of a different sort, for success or failure was to be measured in terms of shares of the market.

EDF's engineers did not have to defend their ideas in an academic arena. Any brilliance or originality in the analysis they developed was of little import. For them the analysis was a question of life and death because the economic future of their project was at stake. No more sophisticated arguments and theorizing! What mattered was to be right: to be able to prove by the very success of their innovation that French society was evolving in the way they claimed it was, borne along by the aspirations of protest movements on which they in turn holed to lean for further support. The rest was of no account. In short, if an engineer-sociologist is to be proved right he or she has to create a new market; success is measured by the amount of profit gained. This, in all its simplicity and toughness, is the test of truth.

For three years the engineers of EDF believed that they were right. Nobody dared interrupt their discourse. Car manufacturers, with Renault in the forefront, kept quiet, terrorized by the future promised them. In order to hold their own, they started to work feverishly on the VEL project. They knew little or nothing about electrochemistry and did not know how to tackle the EDF forecasts that cheap high-performance fuel cells would be available by the end of the 1980s, thus opening up the vast market of private transport. To counter their handicap, they signed contracts with specialized research laboratories in order to acquire the knowledge and expertise they lacked. To begin with, the electrochemists confirmed the optimistic predictions made by EDF engineers. How then could anybody resist a movement that allied consumer aspirations, the wishes of the authorities, and available scientific resources (or rather resources thought to be available in the not-too-distant future)? Nothing could stand in the way of this tidal wave. In addition to these existing forces, another event occurred to weaken still further the position of the traditional motor car: the sudden increase in oil prices, making cars much more expensive to run.

Slowly but surely the tide in favor of the VEL and its society was beginning to turn, or, to use the terms so aptly coined by Hughes (1983), reverse salients began to appear. Things began to go wrong for the EDF engineers. Resistance, of the kind so neatly described by Castoriadis to define reality,[13] got underway. As in guerrilla warfare, it started up spontaneously and unexpectedly in several places. Fairly quickly, the catalysts refused to play their part in the scenario prepared by EDF: Although cheap (unlike platinum), the catalysts had the unfortunate tendency of quickly becoming contaminated, rendering the fuel cell unusable. The mass market suddenly disappeared like a mirage. The VEL, recognized EDF's engineers, needed batteries whose performance was sufficient for the average user, and this sort of battery might be too expensive to produce for a long time to come. In addition, Renault challenged the future of other electrochemical generators identified by EDF. For example, Renault showed that the zinc/air accumulators lauded by EDF's engineers were actually a shaky venture elaborated by a handful of researchers at CGE[14] who were pushing the program without being sure that it was realistic. Furthermore, argued Renault engineers, if zinc/air accumulators were to be used in the VEL, this would presuppose the setting up of a vast network of service stations throughout the country whereby the used electrolyte might be changed periodically. Which industrial groups, they asked, would dare to challenge the all-powerful oil consortiums on their own ground? In contrast to the optimistic view of technological innovation taken by EDF, Renault engineers painted a gloomy picture of uncertain strategies and rival industrial groups with conflicting interests.

The Renault engineers did not stop there. They took their criticism further by showing that what EDT detected as signs of the coming of a post-industrial age was in fact only minor technical difficulties in the current age. According to them, the criticism leveled at the traditional motor car did not change the equilibrium of existing social forces, nor was it a sign of a demand for a new mode of development. It merely expressed temporary and local dissatisfaction with the car industry's lack of dynamism and the poor state of public transport. Pollution could easily be reduced and the reorganization of public transport in cities could be improved, in particular, by using more comfortable and higher-performance buses. They argued that in the space of three years protest movements had quieted down, especially those that had been most virulent in speaking out against the automobile society. Recession was looming large and talk was more of reindustrialization than of post-industrial society.

So it was the Renault engineers, in alliance with the contaminating catalysts and aided by the increasing weakness of the protest movements, who completely rehabilitated the traditional motorcar, although the motorcar underwent some subtle changes in the process (it polluted less, used less petrol, cost less to manufacture, etc.) At the same time, they reconstructed French society (present and future) in a different way. This time it was the EDF engineers' turn to remain silent. They had completely lost their position of strength. In the space of a few months the VEL had become a fiction that no one could believe in any longer. The proclaimed revolution had failed to materialize. EDF's engineers had lost. Their "failure" may turn out to be short-lived, for nobody knows what the future holds.

But in the 1980s, contrary to what the EDF engineers confidently predicted, French society has reaffirmed the traditional motor with its attendant struggle for status, and there is no market for the VEL.

This was a remarkable controversy. The engineer-sociologists of EDF were matched by Renault's engineer-sociologists, who developed a sociology that in its arguments and its analyses was close to Bourdieu's. EDF against Renault is, on another stage and with different stakes and new rules, Touraine against Bourdieu.

The failure of the VEL can legitimately be ignored by sociologists. They have a perfect right to want their analyses judged elsewhere than in the economic sphere. This attitude, as defensible as it might be, seems to me only half convincing. Given the similarity of the controversies, should not sociologists take an interest in the engineer-sociologists, not to take them as models but in order to enrich their own analyses and to diversify their own methods of investigation?

To go along this path, we must leave behind the radical difference that separates sociologists and engineer-sociologists. Sociologists, when they develop, as Bourdieu and Touraine did, analyses that are opposed to each other point for point, can coexist without problems, just as in those pre-paradigmatic situations so well described by Kuhn (1970). For engineer-sociologists this sort of ambiguous situation did not make any sense. Either the VEL would find a market and eliminate competing techniques, or it would become a fiction without a future, thus leaving the road free for the traditional automobile. Both the VEL and the traditional motorcar could not be developed at the same time for the same purpose.

In order to transform the study of technologies into a tool of sociological analysis, I find it appropriate to answer this question: What is the particular faculty that engineers have (which sociologists in this case lack) of being able to evaluate the comparative merits of contradictory sociological interpretations? In order to answer this question, I briefly consider the notion of the actor network, which allows the characterization of the original contribution of the engineer-sociologists: the idea of heterogeneous associations.

Actor Networks

As has been noted in the EDF-Renault controversy, the engineers' projects had mixed and associated heterogeneous elements whose identity and mutual relations were problematic. For example, electrons, batteries, social movements, industrial firms, and ministries had been linked together.

The success of the construction was measured by the solidity and longevity of the heterogeneous associations that were proposed by the engineers.[15] For them, it was not simply a matter of supporting a biased interpretation of French society and consumer tastes. They were attempting to link together fuel cells, electric vehicles, and consumers who were to accept using the VEL as a simple means of transportation despite its rather mediocre performance. The proposed associations, and by consequence the project itself, would hold together only if the different entities concerned (electrons, catalysts, industrial firms, consumers) accepted the roles that were assigned to them. To describe these heterogeneous associations and the mechanisms of their transformation or consolidation, I introduce the notion of an actor network.

The actor network is reducible neither to an actor alone nor to a network. Like networks it is composed of a series of heterogeneous elements, animate and inanimate, that have been linked to one another for a certain period of time (Schwartz Cowan, this volume). The actor network can thus be distinguished from the traditional actors of sociology, a category generally excluding any nonhuman component and whose internal structure is rarely assimilated to that of a network. But the actor network should not, on the other hand, be confused with a network linking in some predictable fashion elements that are perfectly well defined and stable, for the entities it is composed of, whether natural or social, could at any moment redefine their identity and mutual relationships in some new way and bring new elements into the network. An actor network is simultaneously an actor whose activity is networking heterogeneous elements and a network that is able to redefine and transform what it is made of. I show in the case of the VEL that this particular dynamic can be explained by two mechanisms: simplification and juxtaposition.

Simplification is the first element necessary in the organization of heterogeneous associations. In theory reality is infinite. In practice actors limit their associations to a series of discrete entities whose characteristics or attributes are well defined. The notion of simplification is used to account for this reduction of an infinitely complex world.[16]

For example, towns consist of more than public transport, the wish to preserve town centers, and the town councils composed of their spokespeople. They differ from one another with respect to population, history, and geographical location. They conceal a hidden life in which anonymous destinies interact. So far as the EDF engineers were concerned, however, towns could be reduced to city councils whose task is the development of a transport system that does not increase the level of pollution.

EDF's engineers did not need to know more. This definition would remain realistic so long as the simplification on which it was based was maintained. In other words, such simplifications will be maintained so long as other entities do not appear that render the world more complex by stigmatizing the reality proposed by them as an impoverished betrayal: The town council is not representative; living conditions in different neighborhoods cannot be reduced to those in the town center; and the system of public transport is but one aspect of a larger urban structure. The same was true for fuel cells. If the catalysts and electrolytes that were trusted became contaminated or destabilized, the fuel cell, which it was hoped would power the VEL, would become appallingly complex. Instead of being easily mastered, fuel cells were transformed into an apparatus whose ever-increasing elements turned out to be beyond control. A "black box" whose operation had been reduced to a few well-defined parameters gave way to a swarm of new actors: scientists and engineers who claimed to hold the key to the functioning of the fuel cell, hydrogen atoms that refused to be trapped by the cheaper catalysts, third world countries that raised the price of precious metals, etc.[17]

Behind each associated entity there hides another set of entities that it more or less effectively draws together. We cannot see or know them before they are unmasked. Hydrogen fuel cells and zinc/air accumulators were two of the elements that made up the world built by EDF's engineers; however, the controversies that developed in their name rapidly divided them into a series of other elements (much as a watch is dismantled by a jeweler to find out what is wrong with it). Thus simplification is never guaranteed. It must always be tested. The catalyst gave way and the fuel cell broke down, thus causing the downfall of the EDF. As for the catalysts, the electrolytes can be decomposed into a series of constituent elements: the electrons in the platinum and the migrating ions. These elements are revealed only if they are brought into a controversy, that is, into a trial of strength in which the entity is under suspicion. Of course, what there is to say about fuel cells, catalysts, and electrons is also true of city councils or administrations. In the project of the EDF engineers, the city was reduced to the city-council-that-wants-to-preserve-the-city-center-at-all-costs. But to preserve its integrity, the city council must stabilize the elements that hold it together: the middle class electorate that trusts it, the pedestrian precinct that pushes the flow of traffic to the edge of the town center, the urban spread, and the system of public transport that enables the inhabitants of the suburbs to come and do their shopping in the town center.

Such a simplified entity exists only in context, that is, in juxtaposition to other entities to which it is linked. Fuel cells, Renault as a car body builder for the VEL, and users who no longer consider the car to be a status symbol are all interrelated. Remove one of these elements and the whole structure shifts and changes. The set of postulated associations is the context that gives each entity its significance and defines its limitations. It does this by associating the entity with others that exist within a network. There is thus a double process: simplification and juxtaposition. The simplifications are only possible if elements are juxtaposed in a network of relations, but the juxtaposition of elements conversely requires that they be simplified.

These juxtapositions define the conditions of operation for the engineers' construction. In fact, it is from these juxtapositions that the associations draw their coherence, consistency, and structure of relationships that exists between the components that comprise it. If they were not placed in a network, these elements would be doomed. These relationships, which define the contribution of each element as well as the solidity of the construction as a whole, are varied. One must abandon the conventional sociological analysis that tries to adopt the easy solution of limiting relationships to a restricted range of sociological categories. Of course, there may be exchange relationships (the user exchanges money for a VEL), subcontractual relationships (the CGE works for EDF), power relationships (EDF brings Renault to its knees), or relationships of domination. But often the relationships between entities overflow simultaneously into all these categories, and some escape completely from the vocabulary of sociology or economics. How can one describe the relationships between fuel cells and the electric motor in terms other than those of electric currents or electromagnetic forces? Not only are the associations composed of heterogeneous elements but their relationships are also heterogeneous. Whatever their nature, what counts is that they render a sequence of events predictable and stable. Hydrogen feeds the fuel cells that power the motor that ensures the performance of the VEL for which the users are willing to pay a certain price. Each element is part of a chain that guarantees the proper functioning of the object. It can be compared to a black box that contains a network of black boxes that depend on one another both for their proper functioning as individuals and for the proper functioning of the whole. What would the battery be without hydrogen? What would become of consumers without their VELs?

Therefore the operations that lead to changes in the composition and functioning of an actor network are extremely complex. The extent to which an entity is susceptible to modification is a function of the way in which the entity in question summarizes and simplifies one network on behalf of another. If we wish to construct a graphical representation of a network by using sequences of points and lines, we must view each point as a network that in turn is a series of points held in place by their own relationships. The networks lend each other their force. The simplifications that make up the actor network are a powerful means of action because each entity summons or enlists a cascade of other entities. Fuel cells mobilize catalysts, electrons, and ions, which all work for the fuel cell. This, in turn, works for the VEL and the EDF actor network. Through these successive simplifications (which are never as apparent as when they fail) electrons, specialists at Renault, the middle class electorate, and researchers at the CGE have all been enlisted and mobilized. EDF's engineers see and know only fuel cells, accumulators, city council spokespeople, and the public transport authorities. But each of these entities enrolls a mass of silent others from which it draws its strength and credibility. Entities are strong because each entity gathers others. The strength of EDF and the durability of the VEL were built by means of these simplified and mobilized entities. Thus a network is durable not only because of the durability of the bonds between the points (whether these bonds concern interests or electrolytic forces) but also because each of its points constitutes a durable and simplified network. It is this phenomenon that explains the conditions that lead to the transformation of actor networks. It is possible to modify the performance of fuel cells to account for the new demands of users only if the catalysts or electron spin states can be modified in order to increase, for example, the power and longevity of the fuel cell. Each modification thus affects not only the elements of the actor network and their relationships but also the networks simplified by each of these elements. An actor network is a network of simplified entities that in turn are other networks.

Transformation thus depends on testing the resistance of the different elements that constitute our actor network.[18] Is it easier to change the expectations of the users, the demands of the municipalities, the interests of Renault, or the longevity of platinum? This is a practical question that is answered through the continual adjustments that are also negotiated changes. To adapt the VEL by changing this or that aspect of its performance is to act on the actor network, and its success thus depends on the capacity to test certain resistances to their limits, whether these spring from social groups, cash flow, or electrodes to be improved.

An actor network, such as the one described in this chapter, can in turn be simplified. The solidity of the whole results from an architecture in which every point is at the intersection of two networks: one that it simplifies and another that simplifies it. It can be mobilized in other actor networks. For example, the VEL can be linked to the TGV (high-speed train) or to the Airbus, thus forming a part of a new French transport policy. Although simplified into a point and displaced in this mariner, it is still composed of associated entities, and although these entities are susceptible to being molded or shaped, they in turn may transform the actor network of which they form a part.

The actor network describes the dynamics of society in terms totally different from those usually used by sociologists. If car users reject the VEL and maintain their preferences for different types of the traditional motorcar, this is for a whole series of reasons, one of which is the problem of the catalysts that turn poisonous. It is these heterogeneous associations that sociologists are unable to take into account and yet that are responsible for the success of a particular actor network. The post-industrial society that Touraine believes is coming depends in this particular case not only on the capacity of new protest movements to influence the choices of technocrats but also on the way in which the catalysts in the fuel cells behave. Tourainian sociological theory, as with most other sociological theories, remains a clever and sometimes perspicacious construction; but it is bound to remain hypothetical and speculative because it simplifies social reality by excluding from the associations it considers all those entities—electrons, catalysts—that go to explain the coevolution of society and its artifacts. This criticism applies equally well to Bourdieu's interpretation of society. Although his theory happens to work better (explaining the success of the Renault actor network), this is pure luck, for in his explanation of car users' preferences he omits most of the elements that make up and influence these preferences. Although Bourdieu happens to be right and Touraine wrong, this is quite by chance. Although Renault turns out to be right, this is because the heterogeneous associations proposed by the EDF engineers disintegrate one by one: the discovery of a cheap catalyst as a substitute for platinum might have proved Bourdieu wrong and rehabilitated Touraine's sociological theory after all.

A New Methodological Tool

In what way could the analyses and the experiments developed by the engineer-sociologists be useful to sociology?

It was in order to answer this question that I introduced the idea of the actor network, which allows us to measure the distance between the heterogeneous and "impure" sociology of the engineers and the "pure" and homogeneous sociology of the sociologists. In the one case sociological and technical considerations are inextricably linked; in the other they are rigorously dissociated. If EDF and Renault sociology cannot be compared with that of Bourdieu and Touraine it is because its success depends not only on the behavior of traditional social actors but equally on that of catalysts or zinc/air batteries.

One can choose to be satisfied with this declaration and maintain the splendid isolation of academic sociology by underlining the radical difference between it and that of the engineer-sociologists. I would like to suggest now that this defensive position, which seeks to safeguard the orthodoxy, cost what may, is not the only possible one. According to whether one is more or less disposed to transform sociology itself, other more or less radical choices can be envisaged. They all lead to a transformation of the study of technology into an instrument of sociological analysis.

First of all, and this does not in the least imperil sociology, it is possible to use the controversies in which the engineer-sociologists are engaged as particularly powerful tools of investigation. To learn about society, sociologists employ tools that have been developed and tested over years: surveys, interviews, opinion polls, participant observations, statistical analyses, and so on. Another way of learning about society, as shown in this chapter, is to follow innovators in their investigations and projects. This method is particularly effective in cases in which, because they are working on radical innovations, engineers are forced to develop explicit sociological theories. In such cases this method enables sociologists to explore large sections of society (peering over the engineer's shoulder, so to speak). It is in this way that any sociologist, whether or not he or she knows anything about Touraine, could have found in the analyses of the engineer-sociologists of EDF valuable aids to the development of an analysis of the role of social movements in the evolution of consumption.

The study of engineer-sociologists can furnish more than a simple source of inspiration. In effect, the sociology developed by the engineer-sociologists is concretely evaluated in terms of market share, rate of expansion, or profit rate. With the failure of the VEL, EDF's theories about French society and its future collapsed (although perhaps only provisionally). The sociologist has here a powerful tool for evaluating different sociological frameworks of analysis. Engineer-sociologists, then, work for the good of sociology. The sociologists can rest content with following the engineer-sociologists, picking up their analyses and examining the way in which they are refuted or validated by the success or failure of the technical apparatus the engineer-sociologists have helped to bring into being. The results of the test may not necessarily be wholly positive or wholly negative. The case under discussion happens to show a complete reversal of fortune. But in other situations engineers may arrive at a compromise solution and progressively change their sociological interpretations, that is, their associations, and consequently change the shape of the technological devices they develop. In any event sociologists who study engineers shaping technologies have a chance to evaluate the validity of certain interpretations and to follow their successive adaptations in the light of the resistance they encounter.

But the sociologists, if they want, can be still more audacious, can display an audacity equal to that of the engineer-sociologists. They can, and this is the path I urge them to envisage, put into question the very nature of sociological analysis itself. From this point of view the study of technology can play a critical role. Instead of being someone whose ideas and experiments can be turned to the advantage of the sociologist, the engineer-sociologist becomes the model to which the sociologist turns for inspiration. The notion of the actor network then becomes central, for it recognizes the particular sociological style of the engineer-sociologist. To transform academic sociology into a sociology capable of following technology throughout its elaboration means recognizing that its proper object of study is neither society itself nor so-called social relationships but the very actor networks that simultaneously give rise to society and to technology.

As I have noted, the functioning of what I propose to call actor networks is not adequately described by the usual frameworks of sociological analysis. In short, not only does the repertoire of associated entities extend beyond that generally accepted in social science but also the composition of this repertoire does not obey any definitive rules. How can the social elements be isolated when an actor network associates the spin of an electron directly with user satisfaction? How can any interpretation of social interaction be established when actor networks constantly attempt to transform the identities and sizes of actors as well as their interrelationships? The fact that actor networks constantly create new combinations of entities renders this task even more difficult. The notion of actor network is developed in order to handle these difficulties. This notion makes it possible to abandon the constricting framework of sociological analysis with its pre-established social categories and its rigid social/natural divide. It furnishes sociological analysis with a new analytic basis that at a stroke gains access to the same room to maneuver and the same freedom as engineers themselves employ.

Dedicated to understanding the working of actor networks, whose analysis is still to be done, sociology will henceforth find itself on new terrain: that of society in the making. It will also progress resolutely along the path opened by Hughes in his different studies (1983 and this volume) consecrated to technological systems. If, however, we prefer the idea of actor network to that of system, it is essentially for two reasons.

First, the engineers involved in the design and development of a technological system, particularly when radical innovations are involved, must permanently combine scientific and technical analyses with sociological analyses: The proposed associations are heterogeneous from the start of the process. The concept of actor network can be used to explain both the first stages of the invention and the gradual institutionalization of the market sometimes created as a result without distinguishing between successive phases. It is applicable to the whole process because it encompasses and describes not only alliances and interactions that occur at a given time but also any changes and developments that occur subsequently. Certain simplifications become impossible to implement; associations are no longer taken for granted. The actor network is modified under the influence of the forces it seeks, although not always successfully, to enroll; but its structure remains that of an actor network whose development can be traced and followed. The concept enables sociologists to describe given heterogeneous associations in a dynamic way and to follow, too, the passage from one configuration to another.

This leads to the second point I would like to mention, if only briefly. The systems concept presupposes that a distinction can be made between the system itself and its environment. In particular, certain changes can, and sometimes must, be imputed to outside factors. The actor-network concept has the advantage of avoiding this type of problem and the many difficult questions of methodology it raises. For example, how do we define the limits of a system and explain concretely the influence of the environment? To answer such questions precisely, we must develop a formal science of systems, thus possibly depriving the analysis of all its descriptive and explanatory value. Hughes manages to avoid this pitfall by using the systems concept in a pragmatic way.[19] By stressing continually all the connections linking the "inside" and "outside" of the system, he comes close to the actor-network concept. By abandoning the concept of system for that of actor network, I believe we are taking Hughes's analysis—neatly summed up in the ambivalent title of his book, *Networks of Power*—a step further.

Notes

I especially thank Ruth Schwartz Cowan and Gerard de Vries for their sharp criticism, which I have probably failed, most of the time, to answer.

1. For an overview of social studies of technology, see MacKenzie and Wajcman (1985).

2. Several studies have been made to clarify the respective roles played by science, technology, and the market in the beginning and development of an innovation. Put in these terms, the question does not have a general answer. The first reason for this is that it is difficult to draw an indisputable boundary between science and technology. The sociology of science of the last ten years has shown empirically that it is impossible to give a general definition of scientific activity (Knorr-Cetina and Mulkay 1983) and has contested the idea of a noncontroversial distinction between science and technology (Callon 1981b). In addition, for a given innovation it is quite often impossible to outline a genealogy in which scientific and technological contributions that are linked to an innovation can be unquestionably separated. This is what two studies—HINDSIGHT (Sherwin and Isenson 1967) and TRACES (Illinois Institute of Technology, 1968)—have shown.

 Anyway, it is difficult to distinguish market influences from those of science and technology. This is the conclusion of C. Freeman after having reviewed literature pertaining to this question. Following Mowery and Rosenberg (1979), his critique of two models, "technological push" and "demand pull," led him to propose the notion of "coupling," which leaves all possibilities of interaction open and recognizes that uncertainties in the market and sciences are the very motor of innovation. "The fascination of innovation lies in the fact that both the market and the technology are continually changing. Consequently there is a kaleidoscopic succession of new possible combinations emerging" (Freeman 1982, p. 111). Or "the test of successful entrepreneurship and good management is the capacity to link together these technical and market possibilities by combining the two flows of information" (Freeman 1982, p. 111). Freeman correctly notes that "the notion of 'perfect' knowledge of the technology or of the market is utterly remote from the reality of innovation, as in the notion of equilibrium" (1982, p. 111). It is because the innovation is caught between two series of uncertainties, the first concerning the market and the state of society and the second related to the state of knowledge, that it is impossible to describe it other than as an interactive process (Nelson and Winter 1977). Moreover, this point is confirmed by authors such as Peters and Austin (1985) when they seek to identify the organizational forms that favor innovation. Leaning on numerous case studies, they show that innovation is always a compromise that results from a long series of trials, which are at the same time technical and socio-economic. Hughes (this volume) develops this argument in detail. See also Kidder (1982), Jewkes et al. (1969), and Callon and Latour (1986).

3. For this point, see the revealing studies of C. Freeman (1982) concerning research and development of synthetic materials and electronics.

4. This hypothesis is often formed by those who are interested in radical innova-tions. For two examples of this perspective in the fields of economics and history, see two excellent books: Mensch (1979) and Constant (1980).

5. Concerning this point, see the enlightening demonstration provided by Hughes (1983). The cases studies by Bijker and Pinch (1984; also Bijker, this volume), using the notion of interpretative flexibility, also show the impossibility of separating the definition of technical problems from the socioeconomic context to which the inventors associate them. See also Callon (1986).

6. As Woolgar has shown (this volume), engineers are not content with just analyz-ing the society around them. They do not hesitate, if need be, to play the psycholo-gist and propose interpretations of the cognitive capacities of humans.

7. The EDF is a public company that has a monopoly on the production and dis-tribution of electricity. It devotes a large part of its budget to research in the devel-opment of uses for electricity.

8. To study this project, I was able to consult all the archives of different ministries that at one time or another supported the VEL financially. Several interviews were carried out with the different protagonists.

9. This has been analyzed well by Hughes (1983), who shows how Edison conceived the incandescent lamp.

10. In this text the term "battery" is used as a generic term to cover all portable chemical devices for generating electricity.

11. For two contradictory analyses of the May 1968 movement, see Aron (1968) and Touraine (1968).

12. These unforeseen alliances between human beings and animate or inanimate nonhumans have been analyzed in detail by Latour (1984) and Callon (1986).

13. Castoriadis asserts that technology creates what nature is not capable of achiev-ing. How does technology succeed? It succeeds by playing with the differences of resistances that exist within the environment that it uses and transforms, for this environment does not resist in any way and it does not resist stubbornly. Reality is not static because it consists of interstices that permit it to move, gather, alter, and divide; thus there is room to "make." Whether it concerns outside nature, the neighboring tribe, or bodies of people, resistance is regulated. It contains lines of force, veins, and partially systematic progressions. "Technology thus brings about

the division of the world into the following two fundamental regions, which render it human: those elements which resist in all cases and those elements which (at a given stage of their history) resist only in a certain fashion" (Castoriadis 1968). I do not need to be so extreme; I have only to establish a general map of the differentiated resistances that are met by the actors (Latour 1984; Callon and Latour 1981).

14. CGE is a company that specializes in electrotechnology.

15. Concerning the definition and the use of the notion of heterogeneous engineering, see Law (this volume). See also the case of Draper Laboratories studied by MacKenzie (this volume).

16. There is an analogy here with scientific theory. As Hesse (1974) has so persuasively argued, description always entails loss of information and simplification. For a full development of this argument, see Law and Lodge (1984).

17. On the notion of black boxing as a form of simplification, see Callon (1981a) and Law (1985).

18. For a detailed empirical study of the mechanisms of the transformation of an actor network, see Law (1984b).

19. Concerning Hughes's pragmatism in his use of the notion of systems, see the excellent review of *Networks of Power* by Barnes (1984).

II Simplifying the Complexity

Introduction

As we noted in the general introduction, many of the recent developments in the social study of technology consist of detailed empirical analyses of the "content" of various technological artifacts and systems and their environment. This "thick description" results in a wealth of detailed information about the technical, social, economic, and political aspects of technology. In order to contribute to our overall understanding of technology, however, we need to simplify and structure this wealth of information, thus creating order out of the chaos of data. Of course, the construction of models or of "middle-range" theories necessarily oversimplifies the rich texture of each case, but this need not worry us, so long as it is recognized and thus made amenable to critical discussion. A model that incorporates every aspect of the case it deals with would fail to serve its function, being no more than a re-creation or redescription of the original case (Constant 1980; Laudan 1984lb, p. 2). The three chapters in part II deal explicitly with this issue of how to order the results of detailed case studies in this field. Not only are more specific concepts advanced than in the previous section, but there is also more concern with how such concepts may be operationalized. For instance, issues such as how networks and social groups may be identified are addressed. In this section, as is the case throughout the book, the authors provide detailed empirical examples with which to illustrate their arguments.

John Law treats technology as being constituted by "heterogeneous elements . . . which range from people, through skills, to artifacts and natural phenomena." These elements are associated by means of networks. Because for the purposes of this particular study Law has chosen to follow the system-building approach and to focus on the building of the network, various actors play important parts as "heterogeneous engineers." The processes that are at the heart of the matter in Law's model are those of the association of forces by which the heterogeneous elements are built into more or less durable networks. Law's actors can be seen as equivalent to Hughes's system builders.

Henk van den Belt and Arie Rip present an evolutionary model of technological change by extending the sociological themes to be found within approaches developed by the economists Nelson and Winter (1977, 1982) and Dosi (1982, 1984). Kuhn's term "paradigm," as developed in the context of the history of science, has been taken up by these economists as a way of understanding the patterns to be found in the activities of innovating firms. Van den Belt and Rip extend the somewhat mechanistic model concept of paradigm used in this work by introducing Kuhn's notions of exemplar and cultural matrix. The variation and selection processes that are, in the model of Van den Belt and Rip, below the "surface level" often result in a "nexus"—a social institution that carries and shapes the interaction between a "technological" trajectory and the selection environment. They discuss examples of such nexuses as the patent system and patent case law and the test laboratories set up in the synthetic dye industry.

Wiebe Bijker, pursuing the social constructivist approach, elaborates on two particular concepts that he has found useful in the course of his Bakelite case study. The concept of a technological frame refers to the ways in which relevant social groups attribute various meanings to an artifact. The concept of inclusion is introduced to account for the observation that there are varying degrees of interaction within any one technological frame. The social processes underpinning the construction of artifacts, as described earlier in this volume (Pinch and Bijker), are explained in terms of the differing degrees of inclusion of actors within different technological frames. Bijker concludes by outlining a typology of different sorts of technological change that his framework generates.

Besides elaborating sets of rather distinct concepts, the chapters in this part of the volume also discuss various common themes that are central to the research program as a whole.

With regard to the seamless web of technology and society, Law argues for a principle of "generalized symmetry," meaning that "the same type of explanation should be used for all the elements that go to make up a heterogeneous network whether these be devices, natural forces, or social groups." Thus Law does not want to reserve a special explanatory status for social elements. Clearly, this would represent an important departure from the social constructivist approach if by this Law is claiming that the natural exists on its own and outside of social interaction. However, Law remarks that "nature reveals its obduracy in a way that is relevant only to the network when it is registered by the system builders." Thus for Law it would appear that nature is rendered only by the activities of system builders. In the social constructivist approach, the key point is not that the social is given any special status *behind* the natural; rather, it is claimed that there is nothing but the social: socially constructed natural phenomena, socially constructed social interests, socially constructed artifacts, and so on. Indeed, Bijker uses the concept of a technological frame to show how seamless the social constructivist web can be. This point can be further illuminated by taking up the issue of what may count as *explanans* and what as *explanandum* within the two approaches. Law's generalized symmetry leads him to accept the natural as *explanandum* as well as *explanans*. This point is also argued by Bijker. Of course, in Bijker's approach it is the socially constructed natural world that may enter the account as *explanans*. The same issue is briefly discussed from quite a different perspective by Van den Belt and Rip. They question the possibility of combining the radical social constructivist account of technological artifacts with an evolutionary model of the development of those artifacts.

The methodological point of how to demarcate relevant social groups and networks is discussed by Law and Bijker. They both argue that it is not possible to give an easy general rule for identifying networks or social groups; that is, the operationalizations of the general definitions of network and social group will not be the same in every situation. In a sense all analyses must start somewhere, and these authors do not want to be over-specific as to where the starting point should be. Typically, the starting point varies from case to case. In Law's model "the extent of a network is defined by the presence of actors that are able to make their presence individually felt." In Bijker's model "the key element in the identification of a relevant social group is a shared meaning attribution."

The concepts of technological paradigm (Van den Belt and Rip) and technological frame (Bijker) have much in common, despite the differences discussed by Bijker. Perhaps the most intriguing analogy is the role given to Kuhn's notion of exemplar. For Van den Belt and Rip, "exemplar" denotes a cluster of successful heuristics around an exemplary artifact. In Bijker's model, "exemplar" is used to label an artifact that, in stabilizing, structures the technological frames of relevant groups. Because problem-solving strategies are an element of technological frames, the similarity is obvious.

The chapters presented in this part of the book clearly have a close relationship with the preceding three chapters. For instance, the work of Bijker stems directly from the social constructivist approach of Pinch and

Bijker, and Law's chapter arises from his earlier collaborative work with Callon. The themes of paradigms and evolutionary models discussed by Van den Belt and Rip are issues discussed by Pinch and Bijker. The authors of these chapters frequently refer to each other's work, especially to high-light differences between their respective approaches. Without minimizing the importance of these differences, we think that the similarities between the three models are just as prominent. This we consider to be a promising sign of the growing coherence of the field.

Technology and Heterogeneous Engineering: The Case of Portuguese Expansion

John Law

If you want to learn how to pray, go to sea.
—Portuguese proverb, quoted by Diffie and Winius (1977)

How do objects, artifacts, and technical practices come to be stabilized? And why do they take the shape or form that they do? In this chapter I advocate and exemplify an approach to these questions that stresses (1) the heterogeneity of the elements involved in technological problem solving, (2) the complexity and contingency of the ways in which these elements interrelate, and (3) the way in which solutions are forged in situations of conflict. This "network" approach draws on and parallels work by Callon (1980 and this volume) and is developed in relation to secondary empirical material about the technology of the fifteenth- and sixteenth-century Portuguese maritime expansion. In order to clear the ground and situate my argument, I start by commenting briefly on two alternative approaches to the social study of technology.

The first approach is sometimes called social constructivism.[1] This outgrowth of the sociology of science assumes that artifacts and practices are underdetermined by the natural world and argues that they are best seen as the *constructions* of individuals or collectivities that belong to social groups. Because social groups have different interests and resources, they tend to have different views of the proper structure of artifacts. Accordingly, the stabilization of artifacts is explained by referring to social interests that are imputed to the groups concerned and their differential capacity to mobilize resources in the course of debate and controversy. Social constructivists sometimes talk of this process as one of "closure." Closure is achieved when debate and controversy about the form of an artifact is effectively terminated.

The merits of the social constructivist approach are obvious. Many artifacts are, indeed, forged in controversy and achieve their final form when a social group, or set of groups, imposes its solutions on other interested parties by one means or another. The fate of the electric vehicle in France (Callon, this volume) is amenable to such analysis, as are such other cases as the British TSR-2 aircraft (Law 1985), the Concorde aircraft (Feldman 1985), the third airports of London and Paris (Feldman 1985), the bicycle (Pinch and Bijker 1984 and this volume), and aspects of the development of missile guidance systems (MacKenzie, this volume).[2] Indeed, it is easy to think of examples. Whenever there is controversy, the contingent and constructed nature of artifacts becomes manifest, and explanations in terms of differential power and social interests become attractive.

The second approach, which comes from the history of technology and in particular from the work of T. P. Hughes (1979a, 1983, this volume), understands technological innovation and stabilization in terms of a systems metaphor. The argument is that those who build artifacts do not concern themselves with artifacts alone but must also consider the way in which the artifacts relate to social, economic, political, and scientific factors. *All* these factors are interrelated, and all are potentially malleable. The argument, in other words, is that innovators are best seen as system builders: They juggle a wide range of variables as they attempt to relate the variables in an enduring whole. From time to time strategic problems arise that stand in the way of the smooth working or extension of the system. Using a military metaphor, Hughes talks of these problems as reverse salients, and he shows the way in which entrepreneurs tend to focus on such problems and juxtapose social, technical, and economic variables as they search for a solution.

Hughes's study of Edison illustrates both the systemic nature of much technological activity and the importance of the notion of a reverse salient. Edison's problem (his reverse salient) was simultaneously economic (how to supply electric lighting at a price that would compete with gas), political (how to persuade politicians to permit the development of a power system), technical (how to minimize the cost of transmitting power by shortening lines, reducing current, and increasing voltage), and scientific (how to find a high-resistance incandescent bulb filament). That Edison succeeded in resolving this set of problems reveals his success as a system builder, and it also shows that, as Hughes puts it, "the web is seamless"—that the social was indissolubly linked with the technological and the economic.[3]

The social constructivist and systems approaches have much in common. First, they concur that technology is not fixed by nature alone. Second, they agree that technology does not stand in an invariant relation with science. Third, and most important, they both assume that technological stabilization can be understood only if the artifact in question is seen as being interrelated with a wide range of nontechnological and specifically social factors. However, when they specify the relationship between the technological and the social, they start to diverge. Social constructivism works on the assumption that the social lies *behind* and directs the growth and stabilization of artifacts. Specifically, it assumes that the detection of relatively stable directing *social interests* offers a satisfying explanation for the growth of technology. By contrast, the systems approach proceeds on the assumption that the social is not especially privileged. In particular, it presupposes that social interests are variable, at least within certain limits. Although it is true that even on this point the two approaches are starting to reveal some degree of convergence,[4] the basic difference remains: In the end the sociologists prefer to privilege the social in the search for explanatory simplicity, whereas many historians have no such commitment.[5]

In this paper I join forces with Callon and side with the historians in this particular argument. Specifically, I want to suggest that in explanations of technological change the social should not be privileged. It should not, that is, be pictured as standing by itself *behind* the system being built and exercising a special influence on its development. Although it may at times be an important—indeed the dominant—factor in the growth of the system, this is a purely contingent matter and can be determined only by empirical means. Other factors—natural, economic, or technical—may be more obdurate than the social and may resist the best efforts of the system builder to reshape them. Other factors may, therefore, explain better the shape of artifacts in question and, indeed, the social structure that results. To put this more formally, I am arguing, in common with Callon (this volume, 1980b, 1986), that *the stability and form of artifacts should be seen as a function of the interaction of heterogeneous elements as these are shaped and assimilated into a network.* In this view, then, an explanation of tech-nological form rests on a study of both the *conditions* and the *tactics* of system building. Because the tactics depend, as Hughes has suggested, on the interrelation of a range of disparate elements of varying degrees of malleability, I call such activity *heterogeneous engineering* and suggest that the product can be seen as a *network* of juxtaposed components.[6]

As is obvious, this network approach borrows much from Hughes's system-building perspective. There is, however, at least one important way in which it differs from Hughes's approach, and this difference arises from the emphasis within the network approach on conflict. Thus, as the example of the Portuguese, or indeed those of Edison or Renault, reveals, successful large-scale heterogeneous engineering is difficult. Elements in the network prove difficult to tame or difficult to hold in place. Vigilance and surveillance have to be maintained, or else the elements will fall out of line and the network will start to crumble. The network approach stresses this by noting that there is almost always some degree of divergence between what the elements of a network would do if left to their own devices and what they are obliged, encouraged, or forced to do when they are enrolled within the network. Of course, some of these differences are more pressing than others. For the purposes of analysis, however, the environment within which a network is built may be treated as hostile, and heterogeneous engineering may be treated as the association of unhelpful elements into self-sustaining networks that are, accordingly, able to resist dissociation.

This suggestion has an important methodological implication: *It makes sense to treat natural and social adversaries in terms of the same analytical vocabulary.* Rather than treating, for instance, the social in one way and the scientific in another, one seeks instead to follow the fortunes of the network in question and consider its problems, the obduracy of the elements involved in those problems, and the response of the network as it seeks to solve them. As one moves from element to element, no change in vocabulary is necessary; from the standpoint of the network those elements that are human or social do not necessarily differ in kind from those that are natural or technological. Thus the point is not, as in sociology, to emphasize that a particular type of element, the social, is fundamental to the structure of the network; rather it is to *discover* the pattern of forces as these are revealed in the collisions that occur between different types of elements, some social and some otherwise. It is to this task that I now turn.[7]

The Struggle between Cape Bojador and the Galley

In 1291 Ugolino and Vadino Vivaldi set sail from Genoa in two galleys, passed through the Pillars of Hercules "ad partes Indiae per mare oceanum," and vanished, never to be seen by any European again (Diffie and Winius 1977, p. 24; Chaunu 1979, p. 82). In 1497 Vasco da Gama sailed from the Tagus in Lisbon. He too was headed for the Indes by way of the ocean, but unlike the brothers Vivaldi we know what became of his expedition. On May 20, 1498, he anchored in the Calicut Road off the Malabar Coast of southwest India. He entered into negotiations with the Samorin of Calicut about trading in spice. So unsuccessful were these talks that on his second expedition in 1502 da Gama's now heavily armed fleet bombarded the town of Calicut in an effort to force the Samorin into submission (Parry 1963, p. 153). The Portuguese spice trade had begun and with it their domination of the Indian Ocean. I want to suggest that the process that led to this domination can be looked at from the standpoint of system building or heterogeneous engineering. Sometimes the opponents were people, and sometimes they were natural objects. Let me start, then, by talking of galleys.

The galley was primarily a war vessel. It was light and maneuverable, a method for converting the power of between 150 and 200 men into efficient forward motion. In order to reduce water resistance, the galley was long and thin—typically, at least in Venice, about 125 feet in length and 22 feet wide, including outriggers (Lane 1934, p. 3). The hull was lightly sparred, and the planks were laid in carvel fashion, edge to edge to minimize water resistance. The galley was also low. The oarsmen pulled, three to an oar, on between twenty-five and thirty oars on each side. The vessel also carried one mast (possibly more than one, see Landstrom 1978, p. 52), stepped well forward, which carried a triangular lateen sail. This sail assisted the oarsmen, although it was never more than an auxiliary source of power. The ship was steered by means of one or two rudders, and the stern was slightly raised into a "castle." By contrast, the bow was low and pointed, being designed to ram other ships. A typical galley is illustrated in figure 1.

Now let me state the obvious: The galley is an *emergent phenomenon;* that is, it has attributes possessed by none of its individual components. The galley builders associated wood and men, pitch and sailcloth, and they built an array that floated and that could be propelled and guided. The galley was able to associate wind and manpower to make its way to distant places. It became a "galley" that allowed the merchant or the master to depart from Venice, to arrive at Alexandria, to trade, to make a profit, and so to fill his palace with fine art.

The galley is, of course, a technological object. Let me, then, define technology as a family of methods for associating and channeling other entities and forces, both human and nonhuman. It is a method, one method, for the conduct of heterogeneous engineering, for the construction of a relatively stable system of related bits and pieces with emergent properties in a hostile or indifferent environment.

When I say this, I do not mean that the methods are somehow different from the forces that they channel. Technology does not act as a kind of

Figure 1
A galley (Girolamo Tagliente, *Libro Dabaco che Insegnaa fare ogni Ragione Mercantile* (Venice: Raffinello, 1541), 53).

traffic policeman that is distinct in nature from the traffic it directs. It is itself nothing other than a set of channeled forces or associated entities. Thus there is always the danger that the associated entities that constitute a piece of technology will be dissociated in the face of a stronger and hostile system. Let us, therefore, consider the limitations of the galley.

As a war machine in the relatively sheltered waters of the Mediterranean the galley was a great success. As a cargo carrying vessel, however, it had its drawbacks. Its carrying capacity was extremely limited. The features that made it a good war ship—it was slim and low and could carry a large crew that might repel boarders—were an impediment to the carriage of cargo (Lane 1973, p. 122; Denoix 1966, p. 142). Furthermore, the *endurance* of the galley was restricted by the size of its crew. It could not pass far from the sight of land and the possibility of water and provisions. Although the Venetians and the Genoese used galleys to transport valuable cargoes, where reliability was called for, they were replaced in this role by the "great galleys" after about 1320 (Lane 1973, pp. 122, 126).

It must have been in such vessels that the brothers Vivaldi left Genoa in 1291 for what they thought would be a ten-year trip to the Indes (Diffie and Winius 1977, pp. 24–52). Perhaps their galleys were larger than normal, precursors of the great galley. Perhaps they had higher freeboards. But their endurance would have been limited and their seaworthiness doubtful—

one can imagine all too well the consequences of running into a storm off the Saharan coast. And, if indeed the Vivaldis were attempting to row down the west coast of Africa, then they would have had to pass what may be regarded as a point of no return—Cape Bojador, or the Cape of Fear. Chaunu summarizes the problem presented by Cape Bojador:

At twenty-seven degrees north, Cape Bojador is already in the Sahara, so there could be no support from the coast. The Cape is 800 kilometres from the River Sous; the round trip of 1,600 kilometres was just within reach of a galley, but it was impossible to go any further without sources of fresh water, except by sail. In addition there were the difficulties . . . [of] the strong current from the Canaries, persistent mists, the depths of the sea bed, and above all the impossibility of coming back by the same route close hauled. (Chaunu 1979, p. 118)

How brave, then, were the Vivaldi brothers and their men when they sailed their galleys past the pillars of Hercules and out of recorded history! We do not know in what form disaster finally struck. What we can guess, however, is that the galleys, emergent objects constituted by a heterogeneous engineer, were dissociated into their component parts. The technological object was dissolved in the face of a stronger adversary, one better able to associate elements than the Italian system builders. It was a conflict between two opponents, a trial of strength, in which part of the physical world had the final say. Accordingly, it is a paradigmatic case of the fundamental problem faced by system builders: how to juxtapose and relate heterogeneous elements together such that they stay in place and are not dissociated by other actors in the environment in the course of the inevitable struggles—whether these are social or physical or some mix of the two. And it also suggests why we must be ready to handle heterogeneity in all its complexity, rather than adding the social as an explanatory afterthought, for a system—here the galley—associates everything from humans to the wind. It depends precisely on a combination of social and technical engineering in an environment filled with indifferent or overtly hostile physical and social actors.

The Portuguese versus Cape Bojador: Closure and Lines of Force

In the struggle between the Atlantic and the galley, the Atlantic was the winner. We might say that the forces associated by the Europeans were not strong enough to dissociate those that constituted the Atlantic. The heterogeneous engineers of Europe needed to associate and channel more and different forces if they were to dissociate such a formidable opponent and put its component parts in their place. So for over a hundred years Cape Bojador remained the point of no return. Where were the new allies to come from? How might they be associated with the European enterprise?

Three types of technological innovation were important.[8] The first of these took the form of a revolution in the design of the sailing ship in the fourteenth and early fifteenth centuries. The details of this revolution remain obscure, circumstantial, and in any case beyond the scope of this essay, but the result was a mixed-rigged seagoing vessel (figure 2) that had much greater endurance and seaworthiness than its predecessors, one that was able to convert winds from many directions into forward motion. There were no rowers, so manpower was reduced, and it was thus possible to carry sufficient stores to undertake a considerable passage without foraging. This, then, was the first step in the construction of a set of allied entities capable of putting the North Atlantic in its place. The second was the fact that the magnetic compass became generally available in Christian Europe in the late twelfth century. I consider methods of navigation in a

Figure 2
Large late fifteenth-century or early sixteenth-century mixed-rigged vessel (frontispiece from a 1537 Venetian edition of Johannes de Sabrosco's *Sphera volgare nouamente tradotte*).

later section, but here it should be noted that the initial importance of this innovation was that it allowed a reasonably consistent heading to be sailed in the absence of clear skies. Combined with dead reckoning and a porto-lano chart,[9] the magnetic compass took some of the guesswork out of long-distance navigation, and in particular it meant that the sailor did not need to hug the coast to have some idea of his location. This, then, was the second decisive step toward a change in the balance of forces. When new ships combined newly channeled winds with new methods of naviga-tion and consequent knowledge of position, the ground was prepared for a possible change in the balance of power.

What was the decisive third step? To answer this question, we must know a little about the currents and winds between Portugal and the Canaries. It is relatively easy to sail from Lisbon or the Algarve in a south-westerly direction along the Atlantic coast of Africa. The ship is carried along by the Canaries current and is also carried before the northeast trade winds, which are particularly strong in summer. So far, then, the forces of wind and current assist the project of the sailor. It is, however, more dif-ficult to make the return journey for precisely these same reasons. In a ship good at beating to windward, it is no doubt possible to make some north-easterly headway. But this requires frequent tacking, something that was difficult in the square-rigged ships of the day, which could not, in any case, sail close to the wind. Although the wind blows from the southwest for a period in the winter, thus making the return journey easy (Diffie and Winius 1977, pp. 61, 136), at some unrecorded point sailors decided to try to put the adverse winds and currents to good use by beating out to seaward, away from the Moroccan coast, for it turns out that, so long as one has an appropriate vessel, some means of determining a heading, and an appropriate dose of courage, it is much easier to return to Lisbon or the Algarve this way than by the coast. The vessel sails on a northwesterly heading close hauled against the northeasterly trades. It is gradually able to take a more northerly course as the trades are left behind until the westerlies and North Atlantic drift are encountered, when it becomes pos-sible to head east in the direction of Iberia (Chaunu 1979, pp. 111–115). It was the invention of this circle, called the *volta* by the Portuguese, that marks the decisive third step. Ships were no longer forced to stay close to the coast. Cape Bojador, the classic point of no return, was no longer the obstacle it had previously been. The masters could sail beyond it and expect to be able to return.

The *volta* can thus be seen as a geographical expression of a struggle between heterogeneous bits and pieces assembled by the Portuguese system builders and their adversaries, that is, the winds, the currents, and the capes. It traces on a map the solution available to the Portuguese. It depicts what the Portuguese were able to impose on the dissociating forces of the ocean with the forces they had available. It shows us in a graphic manner how the Portuguese were able to convert the currents, winds, and the rest from opponents into allies and how they were able to associate these elements with their ships and navigational techniques in an acceptable and usable manner.

Now we begin to see the advantages and the drawbacks of the systems metaphor in an empirical context. The metaphor stresses heterogeneity and interrelatedness, but it also tends to direct attention away from the *struggles* that shape a network of heterogeneous and mutually sustaining elements. System builders try to associate elements in what they hope will be a durable array. They try to dissociate hostile systems and reassemble their components in a manner that contributes to what is being built. But the particular form that (dis-)association takes depends on the state of forces. Some of these are obdurate: Currents and winds cannot be tampered with, such is their strength. Some of them are manipulatable, but only with difficulty. Here, for instance, the square-rigged ship and navigational practices, although not immutable, were difficult to influence. Others, however, may be more easily altered. In this case the course sailed by the vessels on their return journey was a matter for discretion as a result of the advances in shipbuilding and navigation of the previous 150 years. Here there was, in the most literal sense, new room to maneuver. The course was no longer rigidly overdetermined for the system builder. Accordingly, the *volta* may be seen as tracing the state of forces and measuring their relative strengths in a literal way. It re-presented the state of shipbuilding, the state of navigation, the state of seamanship, and their collision with the forces of nature. The *volta* was the extra increment of force that allowed the new network to be stabilized, for the course was suddenly the most malleable element in the conflict between the Portuguese desire to return to Lisbon and the natural forces of the Atlantic.

The Caravel and the African Littoral: Closure and Adaptation

Africa, as the Portuguese were to discover, does not reduce to Cape Bojador. The capacity to get round the cape and then return to European waters was all very well, but there was more coastline to explore. South of the cape the coastline becomes even more inhospitable until the Senegal River and Black Africa are reached.

For most of this tricky exploration the Portuguese made use of caravels. Although the origins of this type of vessel are shrouded in mystery (Landstrom 1978, p. 100; Chaunu 1979, p. 243; Parry 1963, p. 65; Unger 1980, pp. 212–215), its fifteenth-century features are well known. Weighing less than 100 tons and being perhaps 70 to 80 feet from stem to stern (Parry 1963, p. 65), the caravel was unusual in being a long sailing ship, having a length-to-breadth ratio of between 3.3 and 3.8 to 1 (Diffie and Winius (1977, p. 118) suggest 3 to 1). It was carvel built, quite light and fine in lines, and drew little water, having a flat bottom and little freeboard (Parry 1963, p. 65; Denoix 1966, p. 143; Landstrom 1978, p. 100). It had only one deck and indeed was sometimes even open or only half-decked. There was no forecastle, and the superstructure of the poop was modest, at best containing one room (Parry 1963, p. 65). In the mid-fifteenth century and certainly on the early voyages of discovery, the caravel normally appeared to have been lateen-rigged on all its masts.

We might say that the caravel was well adapted to the context of offshore exploration. Thus we might note (as have many historians, for example, Denoix (1966, p. 142)) that for such a task one needs a vessel that will not blunder into reefs, is light and handy, draws little water, sails well against the wind, and does not require a large crew. All these attributes were true for the caravel, which was indeed well adapted to its task. But what are we really saying when we say this?

The answer to this question can be found in the notion of a network. System builders seek to create a network of heterogeneous but mutually sustaining elements. They seek to dissociate hostile forces and to associate them with their enterprise by transforming them. The crucial point, however, is that the structure of the network reflects the power and the nature of both the forces available and the forces with which the network collides. To say, then, that an artifact is well adapted to its environment is to say that it forms a part of a system or network that is able to assimilate (or turn away) potentially hostile external forces. It is, consequently, to note that the network in question is relatively stable. Again, to say of an artifact such as the caravel that it is adaptable is to note that a network of associated heterogeneous elements has been generated that is stable because it is able to resist the dissociating efforts of a wide variety of potentially hostile forces and to use at least some of these forces by transforming them and associating them with the project. And this, of course, is precisely the beauty of the caravel in the fifteenth-century context in which it was used by the Portuguese. Properly manned and provisioned, it was able to convert whatever the West African littoral might direct at it into controlled motion and controlled return. It was a network of people, spars, planks, and canvas that was able to convert a wide range of circumstances into exploration without falling apart in any of the numerous ways open to vessels when things start to go wrong. Like the *volta*, then, the caravel achieved stability by reflecting the forces around it. It was well adapted because it maintained stable relations between its component parts by associating everything it encountered with that network as it moved around.

Navigation and the Raising of the Sun: Closure and Metrication

Between 1440 and 1490 the Portuguese explored most of the West African coast. As they moved further south and used increasingly larger *voltas*, the Portuguese saw their problems of navigation become more acute. How could they determine their position when they were so far from land? Because the classical European methods of compass course, plain chart, and dead reckoning were of little assistance, this problem was of great concern to the Portuguese. In the 1480s they developed a practical method for the astronomical determination of latitude on board ship. The general idea was that if the *altura*, or height above the horizon, of the sun or a star (normally the Pole Star) could be determined and compared with the known altura of the port of destination, then the ship could sail north or south until it reached that latitude, and then sail, as appropriate, east or west in the certainty of finding its point of destination.

The measurement of the altura was possible with the use of either the quadrant or the astrolabe. Both devices were standard university instruments of astronomy and astrology that carried a great deal of information that was both unnecessary to the calculation of latitude and simply incomprehensible to the layman. On the back of the astrolabe there was, however, an alidade, which was a rule on a swivel with two pinhole sights. The observer held the instrument upright by a swivel suspension ring, peered along the alidade, and measured the altura of the star by reading off the position of the alidade on a scale marked on the rim (figure 3). The quadrant was an instrument with similar functions. It was in the form of a quarter circle, and the star sight was taken along one of the "radii." The artificial horizon was provided by a plumb line suspended from the center of the "circle" and was measured with a scale along the circumference (Taylor 1956, pp. 158–159). In its university and astrological version the quadrant, like the astrolabe, also carried information about the movements

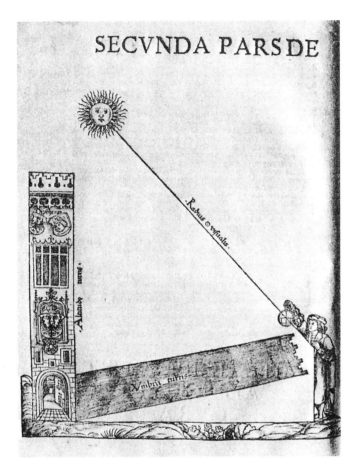

Figure 3
Measuring the altura with an astrolabe (from Sebastian Muenster, *Organa Planetarum* (Basel: Petrus, 1539), 70).

of planets, seasons, and hours. Both of these instruments, shorn of all but their essentials for the measurement of altura, were used by Portuguese explorers, although it seems that the somewhat simpler quadrant was the first to be used by navigators (Taylor 1956, p. 159).

By themselves these instruments were, of course, powerless. The mere fact of sighting a heavenly body through the pinholes of an alidade had nothing per se to do with navigation. That sighting, or the reading that corresponded to it, had to undergo a number of complex transformations before it could be converted into a latitude. The construction of a network of artifacts and skills for converting the stars from irrelevant points of light in the night sky into formidable allies in the struggle to master the Atlantic is a good example of heterogeneous engineering.

I have already mentioned the simplification of the quadrant and astrolabe. This can be treated as the first step in the process.[10] The second stage involved what may be treated as social engineering—the construction of a network of practices that, when associated with the instruments themselves, would lead to the necessary transformations of sun and starlight. This social engineering itself came in three stages. First, in the early 1480s King John II convened a "scientific commission" to find improved methods for measuring the altura. This was made up of four experts: the Royal Physician, Master Rodrigo; the Royal Chaplain, Bishop Ortiz; the geographer, Martin Behaim; and Jose Vizinho, who had been a disciple of the astronomer Abraham Zacuto of Salamanca (Chaunu 1979, p. 257; Taylor 1956, p. 162; Beaujouan 1966, p. 74; Waters 1980, pp. 9–10). The convocation of a "scientific commission" for the purpose of converting esoteric scientific knowledge into a set of widely applicable practices is already remarkable. Even more noteworthy is the fact that these four men, and probably in particular Vizinho, were able to effect that transformation by producing a set of rules for the calculation of the latitude by semieducated mariners. These rules, which form the second part of this experiment in social engineering, took the form of the *Regimento do Astrolabio e do Quadrante,* which was probably available from the late 1480s, at least in handwritten form. The *Regimento* can be read as instructions about how to turn the vessel and its instruments into an observatory—how, in other words, to create a stable if heterogeneous association of elements that had the property of converting measurements of the altura into determinations of the latitude (figure 4).

Even this, however, was not enough. To adopt the new method of sailing, the navigators required a third step: It was necessary to know the latitudes of important coastal features and in particular the major ports and capes. It was, in other words, necessary to generate a *metric* from which the observations might be given absolute north-south meaning and from which the observatory of the ship could be located accordingly. The measurement of important coastal latitudes again involved a major organizational effort. It involved sending out competent observers armed with large wooden astrolabes on the vessels of exploration and having them report back to Lisbon. By 1473 the astronomers in Lisbon had a list of latitudes that reached the equator (Taylor 1956, p. 159), a list that was extended as the century wore on. And it further required that known latitudes be available to mariners, and indeed, a further section of the *Regimento* listed these.

January.			
Firſt.	**Second.**	**Third.**	**Fourth.**
1	2	3	4
1593	1594	1595	1596 Yeere of
1597	1598	1599	1600 the Lord
1601	1602	1603	1604
1605	1606	1607	1608
1609	1610	1611	1612

D. G. M.			D. G. M			D. G. M.			D. G. M		
1	21	50	1	21	52	1	22	56	1	21	57
2	21	40	2	21	43	2	22	9	2	21	48
3	21	30	3	21	33	3	21	36	3	21	38
4	21	20	4	21	23	4	21	26	4	21	28
5	21	9	5	21	16	5	21	15	5	21	17
6	20	58	6	21	1	6	21	4	6	21	7
7	20	47	7	20	50	7	20	53	7	20	55
8	20	35	8	20	38	8	20	41	8	20	44
9	20	22	9	20	26	9	20	29	9	20	32
10	20	9	10	20	13	10	20	16	10	20	20
11	19	56	11	20	0	11	20	3	11	20	6
12	19	43	12	19	47	12	19	50	12	19	53
13	19	29	13	19	33	13	19	36	13	19	39
14	19	14	14	19	19	14	19	22	14	19	25
15	19	0	15	19	4	15	19	8	15	19	11
16	18	45	16	18	49	16	18	53	16	18	56
17	18	29	17	18	34	17	18	38	17	18	41
18	18	14	18	18	19	18	18	22	18	18	26
19	17	58	19	18	3	19	18	7	19	18	11
20	17	42	20	17	46	20	17	50	20	17	54
21	17	25	21	17	30	21	17	34	21	17	38
22	17	8	22	17	13	22	17	17	22	17	21
23	16	51	23	16	56	23	17	0	23	17	4
24	16	32	24	16	38	24	16	43	24	16	47
25	16	16	25	16	21	25	16	25	25	16	29
26	15	57	26	16	3	26	16	7	26	16	12
27	15	39	27	15	45	27	15	49	27	15	54
28	15	21	28	15	26	28	15	30	28	15	35
29	15	2	29	15	7	29	15	12	29	15	23
30	14	43	30	14	48	30	14	53	30	14	58
31	14	24	31	14	29	31	14	34	31	14	39

Figure 4

Tables of solar declination from a navigational manual (Pedro de Medina, *The Arte of Nauigation* (London: Thomas Dawson, 1595), 58).

The new method of navigation proved difficult for most mariners. Only the most up-to-date sailors attempted its practice, and there is evidence that Columbus, among others, understood it only imperfectly. Although the details remain unclear, it appears that in the early sixteenth century, and possibly earlier, classes on navigation were taught to pilots at Lisbon (Diffie and Winius 1977, p. 142). Such instruction, however, was not invariably successful. There were complaints in the sixteenth century that many pilots were inexpert. It seems, then, that in the attempt to create a stable network of elements for the conversion of stars into measurement of the latitude—in other words, in the attempt to convert ships into observatories—it was the mariners who constituted the weakest link. The stars were always there, as were the oceans; they could not be budged. Again, once the instruments and the inscriptions were in place, they proved to be fairly durable. But instruments, inscriptions, and stars were not enough. Part of the association of elements to convert stars into latitudes lay in the practices of the mariners, and it was this element that was the most prone to distortion. It was difficult, although not ultimately impossible, to create a new social group necessary for closure: the astronomical navigator.

So far I have tacitly made the assumption that, when success is achieved, it is obvious. If one arrives at one's port of destination (or for that matter runs aground on the reefs of Cape Bojador), the success (or failure) of the enterprise is readily apparent to all. We might say that in the ultimate analysis it was the capacity of the Portuguese to *return* to their point of departure that marked success. The success of astronomical navigation was that it contributed to that return. Yet, however much final closure depended on the capacity to return, decision making on the voyage would not have been possible without a scale of reference. The success of any course sailed could be measured in the interim only against an entirely man-made metric, a metric that depended on inscriptions and the capacity to interpret those inscriptions. We have, then, the construction of a background against which to measure success—something akin to if not identical with the technological testing tradition described by Constant in the context of water turbine engineering (Constant 1983). The history of navigation can, I believe, be understood as the construction of more (locally) general systems of metrication against which the adequacy of particular courses and navigational decisions might be measured.

The Muslim and the Gun: Dissociation

On July 8, 1497, Vasco da Gama's fleet weighed anchor in the Tagus River and set sail. His four tiny vessels carried 170 men and 20 cannons. They also carried merchandise. Two years later two of the original four vessels returned to Lisbon. The cape route to India had been opened, and spices had been brought back.

The Portuguese encountered various difficulties, which arose in part from the hostility of Muslim traders in India (Magalhaes-Godinho 1969, p. 558). Such merchants organized and controlled the Indian Ocean section of the spice trade. They bought spices in the Calicut bazaars and shipped these, through either the Persian Gulf or the Red Sea, to Arabian ports for further shipment to the Mediterranean and Venice. Not surprisingly the Muslims did not welcome the arrival of da Gama on the Malabar coast at Calicut with enthusiasm. Negotiations went badly between the Portuguese and the Hindu ruler of Calicut, the Samorin. There were many reasons for this, but the most important appears to have been the hostility of the Muslim traders on whom the Portuguese were obliged to depend for translation. The translators spread a variety of hostile rumors about the Portuguese, who were then forced to trade directly with Hindu merchants (Diffie and Winius 1977, pp. 182–183).

Once back in Lisbon, the Portuguese pondered the lessons to be learned. One conclusion that they were quick to draw was that it would be necessary to exercise force in the Indian Ocean. Da Gama's first expedition had carried guns, but more would be needed if the hostility of the Muslims was to be mastered. In fact, the Portuguese had come to this conclusion even before da Gama's return. A much larger and more heavily armed second expedition had already set out; the expedition consisted of thirteen vessels and between 1000 and 1500 men and was commanded by Pedro Cabral. Cabral's orders were clear: He had to install an agent to buy spices in Calicut and was instructed to display force when this was necessary, although he was to refrain from conquest (Magalhaes-Godinho 1966, p. 561). Although negotiations started well, things quickly went wrong again. In response, Cabral put to sea, destroyed a number of Muslim vessels, and bombarded the town of Calicut. The story was repeated with da Gama's second expedition, which, however, used even more force. Together these first three sorties cast the die for Portuguese control of the Indian Ocean over the next few years. Control would have to be maintained primarily by force, as there was no room for both Muslim and Portuguese commerce.

At sea the Portuguese were, at least in the short run, able to exert the necessary military power and choke Muslim maritime trade. Portuguese guns proved better (but not more numerous) than Asian guns. European advances in the technology of gun making had overcome many of the problems that beset the late medieval cannon. In particular, with the development of cast bronze guns, the weight of the cannon had been much reduced, and although still prone to heaviness, they were much less likely to blow up in the faces of the gunners than the welded pieces that preceded them. Again, Portuguese vessels, built for the inhospitable Atlantic, were more solid than those of their Muslim adversaries (Boxer 1953, p. 196). Cipolla puts it this way:

The gunned ship developed by Atlantic Europe in the course of the fourteenth and fifteenth centuries was the contrivance that made possible the European saga. It was essentially a compact device that allowed a relatively small crew to master unparalleled masses of inanimate energy for movement and destruction. (Cipolla 1965, p. 137)

The cannon, the ship, the master, the gunner, the powder, and the cannonballs—all these formed a relatively stable set of associated entities that achieved relative durability because together they were able to dissociate the hostile forces encountered without being dissociated themselves. It is important to note here that some of these hostile forces were physical (the oceans), whereas others were social (the Muslims). Technology, as I have suggested, simultaneously associates and dissociates, and the heterogeneous engineering of the Portuguese was designed to handle natural and social forces indifferently and to associate these forces in an appropriate form of closure.

Having said this, however, it is important not to fall into the trap of technological determinism and assume that it was the technology alone that brought about Portuguese success. As was the case for the caravel, the *volta,* and the practice of astronomical navigation, the durability of the armed warship was a function of a collision between the forces of the Portuguese system builders and those of the seas and, in this case, the Muslims. Thus Boxer (1953, pp. 194–197) argues that the Portuguese "naval and military superiority, where it existed, was relative and limited." It happened that there was no well-armed Muslim shipping in the Indian Ocean. It happened that the Chinese had retired to their coasts. It happened that the Portuguese expeditions were state enterprises, combining the power and organizational ability of the crown with the search for profit. It happened that Muslim merchants traded on their own account and not for their monarchs. It happened that there was little wood available to many of those monarchs in order to build fleets to stop the Portuguese. Under these circumstances the Portuguese were able to dominate shipping in the Indian Ocean. They were not able (and knowing this never sought to) build up sizable colonies on land. There, with the balance of force weighted against them by cavalry and manpower, they risked crushing defeat.

Conclusion

I started by outlining three approaches to the social study of technology. One, that of social constructivism, comes from the sociology of science. I suggested that, although this has many merits, its commitment to a form of social reductionism is unsatisfactory. The second, the systems approach, comes from the history of technology. This stresses the heterogeneity of technological activity and avoids a commitment to social (or technological) reductionism. I argued that this approach, adapted in a way that makes it clear that systems are built, through a struggle, from indifferent or hostile elements, offers a satisfactory model for the analysis of technological innovation. I suggested that "heterogeneous engineers" seek to associate entities that range from people, through skills, to artifacts and natural phenomena. This is successful if the consequent heterogeneous networks are able to maintain some degree of stability in the face of the attempts of other entities or systems to dissociate them into their component parts. It follows from this that the structure of the networks (or systems) in question reflects not only a concern to achieve a workable solution but also the relationship between the forces that they can muster and those deployed by their various opponents. I might, if I were to make more use of the metaphor of force, write of the relative durability or strength of different networks or of different parts of the same network. Thus I have attempted to show by empirical example that, in the collisions among different networks, some components are more durable than others and that the successes achieved by one side or the other are a function of the relative strength of the components in question.

What are the virtues of physical metaphors such as force, strength, and durability? Let me say, first, that I am not strongly committed to these terms. Doubtless other metaphors might serve as well or better. I believe, however, that the strength of the vocabulary lies in its capacity to handle, using the same terms, the various heterogeneous elements that are normally assembled within any system. As I indicated earlier, the method seeks to deal with the social, the economic, the political, the technical, the natural, and the scientific in the same terms on the grounds that (in most empirical cases) *all* of these have to be assembled in appropriate ways if closure is to be achieved. Within any of these (usually distinguished) categories, there may be entities, processes, bodies, objects, institutions, or rules that turn out to have force with respect to the system in question and hence are relatively durable. These may take the form of scientific truths, economic markets, social facts, machines, or whatever. They form, then, a relatively coercive (albeit ultimately revisable) scenery that has to be mastered if a system is to be built. Because, however, durability does not reside in one category alone, I have ignored conventional distinctions among categories, and in particular I have argued that it is not good enough to add the social as an explanatory afterthought. The social (including the "macrosocial") has, rather, to be placed alongside every-thing else if the collisions and closures between forces and entities are to be understood.

Like Callon, I have thus sought to press the principle of symmetry (Bloor 1976) further than is normal in the sociology of science. In the sociology of science this principle states that the same *type* of explanation should be used for both true and false beliefs. It is intended to counter the tendency commonly found in the sociology of knowledge of explaining true beliefs in terms of the way in which they correspond with reality while leaving false beliefs to be explained in terms of the operation of psychological or social factors. The generalized version of the principle of symmetry (Callon 1986) that I have adopted here states that the same type of explanation should be used for all the elements that go to make up a heterogeneous network, whether these elements are devices, natural forces, or social groups. In particular, the principle of symmetry states that the social ele-ments in a system should not be given special explanatory status.[11] The form that these elements take may be, and often is, a function of the technological or natural features of the system. This is a contingent matter, a function of which components of the system are associated most durably and are hence least susceptible to dissociation.

To say this is not, of course, to suggest that it is always the social that is malleable and the technological or the natural that is durable. It is rather to stress that the relationship between them is one of contingency and that it is important to find a way of treating all components in a system on equal terms. But this leads to a further way in which the network approach is distinguished from that of social constructivism. In social constructivism natural forces or technological objects always have the status of an *explanandum*. The natural world or the device in question are never treated as the *explanans*. They do not, so to speak, have a voice of their own in the explanation. The adoption of the principle of generalized symmetry means that this is no longer the case. Depending, of course, on the contingent circumstances, the natural world and artifacts may enter the account as an *explanans*. And in case it is thought that I am giving too much away to realism, let me say that, so long as we are concerned exclu-sively with networks that are being built by people, then "nature" reveals its obduracy in a way that is relevant only to the network when it is reg-istered by the system builders. It is not, therefore, that nature is being promoted to some special status. Rather it is, as I have already suggested, that the social is being demoted. In the network approach *neither* nature nor society has any role to play unless they impinge on the system builder.

This is why, in my explanation of the Portuguese expansion, capes and currents are found alongside vessels and mariners. Once the principle of generalized symmetry is adopted, they cannot be excluded. Indeed, to try to reduce an explanation of the Portuguese system to a limited number of social categories would be to fail to explain the specificity of the *volta*, the caravel, or the *Regimento*. Portuguese views of the sun and the adverse winds are needed to make the explanation work.[12]

A further methodological principle follows from this. It is that the scope of the network being studied is determined by the existence of actors that are able to make their presence individually felt on it. If the system builder is forced to attend to an actor, then that actor exists within the system. Conversely, if an element does not make its presence felt by influencing the structure of the network in a noticeable and individual way, then from the standpoint of that network the element in question does not exist. It is clear that the choice of network on which to focus is therefore crucial. If the focus is on one system, then one pattern will emerge. If the focus is on another system or even on an element within the original system, then a different structure will be seen. Thus the system of Portuguese expansion for Henry the navigator contained elements such as vessels and their masters. A shift in focus from Henry to the master and his vessel would bring a further network of sailors, spars, and stores into focus—a network with its own force that, when placed within the system of Portuguese expansion, acted as a single unit. If the vessel and its master did not play the roles defined for them in the network of expansion, then the elements that make them up might, of course, have become individually relevant in Lisbon and been built into Henry's expansion network. Such adjustment is consistent with, and indeed exemplifies, the original proposition that the extent of a network is defined by the presence of actors that are able to make their presence individually felt.[13]

This also means, of course, that the heterogeneous engineer standing at the heart of his or her network is not in principle analytically privileged. It is true that, for the purpose of the particular study, I have chosen to follow one system-building effort—that of the Portuguese maritime planners. I have done this in order to set practical limits to the analysis. In making this decision, however, I have not committed myself to the notion that system builders are primitive entities that are themselves unamenable to analysis. Just as vessels or navigators are fashioned out of the interaction between networks of forces, so too are heterogeneous engineers. Indeed, the fact that these are in a position to build systems is itself the outcome of a set of interactions among forces of different degrees of obduracy. To put it more simply, the king of Portugal is just as much an effect as a cause: He is the effect of a set of endless transactions that are, in principle, available for analysis. In the present study, I chose, for reasons of simplicity, to treat him as a cause and navigation as an effect, but in another study these roles, or ones like them, might just as easily have been reversed.

In summary, there are thus two closely related methodological principles for the study of heterogeneous networks. The first, that of generalized symmetry, states that the same type of analysis should be made for all components in a system whether these components are human or not. The second, that of reciprocal definition, states that actors are those entities that exert detectable influence on others. Applied to a relatively stable system, we can therefore define the extent of that system or network by the range of actors that operate as unitary forces to influence the structure of the network. In this chapter I have attempted to follow these two principles in an analysis of the Portuguese expansion. In reinterpreting the notions of system, adaptation, and technological testing for a historical case, I hope that I have succeeded in showing the relevance of the approach to the analysis of technological innovation.

Notes

I would like to thank Serge Bauin, Wiebe Bijker, Michel Callon, David Edge, Rich Feeley, Elihu Gerson, Antoine Hennion, Tom Hughes, Bruno Latour, Jean Lave, Mike Lynch, Chandra Mukerji, Trevor Pinch, Arie Rip, and Leigh Star, who all read and commented on earlier versions of this paper. I would also like to thank the University of Keele, l'Ecole Nationale Superieure des Mines de Paris, la Fondation Fyssen, the CNRS, and the Leverhulme Foundation for support and study leave. Finally, I am grateful to the librarian of the University of Keele for kind permission to reprint illustrations from sixteenth-century works held in the Turner Collection of mathematical texts at the University Library.

1. It is fully described by Pinch and Bijker (1984 and this volume). See also Bijker (this volume).

2. I am not suggesting that these authors all use a social constructivist approach but that their material is susceptible to an analysis in those terms.

3. For another study using a systems approach, see MacKenzie (this volume).

4. Pinch and Bijker (this volume) talk of the effects of advertising on the formation of social groups.

5. Although I have made reference to the work of Hughes, the same point can, I believe, be made in reference to Constant. His notion of coevolution (1978) also

seems to represent an attempt to grapple with the interrelatedness of heterogeneous elements and to handle the finding that the social as well as the technical is being constructed. In addition, the analysis of the development of traditions of "technological testability" developed by Constant (1983) can be seen as a study of the way in which a wide range of actors comes to a locally enforceable agreement that certain social/technical relations are appropriate and workable.

6. Arguably we are *all* heterogeneous engineers, combining, as we do, disparate elements into the "going concern" of our daily lives. In the present essay I am concerned, however, only with large-scale, technologically relevant system building.

7. As I have indicated, this approach parallels that of Callon. It also, however, owes much to the work of Latour (1984).

8. What follows is an example of what I call rational reconstruction. See the conclusion of this chapter.

9. The portolano or plain (that is, plane) chart was laid out using wind roses and rhumb lines of constant magnetic bearing.

10. In what follows I have been highly selective with respect to material in order to highlight what I take to be the essentials of the process and to avoid getting bogged down in detail. For similar reasons I have also taken the liberty of reorganizing the chronology of events by talking of the establishment of the latitudes of important points on the coast after discussing the *Regimento*. For a fuller sociological account, see Law (1986a).

11. Similar arguments have been made by Woolgar (1981), Yearley (1982), and Gallon and Law (1982).

12. Having said this, however, I willingly concede that in the present chapter I have sometimes been obliged, because of lack of data on medieval and early modern maritime practices, to make use of a kind of "rational reconstruction" in order to show how nature and society affected the Portuguese analysis of their problems. It should be understood that I use rational reconstruction not for the purpose of epistemological judgment but to try, matter of factly, to work out what appears to have happened in cases in which historical data is lacking. For a more extended discussion of rational reconstruction and inadequacies of data, see Law (1985). It is obvious that this procedure is less than ideal, but unless whole empirical areas are to be denied to us, it is obviously unavoidable.

13. It is clear from what has been said that any network stands at the intersection and (if it is relatively stable) profits from the force contributed by endless other networks that have been simplified into individual units. See Callon (1981a) and Law (1984b).

The Nelson-Winter-Dosi Model and Synthetic Dye Chemistry

Henk van den Belt and Arie Rip

Although economics is virtually unique among the social sciences with its long-standing interest in analyzing the nature and causes of technological change, we do not think that most of its contributions are particularly relevant to the general student of technology.

Some recent "neo-Schumpeterian" theories of innovation and technological change, in particular, those proposed by Nelson and Winter and by Dosi, should be exceptions to this stricture (Nelson and Winter 1982; Dosi 1984). These economists maintain that a major break with the neoclassical orthodoxy of the discipline is needed for coming to terms with the phenomenon of technological change. We think that their ideas and insights may be of interest to other students of technology who do not necessarily share their primary concern for the analysis of competition and industrial transformation.

In this chapter we therefore sketch and expand on the theories of Nelson and Winter and of Dosi, concentrating on their basic ideas about the nature of technological change rather than on the specific formal models developed for purposes of economic analysis.[1] Our strategy is to elaborate and extend the sociological content that is at the heart of the Nelson-Winter-Dosi model (for convenience we use this description to denote the common core of their theories). These elaborations and extensions concern the dynamics of technological development (described in terms of "technological paradigms" and "technological trajectories"), its "social locus," and the interaction with "selection environments."

By avoiding the more narrow issues of economic analysis, we also pass over those features of the Nelson-Winter-Dosi model that might render it particularly attractive or cogent for economists. We therefore hope that, by concentrating on the sociological aspects, our discussion of the model may secure it a wider hearing, particularly among those who are interested in social studies of technology. Unfortunately, some of the central notions of the model, such as technological paradigm and especially technological trajectory, are not entirely free from deterministic overtones that may inspire undue suspicions among those who advocate a social constructivist approach to technology. (See Pinch and Bijker (this volume) for a rather skeptical attitude toward Johnston's and Dosi's proposals to describe technological knowledge in terms of Kuhnian paradigms.) We hope that the following discussion may show these suspicions to be unwarranted. Let us already note at this point, however, that both the Nelson-Winter-Dosi model and the social constructivist approach developed by Pinch and Bijker share a common inspiration, which can loosely be called evolutionary thinking.[2] Perhaps this common inspiration can provide some common ground for comparing and evaluating both approaches.

For empirical illustration we draw on historical materials on synthetic dye chemistry—a branch of scientific technology that was involved in the spectacular rise of the nineteenth-century German synthetic dye industry and which, in due course, was to become the "synthetic everything-else industry" (Lilley 1973, p. 243). Our empirical material is based on an extensive case study but is used here only to illustrate theoretical points. The working of the patent system in connection with the development of (a certain part of) synthetic dye chemistry is the main example of how the interaction between technological development and "selection environments" should be conceived. The example will also help to elucidate the nature of our proposed sociological extension of the Nelson-Winter-Dosi model.

An Outline of the Nelson-Winter-Dosi Model

Neo-Schumpeterian economists reject the mainstream neoclassical approach to technological change. They criticize, in particular, the assumption of perfect rationality and the corresponding "maximization metaphor." Nelson and Winter think it inappropriate to conceive of the behavior of the firm in terms of deliberate choice from a large, well-defined set of production possibilities that extends well beyond the current range of operation.

As an alternative to the neoclassical theory, Nelson and Winter adopt an avowedly *evolutionary* approach. They see firms as more or less loosely structured clusters of *routines*—ways of doing various things and ways of determining what to do. Routines are the organizational counterpart to what is called skill at the level of the individual. Much of the "choosing" postulated by neoclassical theory is actually accomplished automatically by routines.[3]

The set of available routines can be considered the "genetic" make-up of an organization. Just like genes, routines are exposed to the selection pressures of the environment in an indirect way, in this case through competition between firms. The outcome of routines is judged by the market, or, more generally, when government policy and all kinds of institutional arrangements are included in the *selection environment*, which thereby determines indirectly which routines are viable enough to acquire widespread use in the population of firms. The mechanism does not operate exclusively through the differential growth of firms, however; it may also involve *imitation* among firms—in contradistinction to purely biological selection.

For economic and technical evolution the mechanism analogous to genetic mutation is innovation. Because innovative activities may follow search heuristics, they may themselves be patterned, and Nelson and Winter do not hesitate to extend their concept of routine to innovation. This does not imply, however, that the results of innovation are predictable; Nelson and Winter criticized neoclassical theory precisely because it did not do justice to the uncertainties inherent in innovation! Thus firms may possess special kinds of routines, viz. *search routines*, that may stochastically lead to modifications in existing routines or to the introduction of completely novel ones.[4]

Although by now their evolutionary theory seems complete, Nelson and Winter go one step further. They observe that technological development is sometimes even more patterned than is implied by the following of simple heuristics. In those cases a *technological regime* (or a *technological paradigm* in Dosi's terminology) may be said to exist. Their prime example is the Douglas DC-3 model of the 1930s, which guided aircraft design for more than two decades by defining a new technological regime: metal skin, low wing, piston engines. As Nelson and Winter emphasize, their concept is cognitive and relates to "technicians' beliefs about what is feasible or at least worth attempting" (Nelson and Winter 1982, pp. 258–259). Aircraft engineers held strong notions regarding the "potential" of the DC-3 regime, and subsequent innovation essentially involved better exploitation of this potential. Thus, as the example illustrates, technological regimes give rise to *technological trajectories*.

Although these notions of technological regimes and trajectories appear suggestive and potentially fruitful, there is some ambiguity as to where exactly they fit into the scheme of Nelson and Winter's evolutionary theory. Of course, prima facie the notion of a technological regime seems to run counter to the logic of (Darwinian) evolutionary theory, because the emergence of a regime introduces at one stroke a large measure of "directedness" in technological development.[5] But there is still another ambiguity. Technological regimes might be thought of as a kind of master search routines, embodied in firms just like other search routines. In that case they should be considered part of the "genetic make-up" of firms, subject to the normal selection pressures from the environment. In the formal models that Nelson and Winter constructed to simulate the dynamics of Schumpeterian competition (that is, competition that involves innovation rivalry), regimes are treated as part of the *environment* in which firms have to operate (Nelson and Winter 1982, pp. 275–351).

There may be good reasons for not treating technological regimes as (exclusively) embodied in firms. These do not constitute the exclusive social locus of technology. Indeed, we may follow Edward Constant in distinguishing two social loci of technology: firms and communities of technical practitioners (Constant, this volume). The existence of these two social loci raises interesting sociological questions of its own, for example, about their relations and their relative importance. These may vary over time, as can be inferred from Dosi when he writes that establishing a technological trajectory appears to imply a trend toward the increasing incorporation of technology into capital equipment and into complex institutions (such as big firms)—in other words, a movement from "people-embodied" technology toward "organization-embodied" (and "equipment-embodied") technology (Dosi 1984, p. 193). In the semiconductor industry that was studied by Dosi, this process is going on but is still far from completion. In what follows we describe a similar trend in the nineteenth-century synthetic dye industry. Here we note only that the existence of two social loci of technology has implications for the modus operandi of the evolutionary mechanism.

Nelson and Winter nowhere pursue the question of how technological regimes actually come into being, apart from the observation that the original Douglas DC-3 was "the result of the confluence of a number of R&D strands" (Nelson and Winter 1982, p. 257). The question has been examined more extensively by Dosi, although we think his solution is far from satisfactory.

In Dosi's visualization of the process, "selection" with regard to the emergence of technological paradigms is supposed to operate on *notional* possibilities, or, as he says, on "a large set of *possibilities* of directions of development, notionally allowed by science" (Dosi 1984, p. 16). This involves a clear departure from the biological concept of selection. As the French biologist François Jacob asserted, "La sélection s'opère, non parmi les possibles, mais parmi les existants" [selection operates, not on possible, but on existing, living organisms] (Jacob 1970, p. 313). The same point was made by Nelson and Winter when they stated that "selection works on what exists, not on the full set of what is feasible" (Nelson and Winter 1982, p. 142). Dosi's "notional possibilities" sometimes take on a more real existence in the guise of competing technological paradigms, but we seriously doubt the historical adequacy of the assumption of a plurality of competing technological paradigms.

Dosi's view on the relation between science and technology, particularly his *sequence model*, with findings going downstream from science to technology although filtered through socioeconomic criteria, is not helpful at all. In our opinion the *two cultures model* suggested by Barnes and Edge constitutes a preferable alternative (Barnes and Edge 1982, pp. 147–154). In this model the relationship between science and technology (however the two may be demarcated) is not represented as a hierarchical one, with science having "implications" for technology and technology "applying" the findings of science; rather, the relationship is a symmetric one, with both forms of activity possessing their own distinct cultural resources, although both may also, occasionally or more regularly, draw on the cultural resources of the other.

How, then, do paradigms (or regimes) emerge? In fact, there was already the beginning of an answer when Nelson and Winter observed that the DC-3 was the result of "a *confluence* of a number of R&D strands" (our emphasis). Viewing the emergence of an exemplary technological achievement in this way—as a result of a coming together of various lines of development—makes it unnecessary to postulate the complicated kind of selection mechanisms hypothesized by Dosi. In technical evolution, as against species evolution, there is indeed much room for processes of creative combination and synthesis.[6] The process of invention itself was described by A. P. Usher as a "cumulative synthesis of many items which were originally independent" (Usher 1971, p. 50). Perhaps there is nothing special or particularly dramatic in the way technological paradigms come into being. The interesting question may rather be how certain technological achievements (such as the DC-3) come to be recognized as exemplary or paradigmatic. We return to these issues later, in the context of our study of synthetic dye chemistry.

Technological Paradigms, Trajectories, and Selection Environments: Some Refinements

Both the nature of the search process and the interaction with selection environment in the Nelson-Winter-Dosi model require some further scrutiny.

The development of a technology is not always of the kind exemplified by the trajectory of aircraft design following the establishment of the DC-3 regime. Heuristic search processes may exhibit a much less directed pattern; for instance, they may be largely induced by "bottlenecks" in a system of connected processes when and where they occur (Rosenberg 1976, p. 125).[7] And there are other possibilities. Nelson and Winter themselves note that there also exist powerful heuristic guidelines that are not specific to a particular regime but are followed in a wide variety of technologies, giving rise to what they call general trajectories: Examples are mechanization and the progressive exploitation of latent economies of scale.

Thus it can be argued that technological developments proceeding within a technological regime or paradigm present us with a special situation. Dosi draws on the analogy with the Kuhnian view of paradigm-guided development within science to analyze this particular kind of technological development. The analogy can be taken a step further, by slightly improving on Kuhn's formulation.

The occurrence of a technological paradigm can be characterized by the clustering of successful heuristics around an exemplary achievement, such as the DC-3 aircraft. Following Kuhn's terminology, we speak of an *exemplar*. The appearance of an exemplar is a necessary but not sufficient condition for "normal" technological development along a trajectory to occur. In addition, there have to be expectations about the success of continuing work within this cluster of heuristics—expectations that must be embedded in the subculture of the technical practitioners and others involved in the development.[8] Again borrowing from Kuhn's terminology, we can speak of the existence of a *cultural matrix*. The combination of exemplar and cultural matrix forms a technological paradigm, and the further articulation of such a paradigm, partly influenced by the selection environment, leads to a technological trajectory.

Our conceptual reformulation also suggests that the analogy with biological evolution can be only a limited one. In biological evolution, although mutations are random, the selection process is deterministic; that is, there are "well-defined criteria for accepting or rejecting any given mutation" (Elster 1984, p. 6).[9] In societal evolution involving technological development, even the selection process is far from deterministic: Intentions and expectations play a role, and actors can make choices and anticipate reactions of others.[10] More important, human actors also try to influence the reactions of others and change their environments. Thus the assumption of a selection environment that is truly independent of a particular technological trajectory is hard to justify.

Similar issues are at stake in the social constructivist model proposed by Pinch and Bijker. Although in that model "the developmental process of a technological artifact is described as an alternation of variation and selection" (Pinch and Bijker, this volume), it is clear that here too the notion of "selection" does not measure up exactly to its meaning in evolutionary biology. For Pinch and Bijker the meanings given to a particular artifact by various social groups do not just constitute (independent) selection criteria but actually go to make up the content of the artifact. This conception underpins their claim of presenting a *radical* social constructivist account of technological artifacts. Following this sociological version of Berkeley's "*esse* is *percipi*" to its extreme, Pinch and Bijker are consistent only in accepting the conclusion that, when different social definitions are given, there will be more than one artifact; or, as they put it for one of their bicycle examples, "there was not *one* high-wheeler, there was the macho machine . . . and there was the unsafe machine" (Pinch and Bijker, this volume).[11] We think this is one step too far. The claim of *radical* social constructivism, to be sustained, would require the wholesale suppression of the "variation" dimension. For us this would seem too high a price to pay.

Returning to the Nelson-Winter-Dosi model, we observe that the problematic character of the assumption of independent selection is already manifest at the most elementary level; indeed, it is implied by our definition of a technological paradigm. If, as Kuhn also asserts, the success of a paradigm is at the start largely a promise of success (Kuhn 1970, p. 23), then new paradigms must of necessity be protected against the myopia of natural selection.[12] Drawing on the expectations about the success of the heuristics, influence is exerted on the selection environment and a *niche* is created that protects the trajectory against too harsh a selection. On the other hand, when the promise of success largely substitutes for real, practical success (that is, when heuristics can become "successful" by the simple fact of being accepted in the subculture of technical practitioners), the heuristic power of a technological paradigm may be illusory.[13] Whether such a situation obtains or not, may be difficult to determine in practice.

A case in point is the development of the Stirling engine by the Dutch Philips Company. The engine's development illustrates that selection (in a special sense) may be said to be operative even in the absence of commercial success or failure. It is a case of completely protected trajectory: No Stirling engines have been produced for a market, but there has been selection—in this case at the level of expectations. Will there be a demand in ten years' time for clean and silent engines? Instead of a concrete market for realized products of technical development, there is now at work a speculative market of "early promises." In biological evolution there is no analog to this kind of selection.

Even after a technological trajectory has become firmly established, it will still be questionable whether or not an independent selection environment can be legitimately assumed to exist. As Nelson and Winter remark, for market selection environments a relatively clear separation of interests between supplying firms on the one hand and customers and regulatory agencies on the other may generally be presumed (Nelson and Winter 1982, pp. 268–269). But the actors involved in a technological trajectory do not always treat consumer preferences and regulatory rules as if these constituted unalterable structural constraints. Advertising as a means to influence consumer preferences comes readily to mind, but one may also think of the possibility that regulatory agencies become the "captured systems" of the industries they are supposed to regulate (Nelson and Langlois 1983, p. 817).

In our analysis of the synthetic dye industry it will appear that influence on the selection environment often results in a *nexus*, that is, a social institution that carries and shapes the interaction between trajectory and selection environment.[14] Examples are the patent system and patent case law and the application and test laboratories set up in the synthetic dye industry. These examples are discussed in what follows in some detail.

Synthetic Dye Chemistry: Emergence of Heuristics

Before the advent of synthetic dyes, textiles were dyed and printed with colors obtained by isolation or extraction from natural sources, such as the madder root, the indigo plant, and various kinds of dyewood. The application of natural dyes usually involved a series of chemical manipulations, which could be rather complicated, as in the case of alizarin (madder) dyeing. To obtain special shades of color, dye makers used mordants. The technical expertise to conduct these operations resided in large measure with the special professional group of *colorists*, who in some regions (for example, the Mulhouse area in France) were the backbone of the textile printing industry and enjoyed a correspondingly high status (Homburg 1983).

All this was to be changed by the emergence of the new synthetic dyes. The role and competence of professional colorists were to be redefined within a new set of relationships. The center of innovation was to shift from the textile dyeing and printing firms to the newly rising synthetic dye industry.

The beginning of this industry is usually set in 1856 with the legendary discovery of aniline purple, or mauve, by the 18-year-old William Henry Perkin, a student of A. W. Hofmann at the Royal College of Chemistry in London. Perkin's achievement clearly demonstrated that raw materials derived from unpromising coal tar could be turned into magnificent colors. The synthetic dye industry really took off, however, in 1859, with the discovery of aniline red, or fuchsine, which had a much larger commercial success than Perkin's mauve. Aniline red was discovered (or invented) by Emanuel Verguin, a chemist working for the Lyons silk-dyeing firm of Renard Frères, who later patented it. Its success attracted other producers into the synthetic dye field. Aniline red became an exemplar and the corresponding (rather primitive) heuristic was "tinker with aniline" (where aniline is not a pure chemical compound but a mixture of compounds derived from certain coal-tar fractions). In a relatively short time this heuristic resulted in the introduction of a number of red dyes into commerce. In France this occasioned drawn-out patent cases, which were eventually won by Renard Frères.

It is only with hindsight that the successive discoveries of aniline purple and aniline red can be recognized as the beginnings of the synthetic dye industry. Initially, the term "synthesis" was seldom used, and when it was, it did not have its modern connotations. For contemporaries the two new dyes were simply known as artificial colors and put into one class with other recent dyes, such as picric acid, murexide, and French purple (for which we would allow only a semisynthetic status), which were also products of intense innovative activity in the 1840s and 1850s. Furthermore, the processes involved in the manufacture of aniline red were often understood in terms of extraction or isolation of the dye from aniline instead of a straightforward chemical transformation performed on aniline with the help of suitable reactants, thereby suggesting a continuity with the older era of natural dyes. The break with the past was not immediately perceived in its full significance.

The aniline red paradigm was rationalized by Hofmann on the basis of the theory of types (a theory in organic chemistry that was a direct precursor of Kekulé's structure theory of 1859 and his benzene theory of 1865). In 1863 Hofmann proposed a provisional type formula to elucidate the chemical constitution of aniline red, thereby transforming the exemplar into rosaniline and derivatives and giving rise to innovative attempts to develop new dyes derived from aniline red. In general, we think that the role of science is not to provide the technical exemplar but to rationalize it and to articulate the cultural matrix of expectations in which it is embedded.

Although Hofmann's type formula did not represent a definitive solution as to the chemical constitution of rosaniline, it was quite effective in stimulating and directing innovative activity. By contrast, the structural formula proposed by Kekulé in 1866 merely amounted to a translation of Hofmann's formula into his own theory but did not indicate new directions for innovation. Instead of immediately leading to technological applications, the contribution of Kekulé's structure theory must be found in the way it affected the cultural matrix of synthetic dye chemistry. It was only in 1880 that the constitution of rosaniline was finally elucidated by Emil and Otto Fischer in terms of Kekulé's theory. This elucidation was the source of a new sequence of dye innovations.

The attempts at elucidation of the chemical constitution can be seen as the first steps toward a new, more general heuristic: "Look for the *Muttersubstanz*" (that is, the skeleton of the dyestuff's molecule), and if it is found, each derivative of the *Muttersubstanz* may be an innovation.

The heuristic became the central component of a new paradigm of purposeful synthesis, which could already be discerned in rudimentary form in the elucidation of the chemical constitution of alizarin and its subsequent synthesis by Graebe and Liebermann in 1869. The Fischers needed more perseverance in their search for the *Muttersubstanz* of the rosaniline family, but when their efforts finally met with success, the achievement would stand out as a model to be emulated by others.[15] The paradigm of purposeful synthesis was at first directed toward dyestuffs, but the general heuristic turned out to be powerful and adaptable to new areas of application, especially pharmaceuticals.

The heuristics for the elucidation of chemical constitution formed part of both a technological paradigm *and* the developing science of organic chemistry. This coincidence was reflected in the social locus of research. Purposeful synthesis, using the heuristic "look for the *Muttersubstanz*," involved a close cooperation between academe and industry. This may be contrasted with another paradigm of planned synthesis, the azo dyes regime, discussed in the next section, which after its establishment around 1877 became largely an industrial affair.

The anticipatory interaction with the selection environment is especially visible with regard to the customers of the synthetic dye industry, the textile dyers and printers. Originally, these customer industries constituted the main repository of knowledge about dyes and in particular about their application to textiles (compare this with the role of colorists previously described). Suppliers of synthetic dyes had to adapt to their demands and standards. Over the course of time, though, a dramatic reversal of roles occurred. The dyestuffs industry emancipated itself with respect to the textile dyeing and printing industries. The original selection environments were to some extent incorporated into the dyestuffs plants by the establishment of test departments and the performance of application research (at the time, these were social innovations!), and for these purposes colorists from the customer industries were recruited.[16] Testing and application research departments conducted a kind of internal simulation of the selection environment: By anticipating possible selections, they were domesticated and made harmless. In the same movement the selection environment was affected: Textile dyers and printers adjusted their techniques to the requirements of the new products and were supplied with instructions and application manuals. The dyestuffs prospectus of Badische Anilin- & Soda-Fabrik (BASF) was actually used as a handbook for the textile dyer. The emancipation of the synthetic dye industry had transformed its customers into "supplier-dominated firms."[17]

The test departments set up by the dyestuffs plants also provided the infrastructure for the emergence of specific "traditions of technological testability" (Constant 1983) which at a somewhat later stage (after 1900) made it possible for the big dyestuffs firms, through the work of so-called *Echtheitskommissionen* (*Echtheit* means fastness), to establish and impose *standards* of fastness for synthetic dyes. These standards of fastness in their turn defined much more rigorously the nonprice parameters of market competition for existing and prospective new dyes. The establishment of well-defined selection criteria (whose existence is usually presumed in evolutionary thinking) was largely the result of the thrust of the technological trajectory itself.

Test departments and application research are examples of a *nexus* between trajectory and selection environment—a concrete institutionalization (and thus a stabilization) of their interaction (figure 1). In the last section we analyze another nexus, the patent system and its detailed articulation.

Figure 1

Testing facilities for dyeing at the Bayer Company around 1880. These facilities in the next few years would grow into the coloristic department (1887). Photograph courtesy of the Bayer-Archiv.

The Emergence of the Paradigm of Azo Dyes

Alongside the technological paradigm of purposeful synthesis, based on the heuristic "look for the *Muttersubstanz*," another paradigm emerged that was also directed at the synthesis of dyes in a purposeful way but that did not involve a major search for the *Muttersubstanz* (we reserve the description "purposeful synthesis" for the former paradigm). It was based on the exploitation of the so-called *coupling reaction* (or Griess's method), a method capable of wide application for making dyes. The coupling reaction involves the pairwise combination of diazo compounds with aromatic amines or phenols to form azo compounds, that is, compounds with a double nitrogen ("azo") group uniting aromatic rings. Most azo compounds are dyestuffs. When the virtually limitless possibilities of the coupling reaction were publicly recognized in 1877, the synthetic dye industry rushed feverishly into the field of azo dyes. With some justification the subsequent decades can be called the era of azo dyes, for this class of color grew into the numerically largest category of dyes.

How did the paradigm of azo dyes emerge so suddenly? As a matter of fact, it did not. The heuristics emerged slowly; it was only the cultural matrix, the general recognition of the promise of this cluster of heuristics, that can be clearly located in time. We cannot possibly disentangle and describe here all the different strands that eventually came together to form the paradigm of azo dyes, but we can indicate some main lines of development.

Many persons were involved in the process of discovery, with interests ranging from purely scientific to more technological concerns. An important circumstance was the fact that in the 1860s aniline yellow and Bismarck brown, now classified as azo dyes, were available in commerce. These were not, however, recognized as azo dyes, nor were they the result of a conscious application of the coupling reaction; rather they were the outcome of the older heuristic "tinker with aniline." Yet their mere existence encouraged efforts to elucidate their constitution.

Griess's greatest contribution consisted in his discovery of diazo compounds around 1860. One historian of industrial chemistry has expressed his astonishment that these diazo compounds "were not practically developed [for the making of azo dyes] until c. 1877" (Sherwood Taylor 1957, p. 237). The eponymy ("Griess's method") does not seem completely deserved. It may have been a definite handicap that Griess's conceptual scheme did not allow for a sharp distinction between diazo and azo compounds. But many other conditions had to be fulfilled before technological development could occur.

A clear distinction between diazo and azo compounds was the fruit of Kekulé's structure theory. After 1865 he was attempting to complete his theoretical system: Griess's diazo compounds and the two commercial azo dyes had to be fitted in (and there was the additional challenge of the competing conceptual scheme used by Griess). Yet Kekulé did not perceive the simple gestalt of the coupling reaction (as we can perceive it now). Subsequent elucidations of the constitutions of aniline yellow and Bismarck brown led only to the conclusion that there had to exist a third, still unknown dye that was an intermediary between the other two. Attempts at synthesis failed, however.

A first indication of a conception of the coupling reaction (without using the term) can be found in 1875 in a scientific paper by Baeyer and Jäger, who tried to intervene in the continuing controversy between Kekulé and Griess on the constitution of azo compounds. Baeyer and Jäger nowhere indicated, however, that the coupling reaction might be useful for making dyes.

This last idea seems to have first occurred to Otto Witt, a German chemist who was then working for an English dyestuff firm. When he considered the empirical materials relevant for his structure-color theory, on which he was working at the time, his attention was drawn to the "gap" between aniline yellow and Bismarck brown where a suspected third dye should exist. Earlier attempts to fill the gap had failed, but Witt thought of using the kind of reaction described by Baeyer and Jäger. This led to the discovery of chrysoidine (as the intermediary dye was called) in January 1876. The dye was marketed, but its manufacturing process was kept secret (when Witt published his famous structure-color theory shortly afterward, he did not even mention the existence of the dye, although it could have provided useful additional support for his theory).

Independently of Witt, chrysoidine (or a close homologue of it) had also been discovered by Heinrich Caro, the first research director of BASF, using a more complicated route. Caro informed Griess, with whom he often corresponded, about his find. In the spring of 1876 Caro, Witt, and Griess organized a conspiracy of silence: No one should publish the findings.

Meanwhile, in France the pharmacist Roussin had also arrived at a method for making azo dyes, following a much more empirical route. Roussin cooperated with the Paris firm of Poirrier for the commercial exploitation of his azo dyes. An important point was that Roussin used sulfonated reactants from the naphthalene series (naphthols) instead of from the benzene series for the formation of azo dyes. The consequence was that the resultant dyes had valuable properties: a red color (because of the use of naphthalene derivatives) and solubility in water (because of sulfonation). Roussin also maintained the utmost secrecy (he described his findings in *plis cachetés* deposited with the Académie des Sciences).

The game of secrecy was spoiled by Hofmann. In January 1877 he disclosed the secret of the constitution and method of preparation of Witt's chrysoidine, followed in July 1877 by similar revelations on one of Roussin's azo dyes marketed by Poirrier. On the latter occasion Hofmann remarked that "Griess's method" (as he called the coupling reaction!) had opened up a vast domain for the synthetic dye industry, of which the limits were not yet within sight; indeed, it pointed the way to an endless series of new dyes. Hofmann's disclosures opened the eyes of other chemists to the immense possibilities of the field of azo dyes (figure 2). From a social

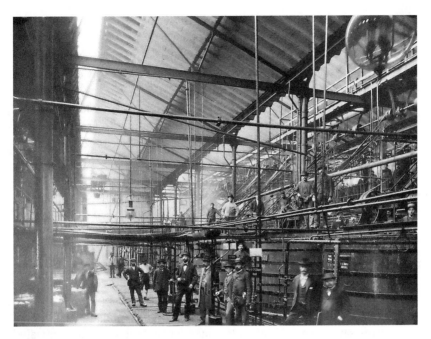

Figure 2
Old azo dye plant at the Bayer Company (before 1901). Photograph courtesy of the Bayer-Archiv.

point of view, this was the definitive establishment of the technological paradigm: The technicians' beliefs were articulated.

Three elements must be emphasized regarding the emergence of this technological paradigm. There was both an exemplary *process* (Witt's method for making chrysoidine) and an exemplary *product* (Roussin's dyes with their valuable technical qualities); both were combined in a cultural matrix through Hofmann's revelations, which also supported the expectations necessary for working within the technological paradigm. Or, as one of the participants in the whole episode, Caro, later declared:

With chrysoidine it was the *process* that was new and promising for the dyestuffs industry [*die Farbstofftechnik*]; with Roussin's dyestuffs it was the new *product* to which textile dyeing [*die Färbereitechnik*] had looked forward for such a long time; with both, it was Hofmann's revelation of their methods of preparation which unleashed the potential energy stored in this domain with the fastness of thought. (Caro 1892, p. 1088)

The Trajectory of Azo Dyes and the Patent System

With the passing of the German Patent Law (*Reichspatentgesetz*) in 1877, a patent system was introduced that represented a selection environment that proved highly relevant for the further evolution of the trajectory of azo dyes. In this section we examine the interaction between the two.

Before July 1877, when the new patent law came into force, there was no effective patent protection in Germany. The practice of secrecy among the inventors of the first azo dyes may have been in no small measure because of this circumstance. Secrecy was common in the German synthetic dye industry, which did not generally favor a new patent law. Under these conditions it was Hofmann, a declared protagonist of patent protection, who set out to demonstrate that secrecy would be of no avail and that only a good patent law could guarantee the undisturbed industrial exploitation of inventions. His disclosures may have served to bring home this lesson to some representatives of the German dye industry who opposed patent protection for chemical inventions. In a polemic with Otto Witt after his first disclosure, Hofmann triumphantly stated: "The time of the arcanists has passed." Indeed, Hofmann's disclosures also demonstrated that in the future no one would dare to put an azo dye on the market without patent protection or else the secret of its constitution and method of preparation would be too easily divined. Thus, with regard to the future chances of the trajectory of azo dyes, the German patent law arrived just in time. Without it, one may surmise that a trajectory of azo dyes would not even have been possible.

Yet the mere existence of a patent law is not sufficient to specify the interaction between technological trajectory and selection environment. There is a widespread tendency to consider legal arrangements as providing some kind of stable framework, within which economic and technological activities can proceed smoothly following their own logic. A justification for this view might be found in Max Weber's opinions on the rationality of law in Western societies, which should somehow secure its "stability," "predictability," and "calculability" (*Berechenbarkeit*) as required by modern capitalist enterprise.[18] General norms do not dictate their application in the particular case, so the qualities mentioned are better seen as a practical social accomplishment.

The German patent system that was introduced with the new patent law of 1877 exhibited some particular features. It provided for a system of *preliminary examination* by the *Kaiserliches Patentamt*, supplemented by an opposition procedure in which third parties (competitors) could raise objections to the granting of patents. A relevant stipulation for chemical inventions was the fact that the law excluded the protection of chemical substances; a chemical invention would be patentable only insofar as it concerned "a particular process" (*ein bestimmtes Verfahren*) for the manufacture of such substances. But the law did not define what was meant by "a particular process." More generally, even the term "invention" was left undefined.

This absence of a legal definition was considered a major source of legal insecurity by many inventors and industrialists. To a certain extent, the gap left by the patent law was filled by Josef Kohler's influential philosophy of invention. Kohler drew a sharp distinction between inventions and constructions; constructions might involve the substitution of "equivalents" in existing processes (doctrine of equivalence). Constructions by themselves should not be eligible for independent patents; only inventions would deserve such protection. Whether or not a certain technological achievement qualifies as a (patentable) invention is, however, not the only question to be decided. Another question concerns the delimitation of the legitimate *scope* of a patent, once an invention is presumed. On this point Kohler's doctrine was also relevant. Kohler held the view that the "subject matter" of an invention, for which a patent was granted, should encompass everything that could be obtained by "constructive" extension of the concrete invention (for example, by the substitution of equivalents). Kohler's doctrine of equivalence seemed to be tailor-made for inventions in the synthetic dyes field because of the ubiquity of such phenomena as homology, analogy, and isomerism in organic chemistry. As a matter of fact, Kohler's theory was eagerly embraced by the dye chemist Otto Witt to clarify the problems of patent law with respect to this domain of technology.

Serious problems of patentability were raised in particular by the class of azo dyes. Inventive activity within the technological paradigm of azo dyes took on a special character. The making of azo dyes was characterized by Heinrich Caro as "an endless combination game." The number of possible combinations was estimated at more than 100 million (taking repeated coupling into account). For the dye industry this class of dyes must have appeared an almost inexhaustible gold mine. Understandably, then, the exploitation of this gold mine gave rise to what Caro called "scientific mass-labor" (*wissenschaftliche Massenarbeit*) in the industrial research laboratories that were largely established for this purpose (figure 3). This also affected the character of inventive activity, which was described by Caro as "construction bound to rules." It was not surprising, then, that the

Figure 3
The scientific laboratory of the Bayer Company in the 1890s. Photograph courtesy
of the Bayer-Archiv.

development of the trajectory of azo dyes raised difficult problems of
patentability.

One of these problems was what exactly the phrase "a particular process"
would mean in this context. At one extreme the coupling reaction in its
generality could be interpreted as a "particular" process, leaving no room
for the patentability of the more specific "processes" within its compass.
At the other extreme one could argue that only a process leading to one
and only one azo dye could count as a "particular" process in the sense of
the law. Somewhere between these two extremes, the Patentamt (or the
courts) had to strike a "reasonable" balance, but where? The actual practice
for those who applied for patents in the field of azo dyes was to try to
cover the widest possible territory by also claiming various classes of ana-
logues, homologues, and isomers besides the substances actually used in
the "invented" process. An average "particular" process would lay claim
to, say, two hundred substances. In 1887 the Patentamt put an end to this
practice by issuing a *sample regulation*: Those who applied for patents had
to hand over samples of the substances for which they claimed protection
to back their claims. This rather arbitrary regulation also cut through the

Gordian knot of the problem of the acceptable scope of a "particular" process. It also earned the hostility of Otto Witt, who condemned the measure for social and theoretical reasons: It put the individual inventor at a considerable disadvantage with respect to large-scale industry, and it was at odds with Kohler's doctrine. Witt's protests had no effect.

Apart from the problem of the acceptable scope of patents, there was also the even more vexed question of whether all the so-called inventions in the azo field really satisfied the requirement of inventiveness. In fact, they were all variations on the same basic theme, to wit, the coupling reaction. By applying that method to new "coupling components," the novelty of the process could perhaps be claimed, but what about its inventiveness? The fact that Caro characterized the inventive activity in the azo field as construction can also provide food for thought in light of Kohler's distinction between invention and construction.

Sometimes, patent applications for manufacturing azo dyes were indeed rejected by the Patentamt for lack of inventiveness, but more often the Patentamt granted patents for inventions of which it was equally doubtful whether or not they satisfied the requirement. At that time, no consistent policy could be discerned.

Some of the doubts concerning the import of the requirement of inventiveness were finally resolved in 1889 by a famous judicial decision in the so-called Congo red case. Congo red was an azo dye belonging to the class of substantive or direct cotton colors; that is, it had the special property that it could dye cotton without the use of mordants. Congo red was, once more, the result of an application of the coupling method. The Congo red patent was acquired by Agfa in 1884. When the dye proved commercially successful, Bayer developed a similar dye (Benzopurpurin) by exploiting a loophole[19] in the Congo red patent, which led to patent litigation between Agfa and Bayer. The legal conflict was settled, however, when both parties realized that neither had a good chance to win the battle. They joined forces to fight a third (much smaller) firm, Ewer & Pick, which was accused of infringing on the Congo red patent. As a countermove, Ewer & Pick initiated legal action to challenge the validity of the Congo red patent (*Nichtigkeitsklage*). In the end the case was brought before the highest German court, the Reichsgericht.

Heinrich Caro appeared as an expert witness. The prospects for the Congo red patent did not seem bright because Caro was unable to discover anything original in the process used to manufacture the dye. He had shown the *title* of the patent to his laboratory assistant and with no other information the boy had been able to prepare Congo red.

Also present at the court session was Carl Duisberg, then a promising young chemist at the Bayer company, later to become the architect of IG Farben. In his memoirs he remembered the following episode of the Congo red case:

During the interval I tried to make clear to Dr Heinrich Caro, who at that time was my opponent, but who became a fatherly friend to me later, that through his point of view he actually delivered the death-blow to the whole chemistry of azo dyes and thereby also brought most of the patents of his own firm to the brink of the abyss. But he did not, after resuming his exposition, withdraw one iota of what he had said earlier. (Duisberg 1933, p. 44)

We think Duisberg is hardly exaggerating: Not only the Congo red patent but also much more was at stake. When all seemed lost for Agfa and Bayer, Duisberg begged permission to speak and to say a few words on what he called the technical details of invention. He took care to make clear to the court what was at stake for the synthetic dye industry. It must have been an eloquent speech, for eventually the validity of the Congo red patent was upheld by the court. At that occasion it formulated what came to be known as the doctrine of the *new technical effect*. The court argued that the process for making Congo red, lacking any inventiveness of its own, would as such not have been patentable. In this case, however, the application of the general method resulted in a dyestuff of undoubted technical and commercial value. Its unexpected and valuable technical qualities more than compensated for the lack of inventiveness of the process. In other words, the court said that, if the requirement of utility is particularly emphasized, it is no longer necessary to look at whether the requirement of inventiveness is also satisfied.

The economic significance of the doctrine of the new technical effect was immense. It saved the *wissenschaftliche Massenarbeit*. The research laboratories of the big companies could continue to turn out "inventions" on a large scale. Lack of inventiveness would be no obstacle to their patentability. Conceptions of inventiveness, derived from individual inventors and their inventive genius, were no longer held decisive in judging azo dye inventions.

Seen in a wider context, the significance of the doctrine consisted in providing a judicial legitimation to the emergence of systematically organized research within industry or, in David Noble's characterization, the rise of "the corporation as inventor" (Noble 1977, ch. 6) along with the concomitant "routinization of innovation" (Schumpeter 1974, p. 132). The organized pattern of inventive activity in the field of azo dyes was only an early example of a larger, more general trend from people-embodied to organization-embodied technology (Dosi 1984). As regards the synthetic dye industry, the new pattern became the rule after 1900, when dyestuff inventions were generally considered to be *Etablissementserfindungen* (corporate inventions). The routinized character of these inventions was graphically expressed by Carl Duisberg (by now a director of Bayer) in one blunt statement: "Nowhere any trace of a flash of genius."

With regard to the azo dyes trajectory in particular, the doctrine of the new technical effect not only safeguarded its future development *tout court* but also reinforced the heuristics oriented toward the demand side. The hoped-for (new) technical effects, such as improved mordant properties or direct dyeing of cotton, became important guidelines. Empirical rules of thumb for the relation between chemical structure and properties became heuristics for micro trajectories within the main paradigm. This involved a closer adaptation to the demand side.

It thus becomes clear that the patent system (with its interpretations) provides a *nexus* between trajectory and selection environment. Both requirement necessary to secure the continuing dynamics of the technological trajectory and considerations of a broad socioeconomic and political character enter into the detailed articulation of the nexus. Patent law is not just a matter of the application of existing legal provisions, it is a social accomplishment that creates new solutions for new situations.

A case in point may be the contemporary situation with respect to biotechnology, in which a process of redefinition and reinterpretation of patent law is going on that in many ways resembles the nineteenth-century developments described here. Of course, such a process does not occur within a legal void. For biotechnology the elaboration of new interpretations also means that the force of the older interpretations that were specifically developed with regard to chemical technologies, that is, their force as "binding precedents," has to be overcome.

Conclusion

In this chapter we have presented a sociological extension of the Nelson-Winter-Dosi model, supplemented with empirical illustrations derived from nineteenth-century synthetic dye chemistry.

We hope that our study represents another demonstration of the utility of viewing technological development as guided by heuristics. In addition, we have emphasized the importance of the cultural matrix of expectations in which these heuristics are embedded. Such a cultural matrix proves to be a strategic factor in the emergence of a new paradigm.

We also think that our notion of a nexus, illustrated by the functioning of test departments and application research and by the working of the patent system, may turn out to be a useful concept to examine the interaction between technological trajectories and selection environments.

Finally, our view of technological development also has implications for policy, but that is another story.

Notes

This chapter is based on a report (in Dutch) about the Nelson-Winter-Dosi model that we prepared for the Directorate of Science Policy of the Dutch Ministry of Education and Science (February 1984). For the historical materials on nineteenth-century synthetic dye chemistry, we draw on an extensive database that was built up in the course of a large research project of the Department of Science, University of Nijmegen (project director W. J. Hornix). We would like to thank the Archives of the Bayer AG in Leverkusen for providing us most kindly with the illustrations used here. We are grateful to Brian Wynne for useful comments on an earlier draft of this chapter.

1. Indeed, we think that these formal models leave out much of what is valuable and interesting in their theories. The basic ideas, which are central to their theories and which are really of a sociological nature, are too easily sacrificed in the building of formal models of competition. Perhaps this is the inevitable price economists have to pay for their professional integrity. We, however, can afford to stick to the basic ideas and to resist the temptation of formalization.

2. More specifically, Pinch and Bijker state that their model (like Constant's) arose out of evolutionary epistemology (Toulmin 1972; Campbell 1974).

3. Nelson and Winter's discussion of routines includes a digression on "routine as truce," which is interesting from a sociological point of view. They observe that routine operation of an organization represents a truce between organization members. The terms of the truce, however, can never be fully explicit, but become increasingly defined by a shared tradition arising out of the specific contingencies confronted and the responses of the parties to these contingencies (Nelson and Winter 1982, pp. 107–112).

4. Perhaps the biological analog to search routines would be "mutator genes" (Dawkins 1978, p. 47).

5. To find a comparable view in biology, one has to look beyond the (neo-) Darwinian mainstream of "modern synthesis" to the theory of the Russian geneticist N. I. Vavilov, who emphasizes constraints of inheritance and channeled variation. His theoretical position has been recently resuscitated by Stephen Jay Gould. For Vavilov's theory and Gould's re-edition of it, see Gould (1984, pp. 134–144).

6. This point was expressed by John Maynard Smith, an aircraft engineer turned biologist: "In the organic world, once two lineages have diverged for some time, they cannot rejoin. In engineering, two inventions, first developed to perform different functions in different kinds of machine, can be brought together in a single machine; the trolley-bus is a "hybrid" between a bus and a tram" (Smith 1958, p. 262).

7. Rosenberg's bottlenecks can be compared to Hughes's reverse salients (Hughes, this volume), except that reverse salients seem much less ephemeral.

8. Sometimes the relevant expectations may be of a broad, even sociological, nature. In the case of the French VEL project (Callon, this volume), the EDF engineers working on this particular project acted on an explicit vision of where French society was going—a vision that was characterized by Callon as a variant of Tourainian sociology. One can also think of the Japanese Fifth Generation project, which seems to have been predicated on, inter alia, Daniel Bell's ideas on the coming of the post-industrial society (Feigenbaum and McCorduck 1984, pp. 35–37). Unlike Touraine for the French EDF engineers, however, Bell became an acknowledged ideologue for the Japanese "knowledge engineers." In our own case the expectations sustaining the technological trajectories were much less sweeping in scope. This circumstance made it easier for us to maintain the distinction between technological paradigm and selection environment, as this distinction was perceived by the participants themselves.

9. But see Gould (1984, p. 334) for recent challenges of the traditional dichotomy between random variation and deterministic selection: "Darwin's strict dichotomy seems to be breaking. . . . Randomness is challenging the determinism of natural selection as a cause of evolutionary change."

10. To cite Jon Elster: "Natural selection always acts as if the environment is parametric, even when it really is a strategic one, whereas man is capable of taking the strategic nature of the context into account" (Elster 1984, p. 18). Actually, this taking into account of the strategic nature of the context implies what John Law has called heterogeneous engineering (Law, this volume).

11. From an ordinary evolutionary point of view, one would simply say that the *same* artifact, by being placed in *different* selection environments and thereby subjected to varying selection pressures, became the starting point for more than one line of development. We would like to stress that, apart from their radical social constructivist claim, Pinch and Bijker's model is quite comparable to evolutionary models in biology. Their multidirectional view, for example, fully accords with neo-Darwinist accounts of species evolution that stress its bushlike rather than its ladderlike character (Gould 1980, pp. 56–62).

12. The myopic nature of natural selection (in biology) is stressed by Jon Elster (1984, p. 9).

13. Weizenbaum (1976, pp. 35–36) offers an interesting discussion of how a particular technique, which for its implementation requires much computation, can be insulated against the risk of manifest failure by "computerizing." On the one hand, "a failure of the technique can easily be explained away on the ground that, because of computational limitations, it was never really tested." On the other hand, when much effort is put into the task of "computerizing" the technique, "an illusion tends to grow that real work is being done on the main problem" and "the poverty of the technique . . . is thus hidden behind a mountain of effort."

14. The creation of a nexus may be considered an example of heterogeneous engineering (Law, this volume), but with the proviso that in our empirical case this kind of engineering operates on a much more restricted scale. We do not think it helpful to conceive of synthetic dye chemistry as consisting exclusively in the building of "systems" or "networks," or the bringing and keeping together of various bits and pieces. Such an analysis would not do justice to the specificity of the technology. On the other hand, there is no denying that a large-scale technology, such as the generation and distribution of electric power, that permeates the whole fabric of society can be usefully analyzed in systems terms, as indeed was done by Tom Hughes (1983 and this volume). But not all technologies are alike.

15. Both Graebe and Liebermann and the Fischers worked under the auspices of Adolf Baeyer, who was undoubtedly the grand master of organic synthesis and particularly known for his synthesis of indigo (which took long years of industrial development to be feasible as an economic process, however).

16. This incorporation of (part of) the selection environment into the dyestuffs firms can be compared with the phenomenon observed by Hughes (this volume) that a large technological system often has a tendency to internalize some of the external factors, thus making for more predictable performance.

17. The term is borrowed from Keith Pavitt: "Supplier-dominated firms make only a minor contribution to their process or product technology. Most innovations come from suppliers of equipment and materials" (Pavitt 1984, p. 356). A graphic illustration of the changed balance of power between the synthetic dye industry and its customers is provided by the *Echtheitsbewegung* in the year after 1900, when the demand for ever faster dyes, made possible by the technological trajectory, was forced on the textile dyers and printers by a massive fastness campaign, or, as a leading representative of IG Farben later called it: "Die (von uns) in die Kundschaft hineingetragene Echtheitsbewegung" [the drive for fastness that (we ourselves) induced in our clientele] (Kränzlein 1935, p. 6).

18. In Weber's view the law should be as calculable as a machine and the ideal judge should be no more than a *Paragraphenautomat* (Weber 1921, pp. 140–142). In an unpublished paper ("The specialization of experts and institutions in the English patent court," 1983) Brian Campbell (of Lancaster and New Brunswick, Canada) uses

Weber's ideas to throw light on the role of experts in patent cases. He identifies the rise of the large corporation and its need for stability of legal patterns as the key factor in the emergence of specialized patent courts and the corresponding change in the kinds of expert witnesses who appear to give testimony (from "generalist" specialists in the art of testifying and patent law to "specialist" specialists familiar with the relevant branch of technology). In our opinion the "stability" of law refers primarily to an important theme in a discursive repertoire (often invoked by parties in patent cases) rather than to a description of a state of affairs. Firms do not have an abstract need for stability but only (variously interpreted) needs for stability in particular situations, which, however, when clashing with each other, may indeed generate "instability." That measure of legal stability which is actually found in practice should be considered a practical, social accomplishment.

19. The term "loophole" is, of course, not a neutral description (it reflects the point of view of Bayer in the legal conflict) and should be avoided precisely because it prejudges the question of the legitimate *scope* of the Congo red patent. The dispute between Agfa and Bayer illustrates the inevitability of *negotiation* for the "proper" delimitation of the scope of patents as intellectual property rights. We do not pursue this question any further here.

The Social Construction of Bakelite: Toward a Theory of Invention

Wiebe E. Bijker

The aim of this chapter is to put forward some theoretical concepts whereby the development processes of technological artifacts can be understood. The approach I suggest extends the social constructivist analysis of the development outlined by Pinch and Bijker (this volume). In the earlier work we proposed a descriptive model that focused on the various meanings attributed by different social groups to an artifact. This allowed us to give a symmetric account of "successful" and "failed" artifacts, and it also had the advantage of incorporating both technical and nontechnical elements in the description. In this chapter I develop the model further by considering aspects of the history of Bakelite.

The chapter is composed of four parts. In the first section I describe the early history of plastics. For my purposes here the emphasis is not on the historical details but on the presentation of some explanatory concepts. In particular, the notion of the interpretative flexibility of an artifact is elaborated on and the concepts of technological frame and inclusion are introduced for the first time. The exact meaning of these terms is the topic of the second section. In the third part some flesh is put back onto the bones of these concepts by following the plastics history into the Bakelite era. Finally, in the fourth section I suggest a more general scheme for a theory about technological development.

A Description of the Plastics History Pre-Bakelite

The history of human use of plastic materials is as long as the history of humanity itself. The Egyptians used resins, natural plastics, to varnish their sarcophagi, and the Greeks made jewelry out of amber. These two applications—varnishes and the production of small solid objects—have provided and continue to provide important markets for the plastics industry. Some of the natural plastics, such as shellac, can be used for both purposes, whereas others, such as rubber, are used for one purpose only.

Until the mid-nineteenth century, the use of plastics had been confined to luxury and fancy goods, ranging from shellac-lacquered scent boxes to ivory jewelry. The vulcanization of rubber, however, created new markets. During the vulcanization process, rubber is heated in the presence of sulfur, which renders it more flexible and durable. This makes the rubber suitable for a wide range of applications. In the second half of the nineteenth century, both rubber and shellac were increasingly used for electrical insulation, especially "hard rubber," also known as "vulcanite" or "ebonite." This rubber was manufactured by mixing a much higher percentage of sulfur with the crude rubber than was done in the ordinary vulcanization process, and it was used for several new industrial purposes for which none of the older natural plastics had previously been employed. Apart from its use as an electrical insulating material, hard rubber was used for the internal coating of chemical apparatus and accumulator storages and for the manufacture of surgical instruments and artificial teeth. Thus, although plastic materials had previously been restricted to the jewelry-wearing upper classes, they now found favor among new social groups. This, however, created a problem.

A Scarcity Problem and Three Variants of Solution

The exotic location of the sources of shellac and rubber led several chemists and industrialists to perceive an imminent scarcity of natural plastics. "We are exhausting the supplies of India rubber and gutta percha, the demand of which is unlimited but the supply not so," remarked the chairman of a meeting of the Royal Society of Arts in 1865 (cited in Kaufman 1963, p. 33). At this meeting Alexander Parkes gave a lecture on his new plastic material Parkesine, which was the first of a series of variants produced in an attempt to solve the scarcity problem by trying to modify nitrocellulose (Parkes 1865b). Another important incentive for these researchers was the long-standing objective of nineteenth-century inventors, to find a substitute for ivory (Friedel 1983).

Nitrocellulose could be produced rather cheaply from paper, wood fiber, or rags. Its importance as an explosive substance immediately caught the world's attention when the Swiss chemist Christian Friedrich Schönbein found a commercially viable production process in 1846. Somewhat later, several chemists and inventors explored the possibilities of dissolving nitrated cellulose in a mixture of alcohol and ether. The "collodion," as this solution was called, was a clear fluid with the consistency of syrup that, when poured out and allowed to dry, resulted in a transparent film. Several applications, such as wound plaster, a means to render fabrics waterproof and a basis layer for photosensitive materials, were successfully developed.

Parkes was the first to attempt to produce "a hard, strong, brilliant material" from nitrocellulose that could be cut and molded (Parkes 1855). The development of a new plastic market for technical applications, besides the traditional market of luxury consumer goods, is exemplified by Parkes's business policy. On several occasions, in presenting his new plastic, Parkesine, Parkes did not distinguish clearly among the different uses to which it could be put: It could be used as a substitute for luxury plastics, such as ivory and tortoise-shell, as well as for industrial substances such as rubber and gutta-percha (a similar substance to rubber but obtained from a different tropical tree). Parkes placed the early emphasis on its use in the production of fancy articles. For example, at the 1862 World Exhibition he presented medallions, buttons, combs, pierced and fret work, inlaid work, pens, and penholders (Kaufman 1963). However by 1866, when Parkes tried to persuade investors to put capital into a newly incorporated Parkesine Company, the prospectus hardly mentioned Parkesine as a beautiful material for making "works of art." Instead, its applications for making carding, roving, and spinning rollers, insulating telegraphic wires, manufacturing tubing, and varnishing and coating iron ships were all stressed (Friedel 1979). Paralleling this shift of emphasis away from the fancy applications, Parkes tried to make his material as cheap as possible (Friedel 1979; Dubois 1972), but this was not enough to maintain the involvement of the new social groups of users. His eagerness to show the applicability of Parkesine to a variety of different purposes meant that he placed less emphasis on finding a dependable chemical formula for at least one specific form of Parkesine. Thus the plastic was not produced with a consistent quality, and a great number of the items sold by the new company were returned as unacceptable because of shrinkage, twisting, and distortion (Worden 1911). In 1868 the Parkesine Company was liquidated.

A second variant of a nitrocellulose plastic was closely linked to Parkesine. The manager of the Parkesine Company, Daniel Spill, attributed the failure of Parkesine to the fact that they had not made their materials white enough. If it could be made whiter, Parkesine would appear as a more credible substitute for ivory. In 1869 Spill founded another company, and with only minor changes in the manufacturing process he continued the production of what by now was called Xylonite. This venture fared no better than the previous one, and it was abandoned in December 1874. Spill had an unshakable faith in his material and established another company in 1875. This time he did succeed in finding a small but rather stable market for what he now also called Ivoride (Kaufman 1963).

A third variant of solution to the scarcity problem of natural plastics was developed by John Wesley Hyatt in Albany, New York. As the popular story goes, Hyatt's research was triggered by the offer of a $10,000 prize for the patent of a material that could he used as a substitute for ivory in the manufacture of billiard balls. Hyatt first tried several well-known plastic compositions, such as wood fiber with shellac. Although this did not result in a satisfactory substitute, an important consequence was Hyatt's acquisition of familiarity with processes for molding plastics under heat and pressure (Friedel 1979). This experience made Hyatt aware of the problems of liquid collodion solutions, such as the ones Parkes and Spill had used: The drying process inevitably caused shrinkage, which made it difficult for such collodion mixtures to be used for molding solid objects. In his own words:

From my earliest experiments in nitrocellulose, incited by accidentally finding a dried bit of collodion the size and thickness of my thumb nail, and by my very earnest efforts to find a substitute for ivory billiard balls, it was apparent that a semi-liquid solution of nitrocellulose, three-fourths of the bulk of which was a volatile liquid and the final solid from which was less than one-fourth the mass of the original mixture, was far from being adapted to the manufacture of solid articles, and that I must initially produce a solid solution by mechanical means. (Hyatt 1914, p. 158)

After Hyatt took out several patents describing such processes, in 1870 a patent was issued that referred to "the use of finely committed camphor-gum mixed with pyroxyline-pulp [nitrocellulose] . . . [rendering] a solvent thereof by the application of heat (Hyatt 1870). Only many years afterward, when he was involved in patent litigation trials, did Hyatt use the term "solid solution" to describe the material produced in one of the first stages of the production process. He used this term in order to bring out the crucial difference from earlier nitrocellulose plastics. His solid solution appeared at the time to be more of a moist mixture:

We conceived the idea that it might be possible to mechanically mix solvents with the pulp and coloring matter while wet, then absorb the moisture by blotting papers under pressure, and finally submit the mass to heat and pressure. (Hyatt 1914, p. 159)

However, his later use of the term "solution" probably added to the perceived importance of the role of solvents in the Celluloid production. Ironically, Hyatt himself did not mention camphor as a solvent, only as an additive. I return to this point concerning the role of solvents later. Together with his brother, Isaiah S. Hyatt, Hyatt founded the Albany Dental Plate Company in 1870. They advertised

a newly-invented and patented material for Dental Plates or bases for artificial teeth, that cannot fail to delight every dentist who desires a better material for the purpose than hard rubber. (*The Dental Cosmos* 13 (1871); cited in Friedel 1979, p. 53)

The dental plates did have various imperfections. Some of them had a strong camphoric taste; some became soft in the mouth (sufficiently so for the teeth to become loose), and plates were found to warp after having been adjusted to the patient's mouth (Friedel 1983). Although these dental plates were far from satisfactory, the concerted effort to produce a material with specific, consistent qualities resulted in the Hyatt brothers' forming the Celluloid Manufacturing Company in order to produce Celluloid in semifinished form (rods, sheets, tubes, etc.). From 1872 to 1880 the Hyatts granted licenses to different companies for the production of Celluloid consumer goods, each company devoting itself only to a narrowly defined market (Friedel 1979, 1983).

Interpretive Flexibility of the Artifact Celluloid as Shown in the Selection Process

Which of the two rival plastics, Ivoride/Xylonite or Celluloid, became dominant? The selection process (see Pinch and Bijker, this volume) was determined to a large extent by a patent controversy. This controversy, between Spill and Hyatt, concerned who had priority in the invention of the use of camphor in the production of a plastic out of nitrocellulose. The debate can be used to show the "interpretative flexibility" of the artifact Celluloid. For Spill, Celluloid meant a mixture of nitrocellulose and camphor that, although prepared in a slightly different way, was essentially identical with his Xylonite or Ivoride. However for Hyatt, a crucial difference between Celluloid and other nitrocellulose plastics was to be found in the fabrication process: He used, he said, a solid solution instead of a liquid solution of nitrocellulose and camphor. Linked to these differences in how these two industrial chemists conceptualized the meanings of their plastics were differences in goals and in the resulting lines of development. Spill mostly valued the use of his plastic as a substitute for expensive natural plastics, as is indicated by the name Ivoride and his emphasis on the need for the material to be white. Consequently, mass production by molding was not his first priority. For Hyatt the goal of constructing a material that could be used to produce a large number of narrowly defined products of consistent quality led inevitably to focusing on the production process and especially on the molding characteristics of the material.

The patent dispute between Spill and Hyatt was resolved by Samuel Blatchford, at that time a justice of the US Supreme Court, and "the most highly regarded patent judge of his time" (Friedel 1983, p. 132). He decided on August 21, 1884, that neither Spill nor Hyatt should be named as the inventor of a camphor-nitrocellulose plastic, because Parkes had already covered that combination of substances in his patents. In effect, this meant a victory for Hyatt, because the judge's decision denied Spill the novelty of using camphor and thus nullified his grounds for litigation against Hyatt. The Celluloid Manufacturing Company succeeded in putting itself on a firm financial base.

The increasing stabilization of Celluloid can be traced by following its use as an intermediate material between cheap but ugly looking plastics, such as rubber, and luxurious materials, such as ivory. For example, the advent of Celluloid brought combs, cuffs, and collars within the reach of social groups that had not been able to afford such luxurious articles until then (luxurious because the original cotton cuffs and collars had to be washed every day and this was such a laborious job that it needed, it seems, servants to do it for you; figure 1).

A Problem with the Artifact Celluloid

Having described some of the processes that led to the eventual stabilization of Celluloid, the next step in the descriptive model is to ask what problems were perceived with respect to this artifact (Pinch and Bijker, this volume). One problem with Celluloid, in the view of certain important social groups, was never solved. This was its flammability. As in the case of the development of the bicycle (Pinch and Bijker, this volume), problems seldom have equal pertinence for all social groups. Thus data about fire and accidents caused by explosions in which Celluloid was said to be involved were interpreted quite differently by different people (Kaufman 1963). For example, it is doubtful whether any chemist would not have thought it the height of folly to heat nitrocellulose under pressure, knowing its explosive character. A professor of chemistry who visited Hyatt's factory warned that if too much heat were applied, the substance would inevitably destroy them, together with the building and the adjacent property!

Figure 1

Advertisement for Celluloid. The advertisers often used anti-Chinese sentiments in the promotion of Celluloid cuffs and collars. Photograph courtesy of the Warshaw Collection of Business Americana, National Museum of American History, Smithsonian Institution, Washington, D.C.

Although Hyatt was skeptical, he was worried enough to put the proposition to the test:

The following day between 12 and 1, when all were out, I rigged up a four inch plank used as a vice-bench, braced it between the floor and ceiling, between the hydraulic press and the hand pump, intending it to shield me from possible harm. I then prepared the mould, heating it to about 500°F knowing it would certainly ignite the nitrocellulose and camphor, and thinking I would abide by the result. The gases hissed sharply out through the joints of the mould, filling the room with the pungent smoke. The mould, press, building and contents were there, including myself, very glad that I did not know as much as the Professor. (Hyatt 1914, p. 159)

However, not many users were convinced by this experiment, and national and local authorities made special safety regulations for Celluloid processing industries (Worden 1911).

Another Artifact and Its Interpretative Flexibility: The Phenol-Formaldehyde Condensation Product

At about the same time that Hyatt was establishing his company for the manufacture of dental plates, Adolf Baeyer in Germany was observing condensation reactions between aldehydes and phenolics. Although he found that under specific conditions chemical compounds that belonged to the group of phenolic dyes were formed, most of the condensation products were resinous and difficult to crystallize (Baeyer 1872).

Many historians of the plastics industry identify Baeyer's condensation product as the first synthetic resin. Having produced the "resin," researchers directed their efforts toward rendering it in an industrial process. This was eventually accomplished by Leo Hendrik Baekeland. For Baeyer himself, however, the reaction product meant something completely different from a synthetic resin. Because the resinous character of the condensation product presented a problem for the usual methods of analysis, Baeyer could not evaluate its importance as a potential synthetic dye. This made the phenol-formaldehyde resin only an annoying by-product that had to be thrown away. A third meaning was attributed to this condensation product by Arthur Michael, who was a student of Hofmann, Bunsen, and Mendeleev and who ended his career as chemistry professor at Harvard University. For Michael the resin did not mean an unpleasant obstacle to synthetic dye research; neither did it mean a potentially useful synthetic plastic. Michael was interested in these synthetic resins for purely academic, biological reasons: He hoped that this research might lead him to a better understanding of natural resins (Michael 1883–1884). He had no interest in potential industrial applications.

Thus the interpretative flexibility of the artifact phenol-formaldehyde condensation product amounts to the existence, in terms of our descriptional model, of three different artifacts: an embryonic plastic material, a potential dye yet to be analyzed, and a method for studying natural resins.[1] However, it was not until the turn of the century that the first artifact came into existence. It is through retrospective distortion that the first artifact is seen to have its origin in 1872. In the next part of this section I address the issue of why the first artifact, Bakelite as it later became known, was not discovered earlier.

Technological Frames and Why No Phenol-Formaldehyde Condensation Plastic Was Constructed

More than a decade passed after the initial observation of the condensation reaction between phenol and formaldehyde, and nobody seemed interested in studying its potential for the production of a synthetic plastic, although at the same time Celluloid's success suggested an attractive market. One might think of the high price of formaldehyde as an explanation for this neglect of the possibility of developing a commercial synthetic plastic. If that explanation is correct, then we would expect the availability of cheaper formaldehyde to lead to a concerted research effort to make a phenol-formaldehyde plastic.

It was not until 1888, after a catalytic process had been developed that enabled formaldehyde to be synthesized directly, that formaldehyde became an easily available material. The dye industry, for example, started to use it in the synthesis of several dyes. We might ask whether there was any trace of a renewed interest in making a synthetic plastic out of the phenol-formaldehyde condensation product at this point.

Indeed, an industrial chemist, Werner Kleeberg, was stimulated by the commercial availability of formaldehyde to take up the study of the condensation reaction. Kleeberg was almost certainly interested in this reaction because, like Baeyer, he hoped to find a new dye. Also for Kleeberg, the "rosarote Masse" meant a substance to be analyzed. And that, again, appeared to be impossible with the available analytical techniques. As a result, Kleeberg concentrated on other formaldehyde reactions that did not produce resinous substances (Kleeberg 1891). Other chemists' interests were triggered by the availability of cheap formaldehyde as well. Otto Manasse and Leonhard Lederer developed, independently, a process to make phenol alcohols (Manasse 1894; Lederer 1894). Probably, both were working for chemical firms, producing raw materials for the synthetic dye industry. These newly discovered chemicals were considered to be of general interest, but they also had commercial value (Lederer 1894). Until then, the production of phenol alcohols had been carried out by the reduction of the respective aldehydes, an expensive and cumbersome process. The abundant availability of formaldehyde suggested another solution: synthesize the phenol alcohols from formaldehyde. Lederer, in summarizing efforts to realize this goal, explained that all such efforts had failed because of the sudden appearance of those "unerquickliche Harze" (awful resins; Lederer 1894, p. 224). Thus, for these chemists too, we can say that the resinous material meant something different. It was no potential plastic to be tamed for molding; nor was it a potential dye to be analyzed for synthesis; it also was not an instrument for studying natural resins; rather it was an uninteresting substance to be avoided because one was after something else.

Apart from further demonstrating the interpretative flexibility of the artifact resinous condensation product of phenol and formaldehyde by adding a fourth construct to the list, the works of Kleeberg, Manasse, and Lederer also indicate that it was not the high price of formaldehyde before 1886 that explains the neglect of the potentiality of that resin as a commercial plastic. Cheap formaldehyde did not lead to the development of a commercial plastic. Another explanation must be sought.

The observation that "they just did not see it" is only a rephrasing of what has to be explained. Why didn't the possibility of producing a synthetic resinous material out of the phenol-formaldehyde reaction figure on the agenda of chemists at the time? Certainly chemists such as Baeyer, Manasse, Lederer, and Kleeberg did not lack commercial acumen. Surely they also would have been familiar with (hard) rubber and Celluloid, if only in their households. Something prohibited the synthetic plastic from becoming an issue for this community of chemists. In order to describe this, I introduce the notion of a *technological frame*.

A technological frame is composed of, to start with, the concepts and techniques employed by a community in its problem solving. (A more comprehensive description of a technological frame is given later, on pages 167–170.) Problem solving should he read as a broad concept, encompassing within it the recognition of what counts as a problem as well as the strategies available for solving the problems and the requirements a solution has to meet. This makes a technological frame into a combination of current theories, tacit knowledge, engineering practice (such as design methods and criteria), specialized testing procedures, goals, and handling and using practice. The analogy with Kuhn's "paradigm" among others is obvious. I return to such analogies in the next section.

If we now apply the concept of a technological frame to the discussion of Baeyer and Kleeberg, it becomes clear why they did not try to modify the phenol-formaldehyde condensation product into a usable plastic. First, they had other goals: the production of new synthetic dyes. But these goals can be changed, especially when large profits are on the horizon. So there is more to it than this. The idea of making a plastic by chemical synthesis simply did not and indeed *could* not occur to them. Chemical theory at that time could not cope with such a substance. Neither could chemical practice: Their daily laboratory practice included all kinds of chemical analysis and synthesis, but the application of pressure and molding techniques were of another world. The technological frame of synthetic plastics was not yet in existence. The same applies to Manasse's and Lederer's not seeing the potentiality of the condensation product: It simply did not fit in the technological frame of their community.

Searching for a Celluloid Substitute within the Celluloid Technological Frame

Celluloid, notwithstanding its success and stabilization in various social groups, still had some important problems. As mentioned previously, for some groups it was quite dangerous because of its flammability; also it was rather expensive because of the price of camphor; and, third, it was not suited to high temperatures, which posed a barrier to many technical applications. This situation prompted several chemists to start searching for an alternative to Celluloid. Other cheaper solvents were tried to replace camphor. All kinds of chemical additives were studied to temper the flammability. And some of these chemists directed their research toward the condensation reaction between phenol and formaldehyde. I argue that this community of chemists had a technological frame that to a large extent was dominated by Celluloid experience.

First, for men such as Smith, Luft, De Laire, Fayolle, and Story, the goal was explicitly to find a Celluloid substitute. Often, they described their products also as "shellac substitutes" or "horn-like substances," but the intended field of application clearly was the same as Celluloid's. Second, most of these inventors did not show any more sophistication with respect to chemical theory than Hyatt had displayed: They made no efforts to say anything about the structure of the condensation product nor about the chemical reaction in detail. Third, their problem-solving strategy focused on finding an adequate solvent. Through the patent litigation trials the choice of the right solvent had acquired the meaning of crucial step in the "invention of Celluloid." And also, much attention was paid to the solvent because of the high price of camphor. This placed the role of a solvent in a central position in the Celluloid technological frame, with respect to both the identification of the crucial problems *and* the problem-solving strategy. Indeed, we see the previously mentioned inventors defining their problem of making a synthetic plastic as how to soften the condensation product, for, having been softened, the material, they hoped, could be treated like Celluloid. Their strategy to accomplish this was to apply all kinds of different solvents at different stages of the reaction. In the words of Baekeland, commenting on Luft's patent:

The whole process of Luft looks clearly like an attempt to make a plastic similar to celluloid and to prepare it and to use it as the latter. The similarity becomes greater by the use of camphor and the same solvents as in the celluloid process. (Baekeland 1909a, p. 322)

However, this strategy did not work in this context and none of these men succeeded in making a commercially viable synthetic resin.

Different Degrees of Inclusion, or How Baekeland Succeeded

Finally, I come to Baekeland. My contention is that Baekeland worked to an important degree within the Celluloid frame and that he worked equally importantly to some extent *not* according to that frame. I want to tackle the descriptional problem by this situation with the concept of inclusion. Baekeland worked within the same frame as Smith, Luft, De Laire, Fayolle, and Story, but he had a lower inclusion in that frame. I describe him as working according to the Celluloid frame because he had the same goal—making a substitute for Celluloid and the natural plastics and varnishes—and because he started to work with the same problem-solving strategy—searching for an effective weakening solvent. But Baekeland did not adhere strictly to the ideas and methods of this technological frame. This relatively low inclusion in the Celluloid frame is linked to Baekeland's high inclusion in another technological frame: electrochemical engineering. For example, after having presented several applications in the traditional field of Celluloid applications, he added:

But its use for such fancy articles has not much appealed to my efforts as long as there are so many more important applications for engineering purposes. (Baekeland 1909d, p. 157)

Obviously, Baekeland intended to focus on other fields of application, overlapping with the Celluloid range of products but distinctly more of an industrial engineering character. When his searching for a weakening solvent did not yield any result, his inclusion in the Celluloid frame was low enough for him not to get stuck on this problem: Baekeland instead started to use one of the familiar problem-solving strategies of the electrochemical engineering frame. He carried out a long systematic investigation to study all the different factors bearing on the reaction. This was the first time that anybody had researched this, despite Baeyer having observed the condensation reaction thirty-three years earlier.

This investigation enabled Baekeland to control the violent condensation reaction. He distinguished three phases in the reaction and, because he could stop the reaction after the first and second phases, he was able to manipulate the molding mass before it changed during the third and final phase of the now well-known (but at that time notorious) thermosetting plastic. The key element in this procedure was formulated in the famous "heat-pressure patents" (Baekeland 1907a, 1907b). Only when high pressure is applied *while* heat triggers the condensation reaction is the production of gaseous products counteracted; otherwise the product is porous and worthless for molding applications.

We leave the Bakelite story at this point in order to return to the notions of technological frame and inclusion.

Technological Frame and Inclusion

Deliberately I have made the concept of technological frame broad enough so as to include such different elements as current theories, goals, problem-solving strategies, and practices of use. ("Practices of use" is to some extent congruent with "existing markets," but it focuses on consumer practices rather than on the economic aspects; an example is given later.) Depending on the technological frame that is described and the purposes for doing so, different elements may require different degrees of attention. For example, the element of current theory in the Celluloid frame is rather empty if we regard Hyatt in his early Celluloid days. As he said aptly at the end of the experiment to test its flammability: "[I was] very glad that I did not know as much as the Professor."

The need to make a technological frame into such a broad concept arises from the requirement that it must be applicable to social groups of non-engineers also. For a social constructivist analysis of technology, it is important not to make any a priori distinction among different types of social groups (Callon 1981b; Pinch and Bijker, this volume). Of course, when describing the technological frame of the social group of dentists with respect to the artifact hard rubber, more details need to be given about the frame elements of goals and using practice than about the element of current theories. Thus a technological frame should be understood as a frame with respect to technology, rather than as the technologist's frame.[2]

The case of the high-wheel Ordinary bicycle (Pinch and Bijker, this volume) provides an illustration. The using practice of the social group of "young men of means and nerve," that is, racing, showing off, and impressing the ladies, constituted the macho machine, whereas the using practice of the social groups of women and elderly men, that is, touring, falling off, and "breaking limbs and bones," constituted the unsafe machine. The macho machine led to a design tradition with larger wheel radius, and the unsafe machine gave rise to a variety of designs with, for example, smaller wheels, backward saddle, or the smaller wheel in front. Thus different using practices may bear on the design of artifacts, even though they are elements of technological frames of nonengineers.

This aspect represents an important difference from most of the related concepts used by other students of technological development. The concepts of technological style (Hughes 1983 and this volume), techno-logical tradition (Constant 1980, 1984; Laudan, 1984a), technological paradigm (Dosi 1982, 1984; Gutting 1984; Van den Belt and Rip, this volume), technological orientation complex (Weingart 1984), and techno-logical regime (Nelson and Winter 1977, 1982; Van den Belt and Rip, this volume) are intended for application to social groups of engineers only. In addition, Hughes's usage of the term "technological style" is primarily meant to account for national differences in technology, which places the concept on a much higher level of aggregation than the intended level for technological frame.

A second feature of technological frame, not yet mentioned explicitly, is at least equally important and differentiates it from most of the other concepts as well. The concept of technological frame is intended to apply to the *interaction* of various actors. Thus it is not an individual's character-istic, nor the characteristic of systems or institutions; frames are located *between* actors, not *in* actors or *above* actors. In that respect frames are similar to Gallon's networks (Callon 1986). Although my usage of the concept of a technological frame draws on other studies in which similar concepts have been developed empirically as well as theoretically[3] the application to technology is, as yet, only tentative. I briefly sketch some aspects of the interactional nature of this concept.

As noted previously, the meanings attributed to an artifact by members of a social group play a crucial role in my description of technological development. The technological frame of that social group structures this attribution of meaning by providing, as it were, a grammar for it. This grammar is used in the interactions of members of that social group, thus resulting in a *shared* meaning attribution (that the meaning of an artifact is shared among members of a social group is, after all, a key element in the identification of relevant social groups; see Pinch and Bijker, this volume). The interactional nature of this concept is needed to account for the emergence and disappearance of technological frames. A technological frame is built up when interaction "around" an artifact starts and contin-ues. Thus the artifact Parkesine did not give rise to a specific technological frame because the interactions "around" it came to an end before really taking off. The opposite happened to Celluloid: Its stabilization was accom-panied by the establishment of, for example, a social group of "Celluloid chemists." The continuing interactions of these chemists gave rise to *and* were structured by a new technological frame. An important element of this technological frame was, as we have seen, the focus on solvents in the chemical process.

In a way, the concept of a technological frame thus results in making less visible the seams of the web that was woven with the descriptional model (see the introduction to part I of this book). On the one hand, a technological frame can be used to explain how the social environment structures an artifact's design. For example, the dominance of the social group of Celluloid chemists resulted in various patents for a phenol-formaldehyde plastic in which the use of a solvent played a crucial role. On the other hand, a technological frame indicates how existing technology structures the social environment. For example, the stabilization of the artifact Celluloid resulted in the rise of specific social groups and technological frames. In this respect an artifact, such as Celluloid in the last instance, plays a role that is similar to Kuhn's exemplar (Kuhn 1970; Gutting 1984, p. 56).

A technological frame structures the interaction of members of a social group. But it can never do so completely: first, because different actors will have different degrees of inclusion in the frame (actors with a high inclusion interacting more in terms of that technological frame and actors with a low inclusion to a lesser extent), and, second, because all actors will, in principle, be members of more than one technological frame, as I have suggested in the case of Baekefand.[4] Also in these aspects—the possibility of various degrees of inclusion in a technological frame and the possibility of being within different frames—the concept of a technological frame differs from the paradigmlike concepts mentioned previously.

The characteristics of the concept of inclusion can be illustrated by contrasting the engineer with a relatively low inclusion in a technological frame with the notorious "marginal scientist" as criticized by Gieryn and Hirsch (1983). There are at least three important differences. First, the marginality concepts discussed by Gieryn and Hirsch are one dimensional. For example, in one study scientists are considered marginal if they recently migrated from another field, whereas in another study "marginal" is operationalized as "being young." The different dimensions yield contradictory results: If Gieryn and Hirsch could choose any single dimension to characterize a scientist, all ninety-eight scientists of their sample would be marginal. In contrast, the concept of inclusion is *multidimensional* because it is related to the multifaceted concept of technological frame. Thus the inclusion of actors in a technological frame can be specified by describing their goals, problem-solving strategies, experimental skills, theoretical training, and so on; then one should go on to indicate to what extent each of these elements is congruent with the respective elements of the technological frame. For example, Baekeland's goals were congruent with the

Celluloid producers' technological frame in that Baekeland intended mass production of plastic articles; his goals were not congruent in that he was focusing on the production of industrial applications, rather than on consumer goods. Second, as I have already indicated, inclusion is not a binary concept: Instead of being either marginal or central, a member of a social group can have different *degrees* of inclusion in the technological frame. This is especially important when we want to pay due respect to the dynamic character of technological development. The degree of inclusion of an actor is not constant but can change in the course of events. For example, Baekeland's degree of inclusion in the Celluloid frame decreased when he switched from the application of solvents to another problem-solving strategy that belonged to the electrochemical technological frame. The third point of difference with the marginality concept was also mentioned previously. Actors will typically be members of different social groups and have (different degrees of) inclusion in various technological frames.

In the next section I again take up the case of Bakelite and follow its history from 1907 on.

The Social Construction of Bakelite

In the orthodox view of the plastics history, Baekeland's patents of 1907 constitute the invention of Bakelite. But as all authors in this volume argue in different tones, an artifact can never be explained as being invented in such a clear-cut way. In 1907 there was not yet a successful innovation Bakelite. The exhibits Baekeland showed during his presentation at the New York Chemists' Club could have proved as illusory as the earlier Parkesine exhibits shown at the 1862 World Exhibition. The first synthetic plastic that would mark the beginning of the "plastic era" still had to be constructed. To understand this part of the developmental process of Bakelite, we are once again helped by the concepts of technological frame and inclusion.

No sophisticated chemical experience seemed to be involved in the Bakelite process, nor hardly any theory (the macromolecular theory to describe this type of process was not developed until the 1920s). That is why Baekeland initially envisaged that he himself could stay out of the manufacturing process; his intention was to issue licenses on a royalty plan for the use of his patents, but

I soon found I was greatly mistaken in this, and that it would have caused no end of disappointment to teach to others chemical details which, to me, seemed rather simple. (Baekeland 1916, p. 155)

This is understandable in the light of the previous discussion. The work of the social groups to which Baekeland intended to turn over the manufacture of the Bakelite molding powders was structured by the Celluloid technological frame. This technological frame focused the interaction within the producers' social groups on, for example, the employment of solvents and the development of new processing machines, such as the compression sheet molding press and the hydraulic planer (both used for the making of thin sheets), and the blow molding press (used to provide new generations with toys, baby rattles, dolls) (Dubois 1972). It did not provide the means to deal with a delicate chemical reaction such as the one between phenol and formaldehyde "in which almost anything may happen but the formation of bakelite" (Chandler 1916, p. 179).

Then Baekeland planned to fabricate the molding powders as intermediate products and to leave the final molding process to the experienced engineers involved in the production of hard rubber, Celluloid, and insulating materials. However, again the Celluloid technological frame posed a barrier:

I found, to my astonishment, that people who were proficient in the manipulation of rubber, celluloid or other plastics were the least disposed to master the new method which I tried to teach them or to appreciate their advantages. This was principally due to the fact that these methods and the properties of the new material were so different in their very essence from any of the older processes in which these people had become skilled. This rather unexpected drawback is so true that even to-day the most successful users of bakelite are just those who were not engaged in plastic before, this simply for the reason that they did not have to divorce themselves from the routine of older methods, and were willing to listen patiently to suggestions from newcomers in the field. (Baekeland 1916, p. 155)

To establish a social group of Bakelite producers, Baekeland had to enlist people from outside the existing plastics-producing groups or those with a low inclusion in the Celluloid technological frame. Thus the social group of Bakelite producers was in the beginning almost totally congruent with the employees of the Bakelite Corporation.

Synchronously with the stabilization of the artifact Bakelite and the formation of a social group of producers, a technological frame came into being. Thus the system of artifact, social group, and technological frame gains *technological momentum* (Hughes 1983 and this volume). These closely interlinked processes can be traced by following the various patent litigation trials and negotiations after 1909. In these trials the meaning of Bakelite for this group of producers was repeatedly made more precise; after the settlement of each patent struggle the losing party became a member of the producers' social group by acquiring a leading position in the

Bakelite Corporation; methods and concepts developed by the other chemists were incorporated in the technological frame of the producers' social group.

Thus the social group of producers was, for example, extended by giving Hans Lebach a function within the Bakelite Gesellschaft mbH, established in Germany in 1910. In 1907 Lebach, who worked for the chemical firm Knoll & Co., had also patented a phenol-formaldehyde condensation product (Knoll 1907, 1908; Lebach 1909). During a heated debate in the *Zeitschrift für Angewandte Chemie*, Baekeland said he was "firmly convinced of the technical worthlessness of this substance" (Baekeland 1909b, p. 2006). This however did not prevent the assimilation of Lebach's process into the technological frame of the Bakelite producers once the struggle was over. This is demonstrated by one of the review articles Baekeland published later on, in which he described neutrally "another indirect method" and plainly acknowledged that "this method was first published by Lebach at the end of 1907" (Baekeland 1912, p. 742). Analogously, the management of the American General Bakelite Company was formed almost totally from the ranks of previous competitors who had been "defeated" in the patent struggles (Redman and Mory 1931), their methods and concepts being partially integrated in the technological frame as well (Thinius 1976).

One of the last important stages in the social construction of Bakelite was the enrollment of two new but increasingly important social groups: the automobile and radio industries. For the radio industry, Bakelite was a good insulating molding material, and, especially for the wireless amateur, it also meant a versatile plate material that could be sawed, drilled, and filed to provide a mounting frame for electrical parts.

Baekeland had, as is also evident from his previous projects, an acute insight into the possibilities of the marketplace. This is illustrated once more in his view of the tasks of an industrial chemist, in which he shows once more that there is much more to a successful innovation than simply producing a new substance:

This question is not just related to the task of creating a certain chemical substance. The subject is much more complicated, because the objective is to manufacture a product in such a way that it can be used reliably for very specific technical purposes. (Baekeland 1909b, p. 2007)

Primarily, Baekeland's efforts were directed toward the production of electrical insulating parts. Electric manufacturing companies, such as Westinghouse Electric Co., Remy Electric Co., and the General Electric Co., were his first customers, buying the molding material from the General Bakelite Company. He worked personally in many plants to help solve the early problems. In establishing these contacts, Baekeland mainly operated at the level of engineers, rather than at management level.[5] His work as an engineer among fellow engineers was efficacious, I think, in stimulating the emergence of a technological frame of Bakelite molders.

By means of the electrical industry, the second important social group of Bakelite users was enrolled—the automobile industry. For the automobile industry Bakelite meant an accurate molding material to produce good electrical insulating parts unaffected by moisture, oil, or other chemicals and able to withstand high temperatures. Kettering's and Bosch's ignition and starting systems were popularizing the motorcar but required insulating parts that needed to be strong and chemically resistant. Subsequently, the use of Bakelite in this industry branched out to nonelectrical parts, such as steering wheels, radiator caps, gear shift knobs, and door handles. Through the initial enrollment of these two social groups, Bakelite acquired a high degree of stabilization by the end of the 1930s in many more social groups.

To finish the story of Bakelite construction I briefly turn to its use in the production of consumer goods. In its meaning of a molding material for electrical insulating parts, Bakelite only partially substituted for other materials; many applications were completely new. Bakelite's meaning as a material for consumer goods (figure 2) is much more ambivalent. Here, the old tension between an imitation material and a material of its own,

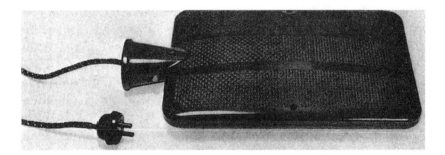

Figure 2
The electric hot water bottle made by R. A. Rothemel, Ltd. Obviously, Bakelite could not be used as an imitation of the (soft) rubber material, but even in the industrial design of products that used to be made of rubber, the imitation is evident. Photograph courtesy of Collectie Becht, Naarden, the Netherlands.

Figure 3
"De Vergulde Hand" soapbox. Form and decoration of refillable boxes were used to enhance the recognition of the product, even more so than the trademark. Photograph courtesy of Collectie Becht, Naarden, the Netherlands.

so intimately tied up with the whole plastics history since Celluloid (Friedel 1983), is prominent. A market survey, carried out in 1938 for the German Bakelite Gesellschaft mbH, nicely illustrates this ambivalence, as does a "retrospective market study" held in 1981 in the Netherlands.[6]

The most important motives to buy Bakelite products were their elegant designs (the material was modern and did not require much maintenance) and their long durability in comparison with porcelain, glass, and clayware. Of course, there were also disadvantages. For example, Bakelite was considered to be rather fragile. Significantly, this view of Bakelite as being fragile was most prominent in industrial areas, where factory workers knew from experience the many intricacies of manipulating Bakelite (they were included in two technological frames!).

Bakelite was often used as packing material, especially for articles that needed to be kept dry (for example, medicine, tobacco, and cosmetics). Many Bakelite boxes were meant to be permanent; one could buy refills. Therefore extra attention was paid to their exterior design (figures 3 and 4). By the end of the 1930s Bakelite was increasingly more accepted as a material of its own. This is reflected in the development of the exterior design. It is possible to detect a general trend from imitative design (for example, the Art Deco style; figure 5) to independent design (for example, the Streamline style; figures 5c and 6). Thus the meaning of Bakelite by the end of the 1930s was very much that of "modern technology," "unlimited possibilities," the "Fourth Kingdom" (besides the kingdoms of minerals, plants, and animals).[7] I would argue that, for a full-fledged account of the history of Bakelite (which was not my objective in this chapter) and for an adequate description of its final stabilization, the social group of industrial designers needs to be given attention. This would bring us rather close to the history of art, rendering the web even more seamless.[8]

Figure 4
Eggcups on a plate with a saltcellar in the center. "That was standing in my mother's best room!" an interviewee recalled when the artifact and photograph were shown to him (Kras et al. 1981, p. 43). Bakelite was also used in luxury products because of its high-tech connotation. Photograph courtesy of Collectie Becht, Naarden, the Netherlands.

Toward a Theory of Invention

By now, the picture has become quite complicated, and I want to conclude by suggesting a way of bringing some order in the chaos of artifacts, relevant social groups, technological frames, and variation, selection, and stabilization processes. As a first approximation I distinguish three possible developmental situations in which an artifact can be at a certain moment of time. These situations are characterized in terms of the concepts of social group, technological frame, and inclusion. In order to make the account more general, I draw on case studies other than Bakelite—the bicycle, the turbojet, and electric power distribution.

First, there is the situation in which no one social group with its accompanying technological frame is dominant.[9] An example of such a situation can be found, I think, in the development of the bicycle, around 1880. Although there were many social groups involved, it is hard to see any one of them dominating the field and structuring with its technological frame the identification of problems and the problem-solving strategies. The second situation is characterized by the dominance of one social group and the corresponding technological frame. Probably, this is the most common situation—"normal technology," to paraphrase Kuhn. The period from 1880 to 1920 in the development of (semi-)synthetic plastics provides an example, with the Celluloid technological frame being dominant. In

Figure 5
(a) Art Deco Philips radio (1933). (b) Functionalistic Ekco radio (1934). (c) Streamline
Sonora radio (after 1945). Photographs courtesy of Collectie Becht, Naarden, the
Netherlands.

Figure 6
(a) Philishave, type 7735, known as the Egg (1948–1951). (b) Advertisement: "The acrodynamical world of the man with his Philishave."

the third situation, two or even more social groups with clearly developed technological frames are striving for dominance in the field. The difference from the first situation is that in that case the many relevant social groups do not yet have distinctive technological frames with respect to the artifact in question, whereas in the latter situation they have. Tom Hughes's analysis of the struggle between the dc and ac systems of electricity distribution offers an example of this third phase.

Having characterized three different phases, the next task is to specify which types of variation, selection, and stabilization processes can be expected to occur in each of the phases. Without being in any sense complete, I briefly discuss some possibilities.

When there is no dominant technological frame, as in the first type of development I have identified, the range of variants that might be put forward to solve a problem is not much constrained. The variation process will tend to be *radical* (see Hughes, this volume). Indeed, in the development of the bicycle around 1880 radically different variants were proposed to solve the safety problem. In the American Star bicycle (1881) the small steering wheel was positioned ahead of the high wheel; Lawson's Bicyclette (1879) had a chain drive on the smaller rear wheel. Thus "radically different" means that all aspects of the bicycle were subject to variation. Hardly any detail of the bicycle was taken for granted, not even the number of wheels (tri- and quatrocycles were constructed) or the method of foot propulsion (besides moving cranks in a circular motion, various lever devices were constructed, requiring a linear vertical motion of the feet). Selection and stabilization of variants will coincide almost totally in this situation. One of the more important stabilization processes in a situation without a clearly dominant social group and technological frame is *enrollment* (Callon and Law 1982). In such circumstances a social group tries to propagate its variant of solution by the enrollment of other groups to organize support for its artifact. One way to do this is by the *redefinition* of the problem (Pinch and Bijker, this volume). If an artifact (for example, the air tire) offers a solution to a problem that is not taken seriously by other powerful social groups, then the problem may be redefined in such a way that it does appeal to them. The problem for which the air tire first was considered to be a solution (the vibration problem) was redefined into a speed problem. The air tire also offered a solution to this problem, and because this problem was important to the racing cyclists, they were enrolled.

In the second development type, when one technological frame is dominant, it is fruitful to further distinguish highly from lowly included actors. Engineers with a relatively high inclusion in the technological frame will be sensitive to *functional failure* (Constant 1980) as an incentive to generating variants. A functional failure may occur when an artifact is used under new and more stringent conditions. Thus Celluloid's flammability presented such a functional failure of this plastic material when its use was extended to applications other than dentures, such as photographic film material. Actors with a high inclusion in the technological frame are bound to generate rather *conventional inventions* (Hughes, this volume)—improvements, optimalizations, adaptations. Thus a large part of the innovative effort of the Celluloid producers was directed toward rendering Celluloid less flammable by finding another solvent.

Actors with a relatively low inclusion in the technological frame interact to a smaller extent in terms of that frame. A consequence may be, as I have suggested in the case of Baekeland, that such actors do not draw much on the standard problem-solving strategies of that technological frame in which they have a low inclusion. Another consequence could be that such actors identify other problems more than actors with a high inclusion in the frame. For example, identification of a *presumptive anomaly* will typically occur among engineers with a relatively low inclusion in the technological frame. A presumptive anomaly, as Constant describes it,

occurs in technology, not when the conventional system fails in any absolute or objective sense, but when assumptions derived from science indicate either that under some future conditions the conventional system will fail (or function badly) or that a radically different system will do a much better job. (Constant 1980, p. 15)

For example, aerodynamical theory in the 1920s suggested a future failure of the conventional piston engine–propeller system for aircraft propulsion. It suggested that proper streamlining would allow aircraft speeds to be increased at least twofold; the propeller would probably not function at the near-sonic speed that would be needed for such aircraft speeds; and theory suggested the feasibility of highly efficient gas turbines. My contention is that especially young, recently trained engineers are in a position to recognize and to react on a presumptive anomaly: They are trained within the technological frame but have low enough inclusion to question the basic assumptions of that frame.

Let me now consider the third situation, in which more than one technological frame is dominant. This is a situation that did not occur in the Bakelite case, at least not in the period I have concentrated on. For an illustration I therefore turn to another case. Around 1890 both the dc and the ac electricity distribution systems were commercially operated, some-times even in the same town (Hughes 1983). The selection process in a situation like this is quite hectic, more so than in the first situation, in which there is no dominant technological frame and when less vested interests are at stake. Arguments, criteria, and considerations that are valid in one technological frame will not carry much weight in other frames. In such circumstances it seems that criteria that are external to both techno-logical frames will play an important role in the selection process. This makes *rhetoric* a fitting selection mechanism in this third situation (Pinch and Bijker, this volume). Tom Hughes described such a rhetorical move in this "battle of currents." A dog is publicly electrocuted: first subjected to direct current of various voltages and then dispatched by alternating current. The objective was to persuade the audience that direct current, as opposed to alternating current, was relatively safe. As Hughes observes, often in such a "battle of the systems" (a competition between two power-ful, equally dominant social groups with respective technological frames) no one wins a total victory. *Amortization of vested interests* is the stabiliza-tion process that will often occur in this situation (Hughes 1983). Of course, the rhetorical closure mechanism may also occur in the second situation, in which one technological frame is dominant. The key feature of this closure mechanism is, after all, that it brings about stabilization by using arguments that do not carry much weight within the actor's own technological frame but appeal forcefully to actors outside it.

I want to emphasize that the situations I have distinguished do not succeed one another in any fixed pattern. For example, Baekeland's first work on Bakelite can be understood, I think, as fitting in the second type of technology development—one technological frame and Baekeland being lowly included. But the subsequent development shows various characteristics that are more in line with the first type of development, in which no one technological frame is well developed.

Conclusion

I have tried to suggest an approach to a theoretical analysis of the develop-ment of technological artifacts that extends the descriptive model intro-duced by Pinch and Bijker (this volume). In the first section on the early history of plastics, two new theoretical concepts, technological frame and inclusion, were put forward. In the second section I discussed these con-cepts in some detail. A technological frame differs in two important aspects from paradigmlike concepts. First, it is applicable to all kinds of social groups, not just to groups of engineers. Second, a technological frame is an interactionist concept. Also, the differences between the concepts of (low) inclusion and marginality were discussed. In the third section these concepts were further illustrated by applying them to the Bakelite case. Finally, I proposed a kind of simplifying scheme to bring some sort of order to the newly created chaos. Three situations were distinguished to characterize the developmental process of an artifact at some stage: no dominant technological frame, one technological frame, and several dominant technological frames. It is stressed that these situations should not be interpreted as forming a rigid scheme of phases through which an artifact successively has to pass. Rather, it is a heuristic device to simplify the description of the "seamless web" of history. In doing so, I found that various concepts developed by historians of technology appeared to be useful. Thus the proposed approach not only brought some order out of disorder but also enabled us to relate different case studies to one other.

Notes

I am grateful to Michel Callon, Ed Constant, Ernst Homburg, Tom Hughes, Stephen Kline, Rachel Landan, Simone Novaes, Trevor Pinch, Jeffrey Sturchio, Sharon Traweek, and my colleagues at De Boerderij for stimulating comments on previous drafts of this chapter. Of course, this substantial help does not make me less responsible for any remaining flaws in the argument.

I would like to thank the Stiftung Volkswagenwerk (Federal Republic of Germany), the Twente University of Technology, and the Netherlands Organization for the Advancement of Pure Research (ZWO) for financial support.

I am grateful to F. A. Becht, Naarden; the Museum Boymans–Van Beuningen, Rotterdam; and the National Museum of American History, Smithsonian Institution, Washington, D.C., for permission to reprint their material in the illustrations of this chapter.

1. The phrase "interpretative flexibility" may lead some readers to think, erroneously, that there is an independent and invariable reality of which only the interpretations may vary. To avoid this misunderstanding, perhaps we should have adopted the phrases "artifactual flexibility" and "factual flexibility." I thank Michel Gallon for his comments on this point.

2. Stephen Kline has suggested that I baptize the concept as sociotechnical frame. Indeed, this describes its connotation more accurately. However, the phrase becomes even more elaborate than the concept of technological frame already is; for that reason I stick to the latter.

3. Similar to some extent are the concepts of figuration (Elias, 1970) and game (Crozier and Friedberg 1977; Van der Meer 1983, 1986; Wilhelm 1985; Wilhelm and Bolk 1986).

4. Wesley Shrum (1984) argued along similar lines in his analysis of technical systems. He refers to Ludwik Fleck (1935), whose "thought style" and "esoteric/exoteric circles" are at the roots of the concepts of technological frame and inclusion (Bijker 1984).

5. This was kindly pointed out to me by Jeffrey Sturchio, who is working on the history of Leo Baekeland and competition in the early American chemical industry.

6. In 1938 the Bakelite Gesellschaft mbH, Berlin, had a market survey carried out by the Gesellschaft für Konsumforschung EV, Berlin. The "retrospective market study" was organized by Intomart Qualitatief BV, Hilversum, the Netherlands. Results of both studies are reported in Kras et al. (1981). One should be careful in generalizing the results of these studies, because differences between using practices in different countries may be considerable (Kaufman 1963). However, it is my impression that these studies of the German and Dutch using practices are at least indicative of the situation in other countries.

7. In 1937 the Bakelite Corporation made a film called "The Fourth Kingdom," in which the production and various applications of Bakelite were shown in much detail. The film starts with a sonorous voice, arguing along the following lines: "Mineral, Vegetable, Animal—the three kingdoms of Nature. They served mankind for ages, but now our modern industrial society finds them insufficient to fill all needs. It has to turn elsewhere; it turns to the fourth kingdom—Plastics" (followed by a crescendo in the symphonic music, of course). I am grateful to Robert Bud for showing me excerpts of this film.

8. Jenkins (1985) also proposes to link the history of technology and the history of art. In his analysis of some aspects of Edison's designs, he gives some intriguing suggestions.

9. Obviously, it is a matter of personal judgment by the historian to decide whether a technological frame is dominant or not. I can offer no quasi-objective measuring instruments for this dominance. In most cases adequate arguments can be given, I think, for the choice of relevant social groups, their technological frames, and their relative importance. For example, the difference between the first and the third situation is often clear. In the third situation, two powerful social groups, with technological frames that can be spelled out easily with respect to the artifacts in question, will have developed their two competing artifacts quite well. In the first situation, any bizarre variant may be considered and may eventually stabilize. For example, Dunlop's air tire became part and parcel of the Safety bicycle without ever having been propagated from the beginning by one powerful social group.

III Strategic Research Sites

Introduction

One of the difficulties facing any study of technology is heterogeneity. Unlike the case of science, in which it is possible to identify communities of practitioners who produce and ratify scientific knowledge, in technology there is a variety of groups involved. Furthermore, it is hard to say that any one group is the crucial one on which research efforts should be concentrated. Among the different participants in any field of technology one can find individual inventors, research scientists, designers and design engineers, production engineers, sales and marketing teams, bankers and financial advisers, lawyers, politicians and state officials, and, of course, consumers—whether individuals, firms, or state agencies. The activities of these groups may take place within one single location, such as a firm or a government laboratory; more often, however, a variety of locations is involved. Technology is such an integral part of modern life that virtually every aspect of an industrialized society intersects at some point with technological issues.

Clearly, part of the task of the emerging new field of technology studies is the identification of research sites at which the complexity of the seamless web is manageable but which at the same time serve to capture key aspects of technological development. We call such locations strategic research sites. In this part several such research sites are offered.

Although the diversity of possible research sites is clear from the following studies, many of the chosen locations do share characteristics that highlight features of the new approach to technology. In particular, the seamless web of technology and society is rewoven by breaking down the all too frequently encountered rigid divisions among different domains, such as between science and technology, among invention, marketing, and consumption, and more broadly between technology and its social impact. As noted previously, the new approach to technology does not recognize such distinctions. Authors are able to move easily among the contexts of university science departments, R&D laboratories, the military, the courts, and the consumer. They go wherever their study takes them, and they do not feel uncomfortable or embarrassed when they cross some sacred boundary. Our picture of technology on one side and society on the other is breaking down. These studies show that society is at work everywhere: within the walls of laboratories, defense establishments, firms, hospitals, and the home.

A number of different technologies are featured in this part. They include a military technology (guided missiles), two medical technologies (drugs and ultrasound), and a domestic technology (the cooking stove). Some authors (notably Edward Constant) embellish their arguments with examples drawn from a range of technologies. Again, we stress that the analytical issues of interest here do not depend on the particular technology studied. All technologies, whether the highly sophisticated guided weapons examined by Donald MacKenzie or the humble cooking stove studied by Ruth Schwartz Cowan, present similar types of problems in terms of producing an account of their development.

In Donald MacKenzie's study of the development of strategic missile technology, Hughes's concept of a technological system is found to be fruitful. Although in Hughes's work the system takes on a physical dimension corresponding to the power lines that spread coextensively with the power system, MacKenzie argues that a seemingly self-contained piece of technology, such as the guided missile, can also be treated as a system. In this case the system includes not only the hardware of the missile and its inertial guidance system but also the strategic considerations embedded in the targeting practices. MacKenzie's study is particularly illuminating because, as in Hughes's own work, it cuts across a variety of social groups and actors. Thus defense bureaucrats, military strategists, and even on occasion US congressmen, as well as the scientists and engineers working at the Draper Laboratory at MIT are all included in the system. This strategic research location thus exemplifies the potential for combining the social, political, economic, and technical factors in the shaping of technology.

Edward Constant's essay, like MacKenzie's, takes up the theme of the systems approach. The main focus of attention in this chapter is a location that has received much previous attention in sociology—the organization. Constant's analysis, however, adds to the previous work by placing the emphasis on the relationships entailed by technological innovation within the organization. His chapter is particularly useful for the choice of strategic research sites because he considers the merits of studying the organization as well as other locations, such as the system and the community of engineers engaged in developing a particular technology. He suggests that a combination of these research locations may prove to be useful.

Ruth Schwartz Cowan places the emphasis on the end of the technological process—the consumer. She shows how consumer preferences in the choice of domestic technology are shaped. By tracing the network of social relationships within which consumers operate, Schwartz Cowan is able to go some way toward explaining consumer behavior and thereby the successes and failures of particular domestic technologies.

In their chapter, Henk Bodewitz, Henk Buurma, and Gerard de Vries analyze aspects of the drug regulatory process. Like Schwartz Cowan, they pay much attention to the consumer. The widespread implications of technology have meant that the modern state has evolved a complex series of sociolegal practices whereby the perceived effects of technology may be delineated and controlled. This is the case particularly for matters relating to drugs, for which several multinational corporations dominate the market, raising concerns in liberal democratic governments about potential health hazards raised by the development of new drugs. The nexus of medical, social, and legal practices, as participants attempt to define the regulatory guidelines on both safety and efficacy to their own advantage, forms the research location for this study. A comparison of these regulatory practices in different national contexts (the United States, West Germany, and the United Kingdom) is provided.

In the final chapter of this part Edward Yoxen describes part of the history of ultrasound imaging. This research site offers him the opportunity to do a comparative study of successful and failed artifacts, and he shows how success and failure can be explained symmetrically. Another theme in this chapter concerns the technology of image generation. In doing this, Yoxen further expands the fruitful convergence of recent science studies with technology studies.

In varying degree, all studies presented in this part are related to constructivist, systems, and network approaches to technology. In such forms of analysis many technical, social, economic, and political aspects of technology are shown to be linked within a system or network, as outlined in part I of this book, However, it is difficult to study the whole system or network, and often authors focus most of their attention on one aspect. For instance, Henk Bodewitz, Henk Buurma, and Gerard de Vries, in their study of the technology of drugs, focus on the nodes in the network, such as regulatory procedures, in which the concerns of the state, the medical practitioners, and the drug companies intersect. Ruth Schwartz Cowan, on the other hand, stresses the need to study the network from within, and this she takes to mean emphasizing the consumers of a technology. Perhaps two pertinent questions to be asked in terms of strategic research sites are whether some parts of the network or system are more crucial to study than others and whether network or system analysis can itself provide researchers with guidelines as to the most salient locations. It seems that at the moment a variety of answers to these questions is possible.

The range of strategic research sites offered here, and for that matter elsewhere in this book, reflects the youthful stage of development of this field. At the moment authors wish to explore just what is possible—such a diversity can only be healthy. Eventually we can expect more systematic comparisons to be made and a consensus to emerge as to where such studies can best be located.

Missile Accuracy: A Case Study in the Social Processes of Technological Change

Donald MacKenzie

Few processes of technological development are fraught with greater significance than the growth in accuracy of strategic missiles. In its current series of flight tests, the MX/Peacekeeper missile has demonstrated the apparent ability to hurl ten nuclear warheads, each with a force many times that of the bomb that destroyed Hiroshima, over a trajectory of 8000 kilometers—and have them land on average less than 100 meters from their targets.[1] Corresponding Soviet capacities are a matter of dispute,[2] and probably lag significantly, but not by enough to have prevented the fear of a Soviet first strike against American land-based missiles from becoming a salient issue in recent American politics. Certainly the world of stable nuclear deterrence that many hoped for in the late 1960s and 1970s may prove a chimera. We approach the last decade of the century with the fear—or in some eyes, the hope[3]—that in crisis it may appear to be to the advantage of one side or the other to initiate a nuclear strike. The growth in missile accuracy and thus in the capacity to destroy even the hardest missile silo or command and control center is only one factor here, but it is a crucial one.

This chapter is an inquiry into the processes of technological development that have led to extreme accuracy in ballistic missiles. My focus is almost exclusively on American missiles. The unclassified technical information on Soviet missile guidance is scant,[4] and the structure of the institutions that produce Soviet guidance technology is unclear. The institutional structure is perhaps the more important point, for the premise of this paper is that technological development cannot satisfactorily be treated in isolation from organizational, political, and economic matters.

To say this, particularly in regard to missile accuracy, is controversial. Many writers have assumed that there is nothing particular to explain, at least nothing *social*, about the growth of missile accuracy.[5] I cannot accept this simplistic technological determinism, for there are related technological areas, notably both military and civil aircraft navigation, in which innovation has been quite evident but technical change has *not* taken the form of the pursuit of ultimate accuracy.[6]

The Technological Systems Approach

To help impose some order on complex issues, I use the technological systems approach developed by Thomas P. Hughes (1983 and this volume). I find it helpful for three closely related reasons. First, Hughes indeed refuses to deal separately with the technological and the social. The subject matter of Hughes's 1983 book is the growth of electricity supply *systems*. He shows that successful system builders *simultaneously* had to engineer technological matters (such as the design of a lamp filament), economic matters (such as the need to compete in price terms with existing gas suppliers), and political matters (such as the legislative frameworks within which electricity supply developed). Artifacts such as Edison's electric light or the Gaulard-Gibbs transformer bear within their design the imprint of the *full* range of circumstances (including economics and politics) within which the system builders worked.

Second, Hughes also refuses to draw an absolute distinction between the micro and the macro, between, for example, the priorities in a particular laboratory and the economic situation of a country. A key bridge between the micro and the macro is Hughes's notions of reverse salient and critical problem. His analogy is military. A reverse salient is something that holds up the growth of a system, as enemy forces may hold out in one particular spot even though in other areas they have been pushed back. System builders typically focus inventive effort, much like generals focus their forces, on the elimination of such reverse salients; they identify critical problems whose solution will eliminate them. Edison, for example, transformed a crucial macrolevel reverse salient in his system, the high cost of copper conductors and the resulting price disadvantage with respect to gas light, into the laboratory-solvable critical problem of the invention of a high-resistance (and thus low-current) lamp filament.

But the transformations are not simply from the macro into the micro: The micro can also be transformed into the macro. System builders transformed surrounding societies so that what was originally a laboratory curiosity, such as the electric light, could spread throughout the land. Herein is a third benefit of Hughes's approach: Successful system builders cannot work with a rigid demarcation between the system and the environment in which the system develops. They continuously seek to mold that environment so that the growth of the system is facilitated, often incorporating what was previously environment into the system, as happened when electrical supply companies came to control the regulative agencies set up to police them.

In reading Hughes's systems approach in this way, I have been influenced by the work of Michel Callon, Bruno Latour, and John Law. Their perspective which is ably presented in this volume and elsewhere (Callon, this volume; Law, this volume; Latour 1983, 1984) makes explicit Hughes's implicit refusal of rigid social/technical, micro/macro, and system/environment boundaries. It emphasizes that systems or networks should not be taken simply as given, as unproblematic features of the world (as they often were in older, discredited uses of systems approaches in sociology); nor should use of the term "system" be taken to imply stability and lack of conflict. Systems are constructs and hold together only so long as the correct conditions prevail. There is always the potential for their disastrous dissociation into their component parts. Actors create and maintain systems, and if they fail to do so, the systems in question cease to exist. The stability of systems is a frequently precarious achievement in the face of potentially hostile forces, both social and natural.

Nor would it be right to view reverse salients as given, independent of the actors involved. Most obviously, to agree on what constitutes a barrier to progress requires agreement on what one is trying to achieve. As I show later, that cannot be taken for granted. More interesting, even those who believe that they are in agreement with respect to goals may not agree on what precisely it is that hinders achievement of those goals. Wisely, Hughes prefers the reverse salient, with its connotation of the unavoidable confusion, flux, and partial information of the battlefield, to more mechanical metaphors such as bottleneck. Only with the wisdom of hindsight (and sometimes not even then) will the nature of the barriers to advance be beyond at least potential dispute. Often, reasonable people will inspect the technological (or military) facts of the situation and disagree. If they agree, it may be as much the result of what they bring to the situation—in training, in assumptions, in interests—as it is the consequence of the effect of the situation on them.

It is worth remarking here that the direction of causation is not simply from reverse salients to critical problems. Solving a problem defined as critical brings with it the expectation of considerable reward, whether in financial terms or in professional status. Thus there is a natural tendency to identify as critical those problems that you are (or will) be able to solve, to see the reverse salient as that particular obstacle to progress that you can remove. To talk of "solutions in search of problems" is a cliché, but it is one that captures accurately an important aspect of technological change. If rewards come from solving problems and if different people have differing capacities for solving different types of problems, then disputes as to what problems most require solution can only be expected. Engineers and accountants, to take an obvious example, differ widely in the type of problem that they can solve competently. They notoriously disagree on whether the reverse salients blocking the growth of a particular enterprise are financial or technological in nature. Similarly, engineers with different skills and types of experience may also disagree on whether, for example, the technological reverse salients are hardware problems or software problems.

Furthermore, the question of what is *possible* arises. Is what is delaying the development of a system a reverse salient to which it is worth devoting inventive effort, or is it simply an intractable natural limitation? Here, too, there is room for disagreement. The form of guidance technology now dominant was, as we will see, held by some to embody a physical impossibility. From the history and sociology of science we know that beliefs about the true nature of the world differ widely and often in socially patterned ways. These beliefs can bear directly on the forms of technical change that are taken to be feasible.

These are not simply matters for the historical or sociological analyst of technology. In my interviews with guidance engineers, on which I partly base this chapter, I found that several of them expressed, obviously in somewhat different words, similar ideas.[7] Interestingly, there seemed to be a rough correlation between how articulate they were on these issues and their degree of worldly success. There is irony here. Technologists often have little time for sociology and other social sciences. Yet successful engineers also know that, to be successful, they have to engineer more than metal and equations. A technological enterprise is simultaneously a social, an economic, and a political enterprise. Successful engineering, to borrow a phrase from John Law, is heterogeneous engineering (Law, this volume). Sometimes, of course, engineers do seek to build their systems only of metal and equations, forgetting the need also to bind in human and organizational allies. These engineers, I suggest, are often those of whom their colleagues say, "X built a brilliant so-and-so, but somehow it never caught on; no one was ever interested in it." Others may be fully aware of the necessity of heterogeneous engineering, yet chafe against it—the spirit of Werner von Braun's classic complaint, "We can lick' gravity, but sometimes the paperwork is overwhelming" (Levine 1982). But whatever attitude is taken, the heterogeneity of engineering is an ultimately inescapable part of technological work. In focusing on heterogeneity, it is worth noting that the social studies of technology are not primarily contributing to the old and ambiguous (though not unimportant) debate on the "social responsibility of the engineer." They are exploring the determinants of what counts as successful engineering.

A Technological Systems Perspective on Missile Guidance

At the most simplistic level, applying a systems approach to missile guidance is absurd. Crucial to the dominant form of missile guidance is that it is *self-contained*. Inertial guidance, as it is known, does not depend on the external inputs typically required by other forms of guidance and navigation. Star sightings, for example, are not needed, a fact that gave early currency to the phrase "astronomy in a closet" as a description of inertial navigation. Inertial guidance or navigation works by measurement of the accelerations experienced by an "inertial measurement unit," and thus by the vehicle—be it missile, aircraft, submarine, or whatever—carrying it. Typically, an inertial measurement unit includes three accelerometers at mutually right angles. These accelerometers are mounted on a "stable platform"[8] held in a known orientation (irrespective of the twists and turns of the vehicle) by a set of gyroscopes and feedback controls. The Advanced Inertial Reference System (AIRS) of the MX/Peacekeeper is one example— an advanced and sophisticated one—of such an inertial measurement unit (see figure 1).

The inertial measurement unit thus measures the accelerations experienced, and, thanks to the gyroscopes, it does so with respect to a known frame of reference. An on-board computer (analog in early systems, now often a powerful digital machine) corrects for changing values of the earth's gravitational field. It then integrates the accelerations to obtain the vehicle velocity and then the vehicle position, given known initial values of these.[9] The system then either provides a simple readout of vehicle position (as in aircraft navigation) or computes corrections necessary to bring the vehicle to a desired target (as in missile guidance).

Additional information *may* be used to supplement inertial guidance. Modern cruise missiles use a radar altimeter to generate an input that is compared with a computer-stored "map." Trident I uses and Trident II will use a sighting of a star to provide additional input. But unsupplemented inertial guidance is, I repeat, self-contained. This has been both a powerful selling point (it cannot be disrupted by bad weather or enemy "jamming")

Figure 1
MX/Peacekeeper Advanced Inertial Reference System. Removal of the two subassemblies allows one of the three third-generation gyroscopes to be seen (left). One of the three accelerometers (MX, Specific Force Integrating Receiver) can also be seen (right). Photograph courtesy of the Northrop Corporation, Electronics Division.

and also, as we will see, the source of the principal objections to its possibility.

In this sense, then, missile guidance is quite unlike the electrical supply systems at the heart of Hughes's most detailed system study (Hughes 1983). Indeed, more analogous to electrical supply are the various systems of radio navigation—sometimes competitors to inertial navigation—based on the receipt of radio signals from either ground-based stations or satellites. They are physical networks, even if the interconnections are electromagnetic radiation rather than wires; a strong military argument against them is precisely this. The degree of their geographical spread is an important issue, as it is for electrical supply networks, but not for an inertial system, for which global coverage is achieved through internal programming. The analogy begins to break down, interestingly, when we move to the more explicitly social aspects of radio navigation systems. It is difficult to establish private property in their output—a problem that also dogged the early pioneers of radio (Douglas 1985 p. 30). Thus, as economic enterprises, they differ from electric utilities, and the distinctive sociotechnical dynamics Hughes associates with the load factor in electrical supply are thus absent even in the more explicitly system technologies of radio navigation.

But the concept of "system" would be impoverished if applied only in a narrow physical meaning. Thinking of a system helps us to understand our subject matter here in at least four ways. First, it reminds us that inertial guidance is self-contained only in a restricted operational sense. To work, an inertial guidance system requires knowledge of the missile's launch point, the precise location of the target, and at least the relevant portion of the earth's gravitation field. Through this form of systemic connection apparently neutral pieces of science gain central military significance, particularly in the fields of geodesy and geophysics. Thus the advent of low-flying cruise missiles has made knowledge of local gravity anomalies of much greater significance than previously (with high-flying ballistic missiles, local gravity anomalies matter greatly only if they are in the vicinity of the launch point, because their effect drops off rapidly with height). Publication of the World Gravity Map has had to be suspended, I was informed, because of this increased military significance of gravity.

Second, an inertial guidance system is indeed a *system*, and missile accuracy is a systemic product. The accelerometers, the gyroscopes, the on-board computer, the gravity map and geodetic data, what happens when the warhead reenters the atmosphere—all these can be argued to be essential to accuracy. Furthermore, the whole can be taken as different from the sum of the individual parts. With the advent of powerful on-board digital computers, for example, it is possible to model mathematically predictable forms of error in the gyroscopes. The mathematics of the different contributions to accuracy, the so-called error budget, thus becomes no longer a simple additive equation in which each component contributes wholly independently to the final figure.

This should not be taken as a mere technical issue, for it directly interacts with issues in the social organization of the institutions responsible for generating and managing guidance technology. With such a complex and expensive technology, a division of labor is inevitable. In the case of US Navy missiles, for example, separate parts of the Strategic Systems Program Office are responsible for the guidance system of the missile and for the navigation system of the submarine that launches it, and wholly distinct civilian organizations (the Charles Stark Draper Laboratory of MIT and the Autonetics Marine Systems Division of Rockwell International) design each. Yet both systems contribute to missile accuracy, and synergistically—the Trident missile's startracker is argued to be able to correct retrospectively for errors in the submarine navigation system. This means that a clear-cut division of responsibility, with wholly separable subgoals for component organizations, can become a difficult social achievement.

This leads us directly to the third advantage of a technological systems perspective on missile guidance. It reminds us that technological issues are indeed simultaneously organizational, economic, and political. This is most clearly seen in the decision-making processes in which the required values of the technical characteristics of missiles, such as their accuracy, are set. Obviously such decisions are technological in the sense that they form a crucial part of the framework in which more detailed design decisions will be made. But they are also organizational, because the health of particular corporations, project offices, even whole branches of the armed services can depend on them, as can a multiplicity of individuals' careers. Typical interests involved often clash. Technologists and organizations in which they dominate typically seek the fullest use of their particular technology and know the likely consequences of the setting of particular requirements for choice of technology. Project managers often have an interest in curbing excessive reliance on innovative techniques and in rejecting overdemanding requirements. "There are two sorts of project managers," one of them told me. "Those who are conservative and those who get fired." Difficult calculations have to be made. Thus in the 1960s the issue was whether it was in the best long-term interests of the navy's fleet ballistic missile program to seek to compete with the air force in counterforce-capable (and thus highly accurate and large-warhead) missiles or to seek a distinct role for navy missiles. On the whole the latter view won out, and the design characteristics of the Poseidon missile in particular, with its large number of relatively small and not superaccurate warheads, reflect this.

In the early years of both the navy and air force missile programs economic considerations were not salient. In a 1955 memorandum on the fleet ballistic missile program, Chief of Naval Operations Admiral Arleigh Burke wrote quite simply, "If more money is needed, we will get it" (Burke 1955, p. 6). But, despite the still considerable insulation of the defense industry from ordinary commercial pressures, the guidance engineers that

I interviewed were fully aware that technological decisions were also economic ones. This has become most apparent with the Small ICBM Midgetman. Economic considerations are obviously extremely important in the choice of guidance system design (an issue currently under active consideration) for this missile, the deployment of which is planned in much larger numbers than the MX/Peacekeeper—perhaps 600 as opposed to 50 to 100.

Setting a design parameter such as missile accuracy is also a political matter. Within the United States the desirability of designing American missiles with the capability to destroy Soviet missiles in their silos has been a matter of fierce dispute, and missile accuracy is central to this capability. On the whole, hawks on defense issues argue for such capability, and doves against it. Particularly in the late 1960s and early to mid 1970s, a congressional lobby led by Senators Brooke and McIntyre opposed technological developments leading to counterforce capability. It is difficult to assess precisely how effective this opposition was, but those seeking greater missile accuracy in those years do seem to have experienced it as a constraint. Furthermore, this debate was not wholly without resonance within the armed services. At least until the mid 1970s, when national policy began to swing decisively to the pursuit of counterforce capability in all strategic missiles, there appears to have been considerably greater enthusiasm for counterforce in the air force than in the navy.[10]

This intertwining of the technological, the organizational, the economic, and the political takes us to the fourth advantage of a technological systems perspective on missile guidance: its emphasis on the need for heterogeneous engineering. Here, the figure of Charles Stark Draper is central (figure 2). Draper's group was not alone in its pioneering of inertial guidance and navigation in the United States. Important early work was also done by the Autonetics Division of North American Aviation (later Rockwell International), by the group of German guidance technologists who came to the United States with Werner von Braun (and whose experience in missile guidance extends back to the 1930s), and, although information here is less clear, by engineers at Northrop.[11] Where Draper stood out among many fine engineers and scientists was precisely in his *heterogeneous* engineering; his only rival in this score was Autonetics's John R. Moore.

Often, the heterogeneous engineering required from those pushing a new technology is the creation of the sense of a *need* for that technology. A radically new device does not find a market ready made: That market has to be constructed.[12] This, however, was not the situation faced by

Figure 2
Charles Stark Draper on the roof of the Instrumentation Laboratory in a simulator
of the Apollo command module. In the background is the Charles River and the
Boston skyline. Photograph courtesy of Charles Stark Draper Laboratory, Inc.

Draper and the early pioneers of inertial guidance and navigation in the immediate post–World War II years. The experience of Second World War bombing had made amply clear the need for accurate navigation, and the military desirability of a device that neither the enemy nor the weather could interfere with was evident. The task, then, was not to establish the need for the technology but its *possibility*. Had Draper and the others failed in this, inertial guidance might now be seen as we see antigravity screens, as a technology for which it is meaningless to speak of there being a need, because it embodies an impossibility.[13]

The idea of navigation by the double integration of acceleration was not new. The earliest reference to it of which I am aware is Murphy (1873),[14] in a discussion of how animals might navigate. But two distinct arguments were raised against its possibility. The first was the practical argument that accelerometers or gyroscopes of accuracy sufficient for useful guidance and navigation could not be built. The second was the objection that inertial navigation would violate physical law, in particular, that the impossibility of distinguishing in a self-contained system accelerations from changes in the gravity vector—a postulate central to the general theory of relativity—rendered the concept futile.

Max Schuler, for example, in a paper later credited with laying crucial theoretical foundations for inertial navigation, described an idea bearing at least some analogy to navigation by integration of accelerations as "ein ganz unmögliches Beginnen" [a quite impossible undertaking].[15] Traditional US gyroscope manufacturers reacted with skepticism when approached to construct gyroscopes of accuracy sufficient for inertial navigation. Draper himself calculated at the time that a hundredfold improvement in gyroscope performance was needed to achieve 1-mile accuracy after an hour's flying time. To achieve what the air force wanted—a system that would navigate a bomber or cruise missile[16] at subsonic speeds to the Soviet Union and that would deliver its weapons with 1-mile accuracy after up to ten hours flight time—would require a thousandfold increase in performance (Draper et al. 1947).

Draper's use of his resources in the struggle to establish inertial technology is fascinating. As a university research group, his MIT Instrumentation Laboratory, known after its separation from MIT in the early 1970s as the Charles Stark Draper Laboratory, had relatively small accumulated financial resources. A constant flow of external funding was necessary. Yet in several ways the university context helped. At that time MIT was (and still is) one of the world's most respected technological universities, fresh from a remarkable phase of successful war work, and association with MIT gave credibility to Draper's project (even if many inside MIT were actually skeptical) and to Draper himself. Draper could accept a plethora of advisory and consultancy roles in the armed services, where an industry figure with an explicit financial interest might have been received with more suspicion. And MIT alumni could be found in many crucial roles in the world of the military and the government. Particularly important were air force and navy career officers who had studied under Draper. Together with those he had worked with during his important Second World War gunsight work, Draper had widespread personal contacts in the armed services. Key contracts could be, and were, negotiated between friends, on occasion over a bottle of whiskey. I seek to imply no impropriety: This was a necessary part of building a technology whose possibility was not established, and it is not clear that under a more bureaucratic system the needed funds could have been raised.

When the moment demanded, Draper could act promptly and with flair. When an attack by physicist George Gamow on the possibility of inertial navigation appeared to threaten continued military funding, Draper countered with a major classified conference reviewing what had already been achieved.[17] He flew across the country to another early conference in Los Angeles in an aircraft navigated by a prototype inertial system, ensuring for himself the conference limelight and cutting the ground from under the feet of anyone disposed to argue the impossibility of inertial navigation. On a later flight, once work in this area had been partially declassified, he took along well-known American television presenter Eric Sevareid. But much of what he did was more routine. On trips to visit his sponsors in the armed services, he would bring along "Doc's dollar bills," wallet-sized graphs showing the performance of his latest instruments (without a scale, to avoid breaching security), to help reassure supporters and convince skeptics that the problems in producing "inertial quality" devices were being overcome.

Sponsors were not the only people who had to be kept "on board" if the new technology was to succeed. Those who worked with the system had to be persuaded often to act in new ways. The laboratory janitor, for example, had to be convinced that if he knocked a test table with his broom, he should report it, or else an unexplained jag might show up in a gyroscope's output. Those who assembled gyroscopes and accelerometers had to be shown the dangers inherent in facial hair and holidays—debris from moustaches and sunburnt skin could play havoc with delicate instruments. Draper devoted a great deal of energy to people such as these whose work under other circumstances might have been taken for granted as unproblematic. And those who handled inertial instruments outside the laboratory also had to be persuaded to act correctly. (A later ploy to achieve this is worth recording. Concerned with "deliberate damage" to inertial-type units, the US Navy Air Systems Command painted them gold, rather than the conventional gray or black, a measure that appeared to succeed! See *Crosstalk* (1980), item 5.)

The Reverse Salients/Critical Problems Dynamic

Had Draper's group limited itself to the activities described in the last few pages, it would have failed (though it would also have failed without them). Work more conventionally recognizable as technological was, of course, equally central. Its nature is well captured by Hughes's reverse salients/critical problems scheme.

From the beginning Draper systematically analyzed the reverse salients in inertial systems. Indeed, he did this in advance of his group's actual construction of a working inertial system, drawing on physics and mathematics. Thus, although different from Constant's notion of presumptive anomaly (defined as the situation "when assumptions derived from science indicate either that under some future conditions the conventional system will fail (or function badly) or that a radically different system will do a much better job" (Constant 1980, p. 15)), Draper's identification of reverse salients in his yet-to-be-constructed system had something of the same flavor to it.

His early analysis (Draper et al. 1947) pointed to the gyroscope, rather than to the accelerometer, as the key. The gyroscope, as I have already noted, was seen as requiring a hundred- or thousandfold improvement for the systems Draper was then considering, whereas a tenfold improvement in accelerometer performance would, Draper argued, suffice for 1-mile-per-hour accuracy. In the early work at the Instrumentation Laboratory the gyroscope was prioritized. Only toward the middle and late 1950s, when attention shifted from bombers and cruise missiles to ballistic missiles, did accelerometer development receive equivalent priority, the accelerometer being analyzed as much more crucial to ballistic missiles than to cruise missiles and bombers.

By 1947 a well-defined and powerful process of gyroscope development (which continues, with vagaries to be explored later, to this day) was already established at the Instrumentation Laboratory. It involved the choice of a specific gyroscope design. Shown schematically in figure 3, this device was the floated single-degree-of-freedom integrating gyroscope. The

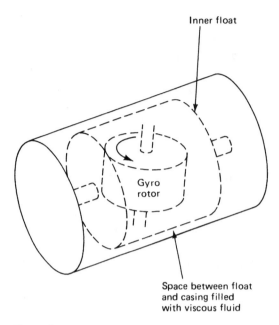

Inner float

Gyro
rotor

Space between float
and casing filled
with viscous fluid

Figure 3
A floated single-degree-of-freedom integrating gyroscope (highly schematized).

design was not simply an abstract choice, although it was defended analytically by Draper et al. (1947), but a representative design with which Draper
had had extensive experience in his Second World War work on gun
control systems. A relatively simple modification, essentially involving
substituting a pendulous element for the gyroscope rotor, turned the
single-degree-of-freedom gyroscope into an accelerometer (Draper 1977,
pp. 240–243). Later, in the 1950s, two more sophisticated basic accelerometer designs, both bearing a strong family resemblance to the single-degree-
of-freedom floated gyroscope, were adopted: the Pendulous Integrating
Gyro Accelerometer (PIGA)[18] and the Pulsed Integrating Pendulous Accelerometer (PIPA).

Adapting to the study of technology a term familiar in the social studies
of science, we can say that the floated single-degree-of-freedom integrating
gyroscope became the *paradigm* for future development. It was the paradigm not primarily in the broad sense of a framework for research but in
the more specific, and "philosophically . . . deeper" (Kuhn 1970, p. 175),
sense of a "concrete puzzle-solution" (Kuhn 1970, p. 175) that serves as an
exemplar and resource for future work. Recently, there has been considerable interest in applying the concept of paradigm to technology (for

example, Dosi (1982)), but, as far as I am aware, only one piece of work (Gutting (1984), although see also Sahal (1981)), has taken seriously the term's more specific meaning. This is sad, because it seems to me that much of the coherence that technological traditions possess comes from shared exemplars. It is thus never an automatic coherence. Sharing an intellectual resource does not dictate only one way of developing it; the idea of a group being "constrained" by a "framework" easily predisposes one to misleadingly mechanical metaphors, such as technological trajectory.[19]

The basic paradigm for Draper's group was subject to successive goal-oriented processes of modification. Although complex and creative in detail,[20] these processes can be described fairly simply. The goal was instruments of ever-increasing accuracy. By "accuracy" I do not seek to imply a self-evident characteristic; rather I mean a whole set of parameters determined by elaborate test procedures, procedures that developed alongside the instruments that were tested.[21] Strenuous attempts were made, involving both empirical and theoretical investigation, to determine the barriers to the achievement of increased accuracy, the reverse salients, and to transform these into solvable critical problems. Although the Instrumentation Laboratory staff was central to the solution of these problems, other parts of MIT were drawn on (for example, the Department of Metallurgy),[22] as were outside companies (for example, in the area of precision ball bearings). Often, too, the Instrumentation Laboratory staff took their "critical problem solutions" outside the laboratory, in the "Route 128" phenomenon. By 1965 at least thirty companies had been founded as spin-offs from the Instrumentation Laboratory, with total sales in that year of $14 million (Roberts 1968; see also Ragan 1980).

At its heart—in the matter of detailed inertial instrument design—this process of "normal technology" was, analogously to "normal science" (Kuhn 1970), simultaneously conservative and powerful. It was conservative in that a contemporary Draper Laboratory–designed instrument, such as the Third Generation Gyroscope used in the MX missile (figure 4), bears a strong resemblance to the basic instrument design of the 1940s, and indeed its ancestry can be traced back directly to this. It was powerful in that the wealth of detail modifications and improvements of three decades of creative work did indeed produce a remarkable increase in performance. Performance figures for the more recent instruments are of course classified, but Draper (1975, p. 26) indicates drift rates between 0.000015 and 0.0000015 degree per hour, which would, if achieved, imply a performance improvement of perhaps five orders of magnitude over the period we are discussing.

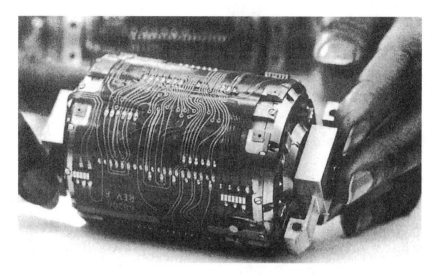

Figure 4

MX/Peacekeeper third-generation gyroscope. Photograph courtesy of the Northrop Corporation.

We are coming close here to the starting point of this paper: the increase in missile accuracy. Although improved gyroscope and accelerometer performance is not the only factor in increased accuracy—more powerful on-board computers, better gravity and geodetic data, and new materials for and designs of reentry vehicles would also have to be mentioned—they are vital factors. With this in mind, and also in view of my earlier point that reverse salients and critical problems are not simply given, it is worth remarking that this process of technological development was carried out almost solely by Draper's laboratory. Only at one other site in the United States, Bell Aerospace Textron, do matters appear similar, although of course it may be that they have also been mirrored at a Soviet analog of the Draper Laboratory.

Let me say clearly what I mean by this. There has been a continuous history in the United States of at least thirty years' development of inertial instruments at at least eight other sites, including Autonetics, Northrop, Sperry, Honeywell, AC Spark Plug Division of General Motors, Kearfott, Litton, and Bell; there are several more sites in Europe. Much of this work does indeed follow a reverse salients/critical problems form. But it differs from that at the Draper Laboratory in two crucial, interrelated respects. First, the reverse salients identified have not been the same. Second, there has been much greater change through time in basic instrument design.[23]

Litton Industries, for example, has moved from floated gyroscopes (especially two-degrees-of-freedom ones), through "dry" (that is, nonfloated) tuned rotor gyroscopes, to, most recently, laser gyroscopes. (The reason I take Bell as a partial exception to this is that their accelerometer design has involved a much more extended period of improvement of a stable basic accelerometer design.)

The more relevant of these two issues here is the first. Why have different reverse salients been identified by different groups? There is an obvious answer and a more subtle one. The obvious answer is that different groups have prioritized different goals and thus have necessarily identified different things as barriers to progress. The central priority given to accuracy in the Draper Laboratory has not in general been considered appropriate elsewhere. Often the structure of goals elsewhere has been to set a threshold value for accuracy (say 0.01 degree per hour gyroscope drift rate) that satisfactory designs must meet but need not surpass, and to concentrate on improving to the maximum possible the reliability, weight, volume, producibility, or cheapness of the instrument. Emphatically, it is not that we are talking here about "low tech" against the Draper Laboratory's "high tech," for the pursuit of these goals has been linked to devices that would normally be regarded as "high tech," such as the laser gyroscope. Rather, we are talking about *different forms* of technical change, informed by if not completely different goals (for reliability, weight, etc. were also of concern in the Draper Laboratory), then at least by different priorities given to different goals.

Explaining why different groups award different priorities to different goals is a complex matter, too complex to enter into fully here. One crucial factor is, of course, simply the structure of needs for instruments with specific characteristics. As discussed in the next section, guidance technologists have been able only partially to influence this, and thus externally defined needs have, over time, created a set of niches, the filling of which demands certain characteristics from the technology. Among these niches we might identify those associated with tactical missiles, civilian aircraft, the bulk of military aircraft and modern cruise missiles, strategic bombers, space boosters, submarine navigation systems, and strategic missiles. Different organizations have come to specialize in different niches—sometimes in more than one but seldom in more than three. Sometimes the original specialization may be close to accidental—a contract won here and lost there—but once specialization has occurred, it is powerfully self-reinforcing as the skills of heterogeneous engineering appropriate for a particular niche develop.

But other issues are also involved. It is hardly accidental that the Draper Laboratory, the one major group not part of a substantial corporation and for most of its history part of a university, came to specialize in the pursuit of high accuracy and in ballistic missile applications in particular. After 1965, when the big American buildup in strategic missiles began to slow down, that market began to shrink, whereas a new market in civil inertial use began to develop and the military aircraft market remained relatively stronger.[24] The civil area especially, and to a lesser but still significant degree the military aircraft area, was seen as a potentially highly profitable field but also as a field increasingly demanding a high "entry price." The incentive of profit particularly predisposed corporations toward them, whereas the barrier of entry price obviously acted as a particular disincentive to groups without the capacity to make a large investment from their own resources.[25] Robert Duffy, shortly to become president of the Charles Stark Draper Laboratory, summed up the logic of the situation as he saw it in October 1972:

I sense a very strong shift in emphasis from performance oriented requirements to one of high reliability, or in commercial terms, low cost of ownership. Although we do not interpret these requirements to be in conflict in all cases, this shift has affected our Laboratory in an unusual way. Our performance orientation has been sharpened rather than decreased, because the market, particularly the commercial market, has drawn the industry itself towards cost/user utility goals, leaving the door to performance somewhat unguarded. (Duffy 1973, p. 3-1)

The more subtle reason for different perceptions of reverse salients by different groups is, as suggested earlier in this chapter, that it would be wholly mistaken to assume that even those who concur on goals will then agree on the barriers to achieving them or on the best means of doing so. An interesting example of this concerns whether or not inertial systems need to be supplemented by external information. In the United States this has been an issue for both submarine-launched ballistic missiles and for modern cruise missiles.[26] I am not concerned here with the debate on whether ballistic missiles *should* be designed to have high accuracies. What is of relevance here is the debate that took place among those who did not question the goal of high accuracy. My interview data suggest that in both cases at least some members of the Draper Laboratory argued that purely inertial systems could provide the requisite accuracy, given sufficient development of the performance of inertial instruments.[27] These arguments were countered, in both cases successfully, by other technologists, in part located elsewhere, who seem to have argued that the key reverse salients were such that further inertial instrument development could not be expected to eliminate them. Thus a debate early in the cruise missile program was described to me as one between the argument for more accurate inertial instruments and the claim that a key issue was the gravity model.[28] Even if the inertial instruments functioned completely without error, then errors arising from incomplete knowledge of the gravitational field over the low, slow flight path of these missiles would prevent desired accuracy levels from being achieved. Hence supplementation of the inertial system was needed (by the TERCOM terrain contour matching system), but with this available, the inertial system need not be highly accurate in itself.

The possibility of this sort of debate, and therefore the negotiable nature of reverse salients and critical problems, should not in the slightest surprise sociologists of science. Much recent work in this field has demonstrated clearly the extent of "interpretative flexibility," the extent to which reasonable people can and sometimes do disagree on what is to be made of any particular set of evidence (see Pinch and Bijker, this volume). It might be argued that this interpretative flexibility is more limited for technology by the fact that technology has to work in a more palpable sense than science does (Mulkay 1979a). But this argument fails for at least four reasons. First, the kind of disagreement I am considering occurs in the design process. The criterion of working is an ex post facto one.[29] Second, what counts as working is problematic. Different groups may have different notions of it (Pinch and Bijker, this volume), and even in hard, esoteric, technical contexts disputes can break out (Robinson 1984; Noble 1984; Wildes and Lindgren 1985). Third, a host of conditions, often including social and economic ones as well as recognizably technical ones, will typically be required for a technology to work, and it may not be self-evident what the cause of failure to work is. Fourth, the acknowledged fact that a device works does not automatically confirm the rightness of every decision taken in its design. Thus Trident I *was* built with a star tracker to supplement its inertial guidance system; it has been tested extensively, and there appears to be no dispute that it has at least met its accuracy specifications, and perhaps even surpassed them (Arkin 1984). Yet in interviews in 1985 I met with the argument (from a highly placed and knowledgeable guidance engineer) that the star tracker was *not*, in fact, contributing to this accuracy, that it was a cumbersome irrelevance.[30]

System and Environment

The third of the advantages of Hughes's systems perspective, listed at the beginning of this chapter, is the way in which it sensitizes us to the fluidity of the boundary between technological systems and their environments, particularly the way in which it raises the question of the extent to which system builders seek to mold their environments to facilitate the growth of their systems. In this final section I examine the extent to which inertial guidance engineers, particularly at the Draper Laboratory, have been able to shape the structure of needs in their surrounding society and thereby have ensured continued support for the pursuit of ultimate accuracy in inertial components.

Two preliminary points are necessary.[31] I in fact found that the extent to which system builders have been able to shape their environment is strictly limited. But had it been otherwise, this would not, ipso facto, have constituted a criticism, for the literature I cited at the beginning of this chapter (for example, Hughes (1983) and Latour (1983, 1984)) shows that this is precisely what many of the great "heroes" of science and technology, such as Edison or Pasteur, did—*had* to do. Shaping needs is not in itself wrong; the moral or political question concerns the character of those needs. Second, the issue is nevertheless of far more than academic interest, for it bears directly on issues of the controllability of military technology. Testifying before a subcommittee of the US Senate in 1969, Jack Ruina argued that "on the issue of guidance accuracy, there is no way to get hold of it, it is a laboratory development, and there is no way to stop progress in that field" (quoted in Gray 1977, p. 4). This argument has certainly been influential, but it does not seem to be correct, certainly for the present and perhaps even for the period in which Ruina spoke. Guidance accuracy is not simply a laboratory development, and even to the extent that it is, it is not immune to external influence. No laboratory development is ultimately self-sufficient. If the environment is not right or is not *made* right by the system builders, any line of laboratory development will lack external influence and may indeed cease altogether.

Charles Stark Draper's own periodization of inertial component development (Draper 1975) allows us to see where this becomes a problematic issue. He distinguishes four generations of inertial systems, classified according to their performance. The first was the gyroscopic aircraft instruments of the Sperry era (see Hughes (1971) for the best description of this period). The second was the inertial navigation technology of missiles of the Polaris era, the Apollo program, and the majority of military and civil aircraft inertial navigators. As we have seen, there was little difficulty in establishing the need for technology of this kind. The issue concerned its possibility, and this issue can be regarded as settled by the late 1950s.

Here then is the crux. Most of the inertial navigation and inertial guidance business has remained content with accuracy levels of this kind (or less than one order of magnitude better), focusing the considerable amount of technical change that has taken place in other directions. For someone who, like Draper, has wanted to drive for greater accuracies, establishing the *need* for such third-generation technology became an issue around 1965.

Draper's major attempt in the early 1970s to create a third generation focused on aircraft (especially civil aircraft) navigation and the growing problem of the air traffic control system. With errors of the order of 1 mile for each hour of flying time, second-generation inertial technology was adequate for transoceanic navigation but could at best be a minor supplement to other systems in the vicinity of airports. Draper himself[32] and others in his Laboratory (see Denhard 1971) argued that more accurate inertial technology could provide a solution to reverse salients in the air transport system, especially delay times.

They did not succeed. Civil aircraft inertial technology has remained firmly second generation in terms of accuracy. Nor, as we have seen, did the attempt to argue for radically improved inertial technology for cruise missiles succeed. Attempts were made by some in the inertial industry, in particular at Honeywell, to use the specific mission requirements of strategic bombers (especially when used as cruise missile launch platforms) to formulate a case for a need for inertial accuracy of 0.1 mile per hour. After lobbying at the level of the Secretary of the Air Force, this succeeded, and a Honeywell system with that specification was included in the G and H modifications of the B-52 bomber. But the success was only temporary, and the accuracy requirements for the inertial system for the B-1 bomber were significantly relaxed and are being fulfilled by a Kearfott design that is a modified second-generation system.[33]

This left ICBMs and submarine-launched ballistic missiles as the only potentially clear-cut market for third-generation inertial components. Furthermore, only missiles with hard-target counterforce capabilities could be designated as requiring this technology. Second-generation systems were acknowledged to be good enough for missiles targeted for city-destroying retaliation alone. So the question arises, Has the perceived military need for such weapons been created by the technologists so as to generate a market for their technology?

It would be fully consistent with a systems approach if we found that this was so. But, to the best of my knowledge, it has not happened like this. I have no evidence of an independently important role played by guidance technologists in pushing US nuclear strategy or targeting policy toward counterforce.[34] It *may* be that my data are at fault here, but I think not, for the guidance technologists have had no need to lobby the air force in this matter: The commitment of the US Air Force to counterforce is deep and long-standing. The commitment of the navy to counterforce is, it is true, arguably neither of these. But in the one clear-cut instance of an attempt there to create a market for high accuracy that I know of, the argument used was oriented toward interservice rivalry rather than directly toward national strategy. In an originally secret paper circulated to the Polaris Steering Task Group, Draper (1959, p. 3) wrote: "Fleet ballistic missiles offer many well-known advantages, but will surely be handicapped in competition for national support unless they can be fired with accuracy levels comparable to those of land-based missiles." Interestingly, this line of reasoning was rejected in the 1960s by the leadership of the Navy Special Projects Office, who, as indicated, did not seek to compete with the air force in missile accuracy.

So the market that third-generation technology has found in MX and Trident II[35] cannot be said to have been simply created by the technologists. The explanation of the dominance of counterforce lies elsewhere.[36]

What of the future? As early as 1975, Draper was looking beyond third-generation devices to a fourth generation, with performance perhaps two orders of magnitude better still, and the annual reports of the Draper Laboratory have reported such devices as under development. Will the funds needed to create such a technology be found (currently they appear to be coming exclusively from within the Draper Laboratory's relatively small independent research and development budget), and would the devices be used if created? A quotation from one of my interviews neatly captures, I believe, the changed problem of heterogeneous engineering:

For a long time, way back, when things weren't as accurate, people would say, and by people I mean our potential customers, and commercial people, they would say, "Gee, if you could get that, we could use that, but that's impossible to do." That was a more standard answer, say [up to] ten or so years ago. Now a more standard answer is, "Gee, we think you can probably do it, but who needs it." That's a lot harder to overcome, you know.

This is the case even in the strategic missile field. It is not that increased missile accuracy has been forsworn, but that there is no longer any consensus that errors in inertial measurement form a reverse salient. Indeed the following interview comment, from Major General John Hepfer, for many years a senior figure in the US Air Force ballistic missile program, indicates pessimism as to the existence of any clear reverse salient that it is productive to attack:

In the sixties . . . the largest contributor was the accelerometer, that was the major error contributor . . . Now, you can take that accelerometer error on down to zero and that [system] error doesn't change because you're in the noise level. The accelerometer is no longer *the* major contributor to the error sources. There are other error sources, and every time we fix one, we find there are others, and so we're sort of at a plateau with probably anywhere from 200 to 500 error sources in there of equal magnitude, and trying to push each one of those down, the guys will do it, they'll identify an error source, and they'll say, now we can work on that and we'll get that out. As soon as you do, you find that that was just one of many in there. . . . We're at the point now where we identify new sources rather than fixing things.

General Hepfer is not alone in seeing matters like this. And compounding these perceived "physical" difficulties in increasing missile accuracy by further refinement of inertial components is an economic concern. The high accuracy of the MX gyroscopes and accelerometers is expensive indeed, perhaps too expensive even for the high-cost world of strategic systems. Says General Hepfer: "The design objective should be (and is being) altered to produce the same accuracy but at a greatly reduced cost." David Hoag, of the Draper Laboratory, in a letter to me denied that a *physical* barrier has been reached but agreed that an *economic* one may have been: "A barrier is hard to recognize distinctly. I believe more accuracy of an unassisted inertially guided weapon system is possible . . . but who can afford it?"

This sense of a barrier may mean that fourth-generation instrument development—laboratory development—may simply cease through lack of funds or through pessimism as to applications, for few engineers wish to build things that will simply be laboratory curiosities.[37] But there would be other consequences of a developing consensus on a (natural or social) limit to practically achievable ballistic missile accuracies by evolutionary development of existing technology. Only a radical departure in technique—most likely some form of terminal guidance in which reentry vehicles, instead of following a free-fall trajectory as normally at present, are guided to their targets—would then be seen as offering hope for substantially increased accuracies.[38] But more than laboratory development would be needed. There is at least some skepticism in military circles about whether terminal guidance techniques can enhance accuracy without military disadvantages such as vulnerability to deception, and any terminal guidance system involves a tradeoff in warhead yield lost because of the additional weight carried in the reentry vehicle. It seems likely that substantial flight testing would be needed before either the United States or the Soviet Union would deploy terminal guidance in their strategic arsenals on any large scale.

This may be a situation of some interest for arms control, for it would indicate that a comprehensive ban on ballistic missile flight testing, or even a well-constructed partial ban, might in fact stem further increases in missile accuracy. If this is helpful,[39] and desirable,[40] it certainly should not be ruled out as impossible.

Notes

The research on which this chapter is based is funded by the Nuffield Foundation. One primary source of information is seventy-five interviews conducted in the United States from 1984 to 1985 with guidance engineers, corporate executives, serving and retired military officers, former officials of the Department of Defense, and others. In view of the subject matter of this particular chapter, it is especially appropriate here to extend my thanks to the Charles Stark Draper Laboratory, the Strategic Systems Program Office of the US Navy, and the Ballistic Missile Office of the US Air Force for the help given by these bodies and the generous cooperation of both present and retired members of them. I am also extremely grateful to those who have commented on earlier drafts of this chapter, especially to Major General John Hepfer (USAF, retired), David Hoag, and John Law. Responsibility for the views expressed here is, nevertheless, my own. I should emphasize that the concerns of this chapter are primarily analytical, and many historically important developments have therefore been treated scantily or not at all. I hope to return to these in greater detail in other papers.

1. See, for example, Robinson (1984, p. 17). I say "*apparent* ability" to signal an important issue that is set aside here: that an accuracy figure is not an unchallengeable given but the outcome of a complex process involving laboratory testing, test firing, theoretical modeling, and mathematical analysis. The extent to which the outcome of the process is a "fact" about the missile is a matter of controversy. See MacKenzie (1989) and note 21.

2. Keller (1985) notes current dispute between the Defense Intelligence Agency and the Central Intelligence Agency over the accuracy of the Soviet SS-19 missile, the DIA estimating a circular error probable of 325 yards, the CIA one of 435 yards. Either figure is probably worse than the current performance of the most closely comparable US missile, Minuteman III, but it has to be remembered that the SS-19, and especially the "heavy" SS-18, carry more reentry vehicles with greater net destructive force.

3. Thus Gray (1982, p. 63) wrote "*In extremis*, the United States has to be able, not incredibly, to threaten central nuclear employment against the Soviet homeland."

Gray identifies success in six programs as central to such a credible threat: "hard-target counterforce," which is my focus here, "strategic anti-submarine warfare," "several layers of ballistic missile defense," "continental United States air defense," "civil defense and industrial hardening," and "robust preparation for societal survival and recovery." Note that what Gray is calling for is these *capabilities*. I am not implying he wishes to see them actually used.

4. Western intelligence estimates of the circular error probable figures of Soviet missiles are, in practice, relatively widely known, and Soviet publications on guidance matters are available; but the publications tend not to indicate the guidance techniques actually used on Soviet missiles. For an indication of how Western knowledge of Soviet missiles is obtained, see Moncrief (1979). Holloway (1977, 1982) is the best source on Soviet ICBM development; see also Berman and Baker (1982), and for theater missiles, see Meyer (1983–1984).

5. This is most explicit in Schroeer (1985). He sees developing computer technology as the cause of the growth of missile accuracy. A better informed view, but one still tending to technological determinism, is Shapley (1978).

6. At the cost of some oversimplification, it can be said that for twenty years the same accuracy standard has reigned: that aircraft inertial systems should have an error of the order of 1 nautical mile per hour of flying time. Systems have become lighter, smaller, more reliable, more producible, and (in real terms) considerably cheaper but not greatly more accurate. For the problematic nature of the 1-mile-per-hour accuracy criterion, see *Crosstalk* (1980, item 13).

7. One published statement of the ideas—significantly, it is an after-dinner speech rather than a formal paper—is Copeland (1981).

8. In some systems, known for obvious reasons as strapdown, accelerometers and gyroscopes are simply fixed to the vehicle body, and the on-board computer is programmed to take account of the fact that the orientation of the accelerometers changes with changes in vehicle orientation as detected by the gyroscopes.

9. Especially in early systems some of these mathematical functions were performed directly in the hardware rather than in a separate computer. Readers interested in a more detailed but accessible account of inertial guidance should turn to Hoag (1971).

10. This emerges both from my interview data and from other sources, such as Sapolsky (1972), Greenwood (1975), Rosenberg (1983), and Kaplan (1984).

11. On the first two groups see Mueller (1960), Haeussermann (1981), and Slater (1966). Hardy (personal communication, 1985) kindly shared with me his recollections of work at Northrop. A historically important aspect of the Northrop work in this period was the ordering of an airborne digital computer, the BINAC, from Eckert and Mauchly. See Stern (1981, pp. 116–136).

12. An excellent study of this is Jenkins (1976).

13. For interest in antigravity screens at around the time I am discussing, see "The Trouble with Gravity" (1950).

14. Credit is due to Claud Powell for bringing this fascinating letter to my attention.

15. Schuler (1923, p. 349). The interpretation of this passage is contentious. See Bell (1969).

16. In this period, bombers and cruise missiles tended to be preferred to ballistic missiles as weapon systems by the US Air Force. The history is complex and in part has to do with the process of establishing the air force as a service separate from the army, a process that was going on in this period. See Beard (1976) for the checkered history of the ballistic missile in the air force and Armacost (1969) for jurisdictional disputes between the army and the air force.

17. Scientific Advisory Board (1949). Gamow's attack was entitled "Vertical, Vertical, Who's Got the Vertical?" Several of my interviewees remembered this event, quite independently, but I have not been able to trace the document. It is, for example, not to be found in the Gamow papers at the Library of Congress.

18. Although different in detailed design, the Instrumentation Laboratory's PIGA was not a new concept. A PIGA designed by Fritz Mueller was used in V-2 guidance systems to detect when the missile had reached the required velocity for engine cutoff. See Mueller (1960) and Haeussermann (1981).

19. These highly condensed remarks are expanded on somewhat in MacKenzie and Wajcman (1985, pp. 11–12).

20. The published account that best captures this process is Denhard (1963).

21. Testing is a topic that should be central to the social studies of technology because it involves what Law (this volume) felicitously refers to as "the construction of a background against which to measure success." See the important studies by Vincenti (1979) and Constant (1983).

22. For accounts of other developments at MIT in this period, see Noble's fascinating *Forces of Production* (1984) and Wildes and Lindgren (1985).

23. There are obviously difficulties in assessing the extent of change through time in design, because assessment involves judging the extent to which two designs are alike. Here I am following what appeared to me to be almost universal judgments among those I interviewed that the laser gyroscope, say, was a radically different device from a mechanical gyroscope, whereas two generations of Draper floated instruments differed considerably less than this.

24. The market estimates I am working with here were kindly provided by Litton Guidance and Control Systems and are consistent with the more informal impressions gained in interviews.

25. In fact, the first entrant to the civil market, Sperry, had to retreat from it, whereas the second, AC Spark Plug Division of General Motors, suffered large losses early on.

26. In the United States it has generally been agreed that land-based missiles do not need such supplementation, unless their launchers are mobile.

27. My information on these two episodes is not as full as I would like, and it may be that the account in this paragraph stands in need of revision.

28. The Draper Laboratory is, of course, far from unaware of the extent to which incomplete information about the gravitational field is a barrier to the achievement of high accuracy. Its solution to this is, interestingly, an inertial device to measure gradients in the gravity field, designed by Milton Traegeser. Such a gravity gradiometer will in fact be used in Trident submarines, although it will be the one designed at Bell Aerospace, not at the Draper Laboratory. For a comparison of the Draper and Bell designs (and a third Hughes design), see Gerber (1978).

29. But the criterion may feed back into the design process either through learning or (interestingly) greater success for those that follow routines that work. See Nelson and Winter (1982).

30. Lest I be misunderstood, I am not endorsing this argument, merely noting its presence.

31. A third point may also be wise. The Draper Laboratory is engaged in a much wider range of activities than the refinement of inertial components. I am talking here only of the fate of this particular technological enterprise, not of the organization as a whole.

32. See the Minutes of the Board of Directors of the Charles Stark Draper Laboratory for the early 1970s, in the Albert G. Hill papers, box 2, files 1/5, 2/5, 3/5, 4/5, 1/3, 2/3, 3/3 (MIT Archives).

33. An area in which accuracy requirements have increased is ballistic missile submarine navigators. For reasons too complex to enter into here, such submarine navigators have failed to provide a market for high-accuracy technology of the *Draper* type.

34. Edward Brooke, who was a senator of Massachusetts, was lobbied during his campaign against counterforce by members of the Draper Laboratory. The Department of Defense has also been lobbied at high levels over organizational matters both at the time of the separation of the Draper Laboratory from MIT and during current debates over legislation that might prohibit as noncompetitive the kind of

contractual arrangement that exists between the Draper Laboratory and the US Navy. Neither of these activities is, of course, in any way illegitimate. Neither amounts to the kind of role described in the text.

35. The MX AIRS contains three Draper-designed and Northrop-built Third Generation Gyroscopes, and three Draper-designed third-generation accelerometers known as SFIRs, or Specific Force Integrating Receivers (see figure 1). The Trident II guidance system will also have three Draper third-generation accelerometers, but the gyroscopes will not be of Draper design, but two Kearfott dry tuned-rotor gyroscopes (a decision, interestingly, probably linked to the supplementation of the inertial system with a star tracker).

36. A full explanation would require consideration of a range of organizational factors, domestic political factors, and factors arising from the situation of the United States as a global power. Let it also be said that counterforce is equally dominant in Soviet war planning. See Erickson (1982) and Meyer (1983–1984).

37. The Strategic Defense Initiative may offer a possible area of application for at least third-generation gyroscope technology, but for precision pointing and tracking, not inertial guidance. Naturally, this is currently an area of considerable interest to the Draper Laboratory. I am unclear as to the extent to which this may alter the prospects for the development of fourth-generation instruments.

38. This will definitely not mean fourth or even third-generation technology, however, because there seems to be a consensus that, if reentry vehicles are to be inertially guided (rather than guided by, say, correlation of radar images with stored "maps"), then the guidance systems should be inexpensive (because one is needed for each reentry vehicle, not simply each missile) but need not be highly accurate (because any errors they involve have much less time to affect missile impact than those generated in the earlier phases of flight).

39. It might not be helpful because missile accuracies are already high. On the other hand, a comprehensive ban on flight tests might erode *confidence* in the accuracy of existing missiles.

40. There has been speculation as to whether *very* high accuracies (zero circular error probable) would permit the replacement of nuclear warheads by conventional warheads in counterforce attacks. I am skeptical. It is worth noting that the considerable increase in the accuracy of US missiles since the late 1960s has *not* been associated with a move to lower warhead yields; if anything, the reverse has taken place.

The Social Locus of Technological Practice: Community, System, or Organization?

Edward W. Constant II

At the risk of doing some violence to the historian's craft, it is probably fair to say that most serious historical treatment of technology falls into one of two broad traditions: intellectual and artifactual accounts that have their origins in classical approaches in the history of science, or in biographical and organizational accounts that count business and economic history as their nearest scholarly kin. Devotees of the first tradition have largely followed Edwin T. Layton's lead and see technology as knowledge possessed by a "mirror-image twin" to the scientific community (Layton 1972). Adherents to the second tradition have focused primarily on entrepreneurship and technological change in the context of economic organizations. Although complementary in principle, the two approaches commonly have emphasized different aspects of the process of technological change and have come to varying conclusions regarding its nature and inspiration. In the first view scientific progress often plays a preeminent role in technological change; in the second view market demand, entrepreneurial creativity, and other economic factors are dominant. On some issues, the importance of patents most notably, the two approaches yield opposite conclusions. For example, for a number of major inventions, from water turbines and Pelton water wheels to steam and gas turbines to turbojets, patents do not seem to have served as either effective information conduits or barriers to multiple or successive inventions (Constant 1980). Yet biographies of inventors and corporate histories clearly demonstrate the central role played by patent positions in shaping both individual career trajectories and corporate business and technological strategies (Hughes 1971; Jenkins 1976; Reich 1980). More recently, the interests of historians of technology, such as Thomas P. Hughes (1983), and of economic historians, such as Alfred Chandler (1977) and Nathan Rosenberg (1982), have converged on the development of large-scale, integrated technological systems and their concomitant organizations.

Current scholarship thus proposes three different social loci for techno-logical practice: the technological community, the complex organization (usually corporate), and the technological system. In this paper I do four things. First, I present a community model for technological knowledge and change and argue that it applies equally well whether community members are defined as individuals or as organizations. Second, I offer a summary of the systems depiction of technological development con-tained in Hughes's *Networks of Power* (1983) and suggest that that approach offers an intellectual vantage point from which the community and orga-nizational perspectives can be better understood. Third, I try to develop an organizational depiction of technology and technological function and try to show how that characterization jibes with the community and systems conceptions. Finally, I suggest at least the shape a synthesis of these viewpoints might take, define areas of consensus, and try to indicate critical issues for further study, the most important of which seems likely to be the role of entrepreneurship in creation of novel, holistic macrosys-tems. The goal here is not finished theory, but an outline of the issues any finished theory of technology or technological change must address. None of this is likely to be as lucid or as conceptually solid as we would like: It represents a reconnaissance of the problem rather than its solution.

Technology as Knowledge: Community

Here I depend largely on my own previous work on turbojets and sum-marize briefly the ideal typical model for community practice of technol-ogy I used there (Constant 1980). In that model technological knowledge is expressed in well-winnowed traditions of practice that are the possession of well-defined communities of technological practitioners. In almost all cases individual membership in a community of practitioners is different from disciplinary commitment: Turbojet design utilizes precepts and people from aerodynamics, mechanical engineering, combustion engineer-ing, metallurgy, and so forth. What distinguishes a turbojet practitioner is adherence to the tradition, not disciplinary training. Such communities may be composed of either individual adherents to the tradition or, equally valid, organizations. Turbojets are designed by a collection of engineers and other specialists, who together constitute an identifiable community of practitioners. Turbojets are designed within and manufactured by a handful of large, complex organizations that are lumped together as an industrial sector, which is simply a way of expressing the same structure of practice at a more aggregate level. The relationship between the two levels is explored at the end of this section. The normal practice of such communities, however they are defined, is the extension and articulation, or incremental development, of the received tradition.

Normally, developmental problems for a tradition of technological practice appear as functional failure, the incapacity to function under new or more stringent conditions. Dynamically stated, the perceived failure is to maintain improved performance or cost or uncertainty reduction trajectories.[1] (These sorts of problems for traditions of practice seem analogous to what Thomas P. Hughes calls reverse salients and subsidiary critical problems in complex, evolving systems, as discussed later.) Solutions to such problems are normally sought within the received tradition through incremental permutation of the conventional system. For example, since World War I, enhanced performance from poppet-valve, gasoline-fueled piston aircraft engines was sought through higher compression ratios, improved materials, improved (internally cooled) valve designs, and new construction techniques (monobloc structure, rather than separate cylinders for in-line engines). Numerous cylinder configurations were tried, and one, the radial, air-cooled type, proved especially successful. Some firms tried to develop diesels, others novel induction systems, such as port induction or sleeve valves. Supercharging, turbosupercharging, and variable pitch propellers, all radical at some level (the implications of which are discussed later) but all well within the piston engine and propeller tradition, drastically improved total system performance. Despite this normal pattern of successful incremental development, however, perception of persistent, recalcitrant failure within any given tradition of practice may provoke a search for more radical alternative solutions: Problems with high-altitude performance during the First World War (ultimately resolved by adoption of supercharging) briefly provoked examination of radical alternative aircraft propulsion schemes.

On rare occasions science intervenes in technological practice directly: New scientific insight may imply that a current system will not work under future conditions or that some radically different system will do a much better job. For a normal technology this situation constitutes what I call a presumptive anomaly: The old system still works, indeed still may offer substantial development potential, but science suggests that the leading edge of future practice will have a radically different foundation. For example, by the late 1920s aerodynamic theory suggested three conclusions: that with sufficient thrust, well-streamlined aircraft should be capable of approaching the speed of sound; that conventional propellers could not operate efficiently at such speeds; and that gas turbine compressor and turbine components designed in accordance with aerodynamic theory should be capable of significantly higher efficiencies than previously thought possible.[2] Together these three insights constituted a presumptive anomaly for the traditional piston engine and propeller combination and led directly to the turbojet revolution. A similar relationship between advances in theoretical science and creation of radically new technologies would seem to apply across a fairly large and significant array of modern technological changes: catalytic cracking in petroleum refining, transistors, magnetic bubble memories for computers, synthetic rubber, nylon, penicillin, multiple innovations in biotechnology, and, of course, nuclear energy. In less than ideal cases many varieties of mixed patterns are possible, as the role of serendipity in the discoveries (but only by well-prepared minds) of nylon and penicillin testifies.

Traditions of technological practice (and their relevant communities) usually also embody higher-level traditions of technological testability, which in turn is composed of both some set of testing technologies and techniques and some set of normative values (Constant 1980, 1983). The tradition of testability sustains and defines, almost tautologically, the specific tradition of technological practice; yet it also normally invokes more general methods and values, such as precise measurement or rigorous testing by means of parameter variation (Vincenti 1979). Incremental change in a tradition of practice rarely evokes higher-level change in testability standards; revolutionary change often does. For example, since Fourneyron successful development of water turbines has depended on simultaneous and parallel development of a whole system of testing and evaluation, including flumes, weirs, and, most important, the Prony brake (or friction dynamometer). Likewise, the testing procedures originally used for reciprocating air compressors could not be sensibly applied to turbo air compressors; the newer devices could not be fairly evaluated using the same techniques or even the same theoretical variable. A change in technology (air compressors) necessitated a change in experimental technique and equipment (Constant 1980, pp. 85–89). The turbojet revolution not only required development of entirely new testing techniques and construction of new (or radical reconstruction of old) test facilities but also necessitated redefinition of the way aero-engine output is measured: thrust for shaft horsepower. Traditions of technological testability, then, linked both to lower-level traditions of technological practice and to a higher-level normative engineering culture, are the major way that communities of practitioners reify the meaning of their tradition of practice for themselves and explain and justify that tradition to outsiders.

In general, the notion of radical change or of technological revolution in a tradition of practice is relative, and probably no issue has caused as much disharmony among students of science and technology as the problem of revolutionary versus incremental change. I suspect that the dispute is not really about the nature of change but stems from a failure to confront the hierarchical structure of all complex technological systems and therefore of virtually all technological practice.[3] Ontologically, systems are composed of subsystems, which are composed of an immense variety of components: An aircraft as system has as a major subsystem, its engine (a turbojet), which in turn is composed of a large number of components (compressor, combustion system, turbine, and so forth).

This intrinsically hierarchical systems structure has a number of implications for technological practice. First, it means that the system is decomposable and can therefore be changed or improved with much greater efficacy (Simon 1969). Subproblems can be isolated and solved independently, subject only to systems interface constraints. Second, this hierarchical decomposability suggests the absolute relativity of all change: Whether a given change is perceived as radical or incremental depends solely on the hierarchical level chosen. A new valve, a new turbine material or fabrication technique may represent a revolutionary solution to a specific subproblem at that level; yet the same change, viewed from the level of the total aircraft system, may appear only as a typical incremental improvement. Third, complex, hierarchical systems imply multiple traditions of practice and multiple communities of practitioners. Each level, each subsystem can be seen as the purview of a distinct community; yet some traditions and some communities may overlap a number of higher- or lower-level systems. For example, gas turbine practitioners are both a distinct community unto themselves and, to a degree, constituents of a larger aeronautical community, but they also design gas turbines for ships and offshore oil production platforms. This layered Venn diagram or fish-scale image of technological communities of practitioners and their traditions, then, goes far in capturing the extraordinary complexity of technological practice.

These considerations in turn justify describing communities of technological practitioners as being composed alternatively of individuals or of organizations. Consider individual members of a given community of practitioners as vectors for a specific replication code, carriers of a powerful set of programs that together constitute the relevant tradition of practice (knowledge, say, of how theoretically to describe, design, fabricate, and test a turbojet engine). Slice open an organization, insert the new vector and its programming, and presto! the organization starts replicating turbojets rather than piston engines, turbosuperchargers, or steam turbines. A piston engine firm, say Pratt & Whitney, has become a jet engine firm. Alternatively, one can reprogram existing software in place; this is commonly called retraining or continuing education.[4] The same characteristics apply for any area of technological practice at any hierarchical level. Clearly, then, complex systems produced by complex organizations involve a multiplicity of communities and traditions, which again goes far to explain the continuous flux observed in real world practice.

Finally, what has been said here about technological knowledge, its relationship to communities of practitioners, and the way it changes probably also applies to the sciences, medicine, accounting, law, railroad traffic management, indeed to any professionalized knowledge system that is both esoteric and dynamic. I return to this conjecture in the third section of this paper.

Technology as System

As implied by the notion of hierarchy, most complex technologies constitute complex systems. At some lower levels some systems can be considered "hardware," for example, a turbojet engine or an airplane. Larger systems are invariably sociotechnical and organizational (they have a large "software" component); their pieces are not necessarily mechanically interconnected: An air transportation system, for example, includes (at least) airplanes, airports, maintenance facilities, ground access, NAV/COM and air control systems, and an immense variety of specialized, hopefully coordinated personnel, plus, of course, some crowd of totally uncoordinated passengers.

A systems portrayal of technology is hierarchically higher and more general than, and thus may be thought to include, the community model discussed in the first section of this paper and the organizational portrayal to be presented in the third section. The systems depiction of technology, therefore, serves to highlight aspects of technological practice of concern in both community and organizational models, areas in which those approaches differ and issues that neither successfully addresses. Perhaps the boldest and most comprehensive recent examination of large-scale sociotechnical systems is Hughes's *Networks of Power* (1983). Let me briefly summarize what I see as the model contained in Hughes's work; I then try to show how that model fits and how it goes beyond the ideas presented thus far.

Vastly oversimplified, Hughes's model consists of three interrelated structures. First, it contains a temporal stage model for the development of technology. In this case electrical technologies evolved through three distinct phases: electric lighting systems, universal lighting and power systems, and large regional power systems. Second, according to Hughes, each stage in the development of these technologies is characterized by specific reverse salients in the advancing technological front; the reverse salients are in turn decomposable into subsidiary critical problems that attract the attention of relevant practitioners. At each stage these critical problems elicited the emergence of characteristic types of problem solvers: inventor-entrepreneurs (Edison), manager-entrepreneurs (Insull), or engineer-entrepreneurs (von Miller). Third, Hughes argues that each developmental phase produced a specific "culture of technology," composed of distinctive values, ideas, and institutions. Some values are quite general—technical efficiency, for example—and are shared with much of engineering. Yet for each phase such general values have system-specific referents: the importance of the load factor as the key to efficiency in large-scale systems, for instance. Such cultures of technology also comprise specific organized knowledge that, by the beginning of the twentieth century, constitute a virtual science of technology. The technology itself, its systematized knowledge, and its culture are embodied in a variety of economic organizations and social institutions. It is this culture of technology, expressed both in large-scale organizations and institutions and in the career commitments of individual practitioners, that creates technological momentum, the propensity of technologies to develop along previously defined trajectories unless and until deflected by some powerful external force or hobbled by some internal inconsistency.[5]

Yet this whole rich and complex model for the character of technological change does not imply technological autonomy: It is underdetermined by factors internal to the technology. Hughes argues persuasively that it is the interaction of these properties of technology with a vast array of geographical, economic, political, and historical contingencies that yield observed, unique technological styles.

For all its richness, Hughes's model leaves several issues unresolved. First, it is not specified how reverse salients are parsed into (solvable) critical problems. It would seem likely that such parsing is an iterative process, with early attempts made to define subproblems within existing grammars, that is, along existent community boundaries or, for technologies already contained within organizational structures, along the lines of functional departments. To the extent that such problems fall within the domain of well-organized knowledge and existent social groups, they are more likely to be easily and efficaciously solved. Recalcitrant reverse salients may not parse easily or at all, or intractable problems may not fall within existing community or organizational boundaries. Such salients or problems may require multiple passes at decomposition and may necessitate extensive search among multiple communities, organizations, and institutions before an appropriate solution is found (Kline 1985). For resolution some reverse salients, such as high-voltage transmission, require new fundamental knowledge in engineering science, whereas some large problems give rise to wholly new (but usually synthetic) disciplines, such as aerodynamics, as well as to new related institutions, the National Advisory Committee for Aeronautics (NACA), for example.

Second, how systems and functions within systems are divided among organizations is not specified in Hughes's model. Clearly, successful parsing of systems and their problems entails organizational and institutional development, which in turn underlie technological momentum. Alfred Chandler, of course, would argue that broad encapsulization of systems within single, large, vertically integrated organizations occurs when the nature of the underlying technology is such that throughput rates reach a level at which managerial coordination is more efficient than market or interorganizationally negotiated coordination (Chandler 1977). Nathan Rosenberg, in contrast, suggests that under appropriate technological circumstances (technological convergence or economies of scale in a highly specialized function) vertical disintegration is more likely (Rosenberg 1963). As Hughes's depiction of the central role of consulting engineering firms (Stone & Webster, Charles Merz) in the creation of large systems demonstrates, however, the variety of possible organizational arrangements seems to increase as a function of system scale and complexity. Finally, as Hughes's earlier biography of Elmer Sperry illustrates, individual inventor-entrepreneurs can have personal styles that carry them through a succession of technologies and systems (Hughes 1971).

Clearly, the systems perpective on technological practice says a great deal about the forces driving, and the developmental paths of, such systems. Yet those insights still require integration with community and organizational approaches.

Technology as Function: Organization

Despite what has been said so far, people rarely buy either technological knowledge or technological systems in their entirety. In general, people buy artifacts (widgets) or the output of complex systems (electric light), both of which embody and require an immense amount and variety of technological knowledge.[6] Purchase or use of almost any modern technology is mediated by the complex organizations that are required to integrate the knowledge and resources necessary to produce and distribute the artifact or service. This custom of buying an organizationally mediated function, not knowledge or a system, has two related implications.

First, technological knowledge is never pure; it may exist as pure Platonic form, but if it is to be expressed as a tactile, functional artifact, it must be mixed with both base matter and other varieties of technological knowledge. As noted in the first section, this circumstance implies the cooperation of adherents of multiple communities of practitioners in the creation of any complex system. Clearly, integration of this multiple expertise in turn implies complex organization. John Kenneth Galbraith calls this technically required organization a technostructure and describes its necessity in essentially cognitive terms:

The real accomplishment of modern science and technology consists in taking ordinary men, informing them narrowly and deeply and then, through appropriate organization, arranging to have their knowledge combined with that of other specialized but equally ordinary men. This dispenses with the need for genius. . . . No individual genius arranged the flights to the moon. It was the work of organization—bureaucracy. . . . Finally, following from the need for this variety of specialized talent, is the need for its coordination. Talent must be brought to bear on the common purpose. (Galbraith 1972, p. 61)

Galbraith thus clearly identifies the two central requisites for technological function: specialized knowledge (which was identified in the first section with communities of technological practitioners) and organization.

The second implication of buying function and its relationship to organization is more subtle. That organization is required immediately invokes a vast array of organizational variables in the performance of function. First, as James G. March and Herbert Simon have so eloquently shown, organizations have an extensive set of behavioral characteristics, such as differentiation of subunit goals, merely as a consequence of being organizations (March and Simon 1958). Managing intraorganizational interfaces, say among functionally differentiated departments or between hierarchically higher and lower managerial levels, is only the problem of technical systems interface written differently. Second, the necessity of social organization redefines technological function. I not only want a car that is built in accordance with the basic principles of physics and in light of an empirically verified, well-winnowed tradition of automobile design, but I also want it screwed together right. Quality, reliability, service, style, economy of operation, expected resale value—all matter and all are to a major degree a function of organizational decisions and efficacy rather than straightforward technical solutions to straightforward technical problems.

These considerations lead to the vast variety of faces that the same basic technology can wear. A Honda Civic and a Lincoln Town Car clearly are based on the same great international pool of technological capabilities, the same broad traditions of practice in internal combustion engine design, transmission design, tire fabrication, robotics and machine tools, factory organization, and so on. Each of those traditions of practice is represented by individual community members employed by Honda and by Lincoln; the technological knowledge so represented is embodied in the complex, functionally differentiated organizational structure of the two firms. Yet the two cars as artifacts express radically different organizational appreciations of how automotive technology ought to go about performing its function and about how the organization itself should go about providing that function. The two varieties represent the same generic technology, but their respective parent organizations have adapted that technology to "fine track" a relevant organizationally defined and perceived environment.

Technology as function, then, is embodied in and mediated by organization. If technology as knowledge finds its home in communities of practitioners and their associated professional societies and educational programs, then technology as function has as its locus complex organization. Knowledge and function, community and organization, of course, are the building blocks of the "culture of technology," which Hughes finds necessary to sustain complex technological systems and their momentum.

How then might these relationships between community knowledge and organizational structure be formalized? And how do they connect with large-scale systems? Specifically, given that complex organizations are necessary to performance of complex technological functions, how does the internal differentiation (or departmentalization) of such organizations map onto communities of practitioners? Why is hierarchical, bureaucratic integration of specialized community knowledge so universal and so effective? What dynamic do such organizations themselves express? Possibly the richest source on the structure and functioning of the modern business enterprise is Alfred Chandler's *Visible Hand* (1977); perhaps his insights, combined with what has been said thus far, can be used to answer such questions and to begin building a fruitful ideal typical model for organizational function.

Chandler sees the rise of large business organizations not primarily as the result of a process of internal differentiation and specialization, but rather as the result of a process of successive inclusion. Large enterprises internalized separate steps in the sequence of production or distribution that traditionally had been performed by separate business units whose activities were coordinated by market transactions. In place of market coordination, large, vertically integrated, functionally departmentalized, multiunit business enterprises replaced direct control by bureaucratic managerial hierarchies. Chandler argues that these large, hierarchical organizations have come to dominate most economic sectors because they are more efficient: The "visible hand" of management can coordinate the flow of production (or the flow of goods through the channels of mass distribution) more effectively than the market can.

There are two complementary keys to this managerial efficiency. One is the internal division of managerial tasks among hierarchically organized, specialized middle managers who populate differentiated functional departments. The narrowly but deeply informed specialists of whom Galbraith speaks (and who are thereby adherents to well-defined traditions of practice) are not randomly distributed through the organizational structure but are grouped into functional subunits. The other key to efficiency is reduced information or transaction costs. Bureaucratic management of throughput has lower information costs (because information flows are routinized and because problems of access, credibility, and negotiation are minimized) than does a functionally equivalent series of market-mediated or negotiated exchanges (Chandler 1977; O. Williamson 1979, 1981). Chandler maintains that large, hierarchical firms thrive not because of economies of scale associated with mere size (which he argues are trivial) but because of radically enhanced rates of throughput, which of course require precise coordination and control.[7]

Historically, Chandler locates the origin of modern managerial hierarchy in the application of engineering methodologies or modes of thought to the operating problems of mid-nineteenth-century American railroads (Chandler 1977, p. 95). Functionally departmentalized, multidivisional forms of organization as well as effective cost accounting and the practice of separating line and staff functions were invented then, and the first large cadres of specialized middle managers appeared. From the railroads, these organizational motifs were rapidly adapted to a succession of industrial sectors. In each case, that adaptation depended on linkage of at least three sets of innovation, which together invariably permitted radically increased throughput: transportation and communication (initially railroads and the telegraph), plus those innovations specific to the industry in question. The industry-specific sources of growth almost always were composed of both technological and organizational components: Duke used the Bostick cigarette machine and creation of a direct marketing force; McCormick used the reaper and his invention of the franchise dealer network; Singer used the sewing machine and his direct sales stores; Swift used refrigerator cars and local distribution cold storage warehouses. Whether a given product or service became the province of vertically integrated producers or mass distributors or retailers depended on the relative capital and energy intensity of the production process, on where the advantages of managerial coordination and high throughput velocity could be reaped most easily (Chandler 1977, pp. 238–241). Ironically, given the central role Chandler attributes to technological change in driving organizational innovation, he treats the technology itself almost as a *mechana ex dea*. In all likelihood, because Chandler's primary interest is organizational change, he has little to say generally about the creation of new technology or specifically about research and development and its integration into the firm.

Nevertheless, Chandler's richly informed portrayal of organizational development does offer a possible path to synthesis of community and organizational depictions of technological knowledge and function and to integration of those insights with aspects of the systems perspective. Figure 1 presents a diagram of a modern industrial organization.[8] This idealized organization is hierarchical, functionally departmentalized (production, assembly, and so on), and separated into line (across the bottom of the figure) and staff units. Ideally, each module or department can be considered the location or address of some specific set of technological community knowledge (each ranging from the explicitly formulated to the utterly tacit) necessary to perform that module's function within the organization.

More abstractly, each module or department could be considered coterminous with that organization's "chunk" of a specific community of practitioners (represented by individual adherents to a broader technological tradition who happen to work for the organization). For example, if a manufacturer has a design department (or module) consisting of people who design turbojets, that module does three things: It performs the design function for the organization; it contains adherents to the broader turbojet tradition, and it makes the whole organization a turbojet company—a member of the turbojet community, with community now defined in

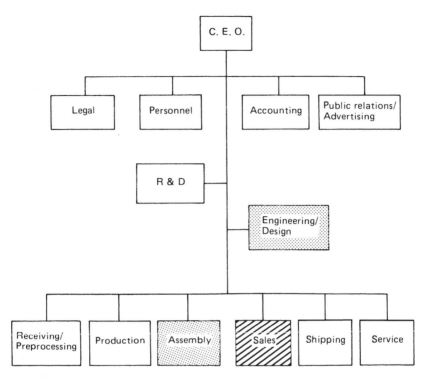

Figure 1

Idealized portrayal of functional organizational modules. Engineering/Design and Assembly are shaded to indicate substantive change (such as the emergence of flush riveting) in those modules. Sales is an especially permeable module.

organizational rather than individual terms. Similarly, an accounting department would certainly do the first two things: provide accounting services for the firm and offer access to the broader traditions and practices of the accounting profession. The same could be said for any other department or module that contains individuals with esoteric or specialized knowledge.

Each module in the organization, then, is a locus for potential change, both for performance in that module and for practice in a broader represented community. Knowledge can simply change autonomously: People learn by doing. In addition, as noted in the first section of this chapter, by pulling the R&D or engineering design module and substituting another or by reprogramming the existent module, one can radically transform the whole organization's replication code: turbojets for piston engines. Fur-

Figure 2
Idealized product life cycle. Implication of strictly linear development is not valid. Rather, movement back and forth along a spectrum of activities and movement up and down through a systems hierarchy are the norm.

thermore, by plugging in a product lift-cycle loop (figure 2) linking R&D, engineering design, production, and sales, one can more nearly capture the dynamic disequilibrium characteristic of a technological open system. This figure is also schematic: The evolution of technological systems in fact implies a continuous movement back and forth along this spectrum of activities, as well as up and down through the hierarchical structure of the system itself (Hughes 1976a; Kline 1985; Hummon 1984).

Each of these idealized modules also represents an environmental interface, a semipermeable membrane through which the organization both receives information and acts on the external world. Each module, then, is capable of internal development and of exchanging information with the outside world, most often with the specialized communities of practitioners to which organization members belong. It is within these modules or through these modular interfaces that in the ideal case reverse salients or critical problems, or functional failures or presumptive anomalies, are identified and resolved. It is in these modules (as well as in the individual practitioners who inhabit them, in the organization as a whole, and in relevant communities of practitioners) that technological frames might be thought to exist.[9] The larger organization (and its upper management especially) spends most of its time and energy trying to combat the centrifugal tendencies common to all such sociotechnical systems: If you want to picture the visible hand of managerial coordination, visualize a committee meeting. Clearly, the existence of social structure is not in any sense an explanation in and of itself; rather, the structure's existence is explained by the Galbraithian necessity of utilizing and integrating the knowledge of well-inculcated specialists to perform technological functions and by the centrifugal tendencies of all organizations.

This idealized modular (or departmentalized) conception of an organization thus offers a means of portraying the way changes in community knowledge, organizational function, and systems development are related. The effect on an organization of changing technological community knowledge, specifically when induced by changing scientific knowledge through presumptive anomaly, was discussed in the first section of this chapter. The modules most affected there were R&D and design; they were the principal loci of organizational change (neglecting, for the sake of simplicity, changes in production machinery, materials, and so on). Radical change in any module can similarly transform the organization in major ways. Vincenti has discussed the efflorescence of flush riveting in the aircraft industry, almost from the shop floor (Vincenti 1984). Transfer of cost accounting techniques from the railroads to the steel industry had revolutionary consequences. Marketing recognition of new trends in family formation or enunciation of new legal doctrines can radically transform corporate strategy and structure. The simple act of patenting can have an immense variety of different purposes, depending on how different individuals or groups within the organization (representing a module or set of modules) perceive the relevant legal, economic, and technological environment (Hughes 1971, 1983; Jenkins 1975; Reich 1980, 1983; Leslie 1980). Sales, of course, is the most persistently permeable of these modules. It is there that firms most directly act on their environment, and it is there that most persuasive news of impending environmental elimination is felt: Even in Chandler (but not in Galbraith) the market still selects. In short, viewing technology as a function embodied in such modular organizations offers a way of systematically building up the observed complexity of actual behavior, contemporary and historical.

Intimations of Synthesis

As should be obvious by now, the three conceptions of technological practice—community, system, and organization—are not really contradictory. In each mode of analysis the relevant social groups have mechanisms for self-identification, persistence, and development, both socially and technologically. For communities of practitioners that mechanism is tradition of practice. For systems it is momentum. For organizations it is technological function. All have subsidiary processes for problem recognition, definition, and solution. For communities of practitioners those processes include functional failure and presumptive anomaly. For systems they are reverse salients and critical problems. For organizations they are a variety of environmental interfaces and internal developmental imperatives. Figure 3 offers one way of visualizing the structural relationship among communities of practitioners, organizations, systems, and the host culture.[10]

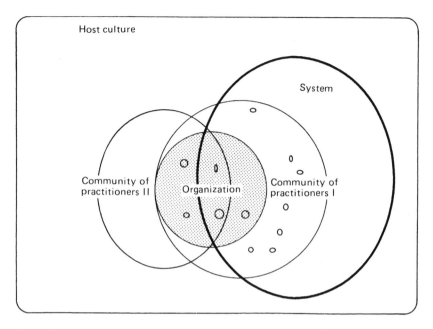

Figure 3
Relationship of an organization to communities of practitioners (I and II), the system, and the host culture. The small circles represent individual community members, or vectors.

Despite these commonalities and connections among community, systems, and organizational conceptions of technological practice, their goals and purposes remain paradoxically both similar and different. The core purpose of the community of practitioners model is to explain technological development and revolution, conceptual or cognitive novelty, invention, and innovation. By and large, organizational portrayals seek to explain technological change, innovation, entrepreneurship, and economic and organizational growth. The systems perspective seeks to explain the emergence and development of large-scale sociotechnical systems— especially invention and entrepreneurship. The modular depiction of practice offered, integrating as it does the processes of problem recognition, definition, and solution selection central to both community and organizational models, goes far in explaining middle-level technological change, normal or revolutionary. What the modular portrayal does not adequately capture is creation of novel macrosystems or the process that might be termed grand entrepreneurship.

This common but differently focused concern with novelty and entrepreneurship highlights the contrast, and the complementarity, among the three notions of technological practice and points to some major unresolved problems for each. Entrepreneurship itself in fact commonly denotes at least three distinct types of activity. First, it is often carelessly used to mean simply economic risk taking—starting a new business, for example—and in this usage it does not necessarily imply either the use of a new technology or the novel use of an existing technology. This type of entrepreneurship is of little interest in the present context.

Second, entrepreneurship more often means both technological and organizational innovation—bringing to market a new product or using a new process within either a preexisting (but changed) organization or a new organization. Radical technological change may or may not imply creation of wholly new organizations: Turbojets ultimately were manufactured solely by preexisting firms; amateur photography or Xerography created new firms. To a degree, Hughes's depiction of the way entrepreneurs (of whichever of his types) recognize reverse salients and define and solve critical problems corresponds to this conception of entrepreneurship. This class of entrepreneurial activities also fits more easily into the idealized modular conception of technological function: As technological systems evolve through time, different modules "light up" as the critical research and development sites, partly because they are where the action is in terms of internal technical development and partly because they may be an especially critical environmental interface (see figure 1). Dynamically, it is through the agency of this type of entrepreneurial activity that technology and organization coevolve.

A third type of entrepreneurship is more grandiose and much closer to what really attracts both Chandler and Hughes: the creation of novel forms of organization and of attendant ideas, such as load diversity, necessary to deploy and exploit a new technology. This rather extraordinary type of entrepreneurial activity, the topic of so much of the literature in business and economic history, is perhaps irreducibly Schumpeterian and creative, a breaking out of both existent community knowledge and an existent organizational repertoire. It is the punctuation in long-run equilibria. Hughes indeed pursues this type of entrepreneurial activity to the creation of true macrosystems, whose critical problems become largely political and institutional. Ironically, at this level Hughes begins to see vertical disintegration (Rosenberg 1963) of entrepreneurial functions to consulting engineering firms or holding companies that are not only technically expert but also politically adept. These firms, in turn, become functionally departmentalized, as the latent functions of entrepreneurship itself become manifest—the modular model reemerges at a much higher hierarchical level within a sociotechnical macrosystem (Hughes 1983, pp. 386–389).

It should perhaps be reemphasized that the evolution of both the managerial systems depicted by Chandler and the sociotechnical systems portrayed by Hughes have first-order technical foundations. As noted, Chandler defines the critical element in the rise of the visible hand as throughput velocity—the capability of managerial coordination to increase the rate of flow of goods through the stages of production, thereby enhancing the efficiency of both capital and labor inputs. Chandler attributes this enhanced managerial efficiency to a whole array of technological and organizational innovations. Hughes carries the analysis a step further. He sees electrical systems evolving through the continual interplay of three elements: first-order (hardware) technological change, which may autonomously, in the nature of the technology, be biased toward economies of scale; organizational innovation (creating new organizations); and, most notably, the invention of systems concepts, which both shape the direction of technological invention and define the context of organizational development. Hughes, to a substantial degree, thus sees both technological and organizational innovation as derivative of higher-level invention of systems concepts.

In summary, then, seeing communities of practitioners (defined either as firms or organizations) as the social locus of technological knowledge, organizations as the social locus of technological function (with the modular conception of function used to portray the way knowledge and function are integrated in complex organizations), and sociotechnical macrosystems as the broader dynamic and holistic structural context of both may offer a way of reconciling the insights of much of the history of technology, business, and economics. This approach also highlights but does not adequately depict the creation of macrosystems and related grand entrepreneurship. Nor do these perspectives adequately capture the interaction between cultures of technology, organizational cultures, and society at large. These problems, defined and focused by a more secure understanding of technology and the way it changes, should rank high on the agenda of future research.

Notes

1. Whether or not improved performance and uncertainty reduction amount to the same problem is a matter of some contention. See Vincenti (1986).

2. Walter G. Vincenti, in his "The Davis wing and the problem of airfoil design" (1986), discusses the extraordinary case of laminar-flow airfoils. Laminar-flow airfoils were originally designed according to the precepts of subsonic aerodynamic theory to reduce profile drag at moderate subsonic speeds. The inadequate standard production techniques, however, prevented their achieving that purpose in practice. Nevertheless, quite serendipitously, those airfoil shapes proved well suited to an environment unanticipated at the time of their inception: high subsonic speeds. Yet the scientific brilliance and novelty of the design technique that underlie either application remains undiminished.

3. Michael Polanyi (1968, p. 1310) defines a hierarchy as a structure in which, "each level . . . relies for its workings on the principles of the levels below it, even while it itself is irreducible to these lower principles." As Polanyi's article testifies, a hierarchy is an idea that is both compelling and fraught with controversy. I am indebted to Stephen J. Kline for calling my attention to this article as well as for stimulating discussions on the nature and importance of technological hierarchies.

One suspects that the same argument applied here to hierarchies of technology and technological knowledge (and communities so defined) applies also to the structure of scientific theory and practice and that issues of incremental versus revolutionary change in science are isomorphic with the same problems in technology.

4. This notion developed from conversations with Walter Vincenti and Virginia Dawson.

5. See Wiebe Bijker (this volume), especially the later sections, on the critical role of social groups in generating and sustaining technological momentum. Although not clearly spelled out by Hughes, technological momentum clearly implies both size (or mass) and rate of change (or velocity) components. Presumably, the greater the size of a technology (as measured by extent, diffusion, scale, complexity, or investment) and the greater its rate of change, the greater its momentum. A large, static technology will have great inertia and, by definition, will not change: momentum at rest. More important, a rapidly changing large technology will have great momentum and great impetus for continued change: It will generate reverse salients and critical problems at a fierce rate. Although the connection is not made by Hughes, the work of Jacob Schmookler offers some empirical support for this interpretation. Oversimplified, Schmookler found that rates of patenting (his surrogate for invention, or at least inventive activity) varied directly with but lagged slightly behind rates of investment in a given industrial sector (Schmookler 1966; see also Rosenberg 1972, pp. 39–42). Schmookler's rate of investment might be regarded as a measure of the rate of growth of a system or technology and therefore of the rate at which it is likely to encounter reverse salients; his rate of patenting can likewise be seen as a measure of the rate of critical problem solution, which is some (unspecified) transform of critical problem definition. See also Hughes's own discussion of momentum and that of Donald MacKenzie, both in this volume.

6. People also buy services, which usually entails use of some mix of artifacts and knowledge by the performer of the service, but the general point remains the same.

7. One suspects that the distinction between technological economies of scale and higher organizational throughput is often only analytic. For example, if a firm were to replace a 36-inch (width) sheet steel rolling mill with a 48-inch mill without increasing the labor input or capital investment, it would increase its potential throughput by fully 33 percent. Yet that enhanced capacity reflects purely the economics of scale associated with the 48-inch mill technology. Of course, the organization must adapt to (that is, sell) the increased output.

8. Why an idealized schematic is appropriate should be obvious from examination of an ostensibly more realistic organization chart. See Chandler (1977, pp. 394–395).

9. See Bijker's discussion of technological frames (this volume).

10. This diagram is adapted from Meryl R. Louis's depiction of transorganizational culture (Louis 1981).

Regulatory Science and the Social Management of Trust in Medicine

Henk J. H. W. Bodewitz, Henk Buurma, and Gerard H. de Vries

Over the past two centuries treatment of the sick has become the monopoly of the medical profession, supported by an increasingly complicated medical technology and protected, regulated, and in large part paid for by public agencies. Folk medicine, magic, and quackery have waned. In all industrial societies social systems of medical care have evolved. Such systems may be conceived of as densely populated networks of heterogeneous arrangements and dependencies. Nodes of these networks include hospitals, pharmacies, insurance companies, government departments, university faculties, the multinational pharmaceutical industry, and, of course, doctors and their patients. Also part of these networks is the various national drug regulation agencies, on which we focus in this chapter to study part of the processes from which an important segment of modern medical technology—drugs—evolves.

In all industrial nations national drug regulation agencies have been installed to register, control, and regulate the entrance into the open market of all chemical substances labeled as drugs for particular diseases. Such regulation agencies may be conceived of as instances of a special kind of complexity-reducing nodes, which tend to emerge in social networks that become increasingly complex. They may be seen as the social system's reaction to system complexity (Lehmann 1973, 1984). In these newly formed nodes, actors characteristically take a bird's-eye view of activities and relations elsewhere in the network. In drug regulation agencies considerations are assembled concerning the production and use of modern drugs, which are nowadays usually industrial products with (compared to their predecessors) often extremely rapid action and high potency. For that reason they present unique research sites for observing important aspects of the social construction, distribution, and acceptance of drugs. In these institutions information and influences from different corners of the network are assembled, and both intended and unintended consequences of actions are discussed and negotiated. One finds a microcosm in which the ideas, values, and professional attitudes that operate in the network as a whole are reflected and in which interests, expectations, and insights from a variety of groups become linked.

Typically and often because of conscious action, in complexity-reducing nodes a collective background is generated; this background serves as a shared framework for orientation and appraisal in the network at large. Imposed as a normative guidance (for example, through legislation) such backgrounds determine relevance structures and affect the perception of critical problems and reverse salients (Hughes, this volume) by the various individuals and institutions taking part in the networks. In drug regulation agencies an essential ingredient of the shared backgrounds is constituted by modern medical science. In one way, this just reflects the importance of science in the social system of medical care as a whole, as modern medicine serves as a rich resource for the development of new drugs, delicate instruments, and therapies and also helps to legitimize often painful interventions. No less important for its role in regulation agencies, however, science also stands for the promise that a rational, neutral, and objective assessment of drugs is possible—an assessment that is legally defensible and satisfactory to all parties affected by the regulations of drugs: the industry, consumers, medical professionals, and laymen.

Drug regulation agencies form an important bottleneck before diffusion of drugs can take place. Every change in the composition, production, and description of the drug and/or its potential use has to be approved by these agencies, which make the final decisions regarding the indications for which the drug may be prescribed, the content of the package insert (the information leaflet accompanying the product), and eventual advertisements. The agencies also determine the drug's availability: in hospitals and/or in pharmacies; over the counter or only by prescription; prescribed by general physicians or only by a medical specialist. It is this registration process that formally constitutes certain chemical substances as (legalized) drugs.

To apply for registration, a pharmaceutical company has to submit extensive documentation to demonstrate that its product meets criteria set by national legislation. In most countries these criteria imply proof of chemical-pharmaceutical quality and efficacy and safety of the drug. Partly as a consequence of differences in medical tradition and political and legal frameworks, the standards and regulation practices within which these general criteria are operationalized may vary considerably in different countries. In general, the application file has to include data from preclinical as well as clinical research on both healthy volunteers and diseased persons. In some countries, notably the United States, the start of the clinical research phase also has to be approved by the national regulation agency. In other countries such intermediate decisions are not required.

Regulation of drugs, of course, means different things to different participants in the social system of medical care. For the pharmaceutical industry regulation primarily presents an important barrier that has to be overcome before a product can be marketed. Much of the industry's research effort is directed toward determining as early as possible presumptive problems in regulation of its products. For consumers regulation is a guarantee of the overall quality of the commodity perceived as essential for their well-being. For the medical profession regulation is an additional safeguard for public health and, by distinguishing prescription from over-the-counter drugs, a support for the social position of medical professionals as specifically qualified experts. Unsurprisingly, given its obvious importance for modern medicine, regulation policy has become the subject of heated debate.

Unfortunately, demands of (industrial) secrecy block (especially in Europe) direct analyses of processes of consensus formation and decision making in regulation agencies. Therefore we have to rely on more indirect approaches. In this chapter we proceed along two lines. First, we analyze contemporary international debates about regulation policy. As we demonstrate, the ways in which the rationality of regulation is questioned reveals important aspects of the framework that is used to assess the acceptability of drugs. Second, we compare the emergence of drug regulation in the United States and in some Western European countries to show the ways in which different groups were involved in the development of the respective institutions. Our emphasis is on the role of scientific considerations in the development of a collectively shared and widely trusted framework for assessing the chemical-pharmaceutical quality, efficacy, and safety of drugs.

Of course, there is much more to medical technology than science: Economics, nonscientific traditions of thought, and social and political factors (not to mention fraud, shortsightedness, and other human imperfections) all help to shape its evolution. However, patients' trust in their doctors and medical experts' confidence in the reliability of medicines supplied by pharmaceutical industries are important prerequisites for the smooth functioning of social systems of medical care. Because doubts about the rationality of a certain medical practice are often expressed as doubts about the scientific basis of that practice, rather than, say, its ethical rationale, the framework on which this mutual confidence and trust rests in our societies happens to include science as one of its main pillars. The reception and use of medical technology are thus bound up with assessments of its scientific basis. An analysis of the role that these considerations play may thus help to reveal closure mechanisms for the networks in which medical technology evolves. Moreover, such analysis may contribute to bridging the gap between the sociology of science and the social study of technology (Pinch and Bijker, this volume).

Debates about Regulation Policy

In most countries drug regulation agencies arose in the early 1960s. Since the mid-1970s extensive discussion of regulation and deregulation have commenced. By that time the demand for chemical-pharmaceutical quality had been laid down in generally accepted and highly standardized rules of good laboratory and manufacturing practices. Hence discussions about (de-)regulation were restricted to the demands of efficacy and safety. In these discussions, which took place—and still take place, in a variety of forums, representatives of different groups participate. As a first approximation four such groups may be distinguished, each occupying a specific position in the social system of medical care and each expressing a specific perspective on drug regulation practices: (1) the general public, (2) the pharmaceutical industry, (3) representatives of drug agencies, and (4) medical scientists and practitioners. We begin by briefly reviewing their respective positions.

Although sociologically "the general public" is of course a cloudy category, its viewpoint may be considered to be rather straightforward. As expressed by means of the media by, for example, consumer and health organizations and, perhaps even more important, attributed by other participants in the debate, the general public appears to stress the rational nature of drug regulation and its scientific basis. To the layman it seems to be "a justifiable expectation that by now the evaluation of medicines has become a highly scientific, rational activity, capable of providing definite answers regarding the safety and efficacy of new medical products" (Griffin and Long 1982). In this quest for safety consumer organizations have delayed the approval of some drugs. For example, in answer to the pharmaceutical industry's complaints that the relatively late introduction of beta blockers on the US market created a "drug lag," consumer organizations said that all that could be shown to exist was a "deathlag" and that the introduction of beta blockers was rightly delayed so long as safety could not be guaranteed absolutely (Daniels and Wertheimer 1980).

In contrast, although agreeing on the need for regulation in general terms, the pharmaceutical industry challenges the unlimited growth of standards. In regulation, it is contended, "there has been a steady trend to more 'scientific' and ritualistic interpretations of the criteria, which may have gone beyond the point of medical realism" (Wardell 1981). Procedures are said to have become ritually routine tests of limited value, governed by a desire for regulation rather than rational thought. The expansion of regulation over the years is now perceived as a serious threat to creative research and development and is thought to be responsible for a decrease in the amount of new drugs appearing on the market in recent years.

Against this criticism, representatives of drug regulation agencies tend to maintain the scientific nature of their assessments. Some of them, however, also acknowledge that there are still serious problems. One former UK commissioner referred to such problems when he conceded that "there is no such thing as a regulatory science. [It] is an art and a very difficult one, one of the troubles [being] lawyers and the legal systems within we have to work in different countries" (Wade 1977). Other problems are, for example, related to the question of whether nonscientific factors (for instance, medical need and cost) should be explicitly included in the evaluation of a new drug's merits. Advocates of this viewpoint to the urgency of a neat tuning between the agency's perception of an approved drug's use and its later role in actual therapeutic practices. Other discussants, however, stress the dangers of adding such nonscientific factors, making regulatory agencies subject to extraneous considerations and possibly causing them in due course to compromise scientific standards and procedures (Dukes 1979; Wardell 1981).

The fourth group we have distinguished, medical scientists and practitioners, played an important role in early discussions about drug regulation. In later years, however, their role has been reduced, and in contemporary debates medical practitioners and scientists almost exclusively participate as spokesmen of one of the aforementioned groups rather than voicing an independent professional view.

Even from this brief review it is obvious that, no matter how present regulatory practices are evaluated, for all participants science provides the yardstick for assessing their rationality. Not surprisingly, the image of science that is adhered to in this context is unmistakenly the standard, empiricist-oriented "received" view (Suppe 1974; Mulkay 1979b, ch. 1, 2). Although now abandoned by most philosophers, historians, and sociologists of science, this view nevertheless continues to serve as an important normative ideal in public debates. One of the reasons behind this, no doubt, is that the empiricist image of science seems to be tailor-made for legal and juridical considerations, promising to all concerned that, because of the rational rules of scientific method, objective assessments of the facts by experts are indeed possible and that the separation of questions of fact and of value can be achieved. That the ideal is never completely reached is not considered as a reason to question the underlying assumptions about scientific rationality. It is the contingencies of actual practice that transform the nature of the activities of experts into an art, rather than into what it is supposed to be—a science.

Although widely used in the debates on drug regulation, the description of the craft of regulators as an art is not of much help. It hardly implies more than that these activities do not yet meet empiricist standards of rationality. It neither casts light on the specific characteristics of contemporary regulatory practices nor helps us to envision how procedures might be improved to achieve a more scientific basis for regulation. It leads us only into the blind alley of the science/art dichotomy.

A more adequate perspective on the role of science in drug regulation may be derived by focusing on an asymmetry that appears in the debates on regulation policy. Whereas standards of *efficacy* for drugs are widely accepted as unproblematic, standards of *safety* for drugs are consistently contested and disputed. Moreover, a second look at the science/art dichotomy used in the debate suggests that it is particularly those practices that relate to assessment of safety that are labeled as art, whereas assessment of efficacy generally is considered to be a matter of scientific judgment (Liljestrand 1977).

For proof of the *efficacy* of new drugs, controlled clinical trials using double-blind methodology are generally accepted as standard. This methodology is one of the main contributions of clinical pharmacology, a young discipline in the medical sciences. According to its rules patient populations are divided into two groups. One of these groups is prescribed the drug that is to be tested; the other group receives a placebo for the duration of the test. During test runs, neither the patients nor the researchers who measure effects know which individual received what. Decoding takes place only after completion of tests. Except for ethical doubts concerning the use of placebos instead of, for example, drugs that have already been in use with known positive effects and a number of objections that arise from alternative medical traditions (for example, homeopathy) and that are related to discussions in West Germany (to which we return later), double-blind methodology is not under attack. Once a number of pitfalls in the measurement of clinical effects are eliminated, its contribution to the reliability of efficacy testing is largely uncontested.

In contrast to the situation relating to efficacy tests, there is no comparable, generally accepted methodology for testing the *safety* of drugs. Preclinical studies, that is, studies in animal pharmacology and toxicology, turn out to have low predictive value; their results are not conclusive in the clinical stage either for proof of safety or for the diffusion and action of the drug in the human body. In reacting to these disappointing facts, regulation agencies in most countries have consistently raised the standards for preclinical research, extending the time scale on which effects have to be monitored, demanding tests on an increased number of species, and in particular focusing on specific problems, such as carcinogenicity, mutagenicity, and teratogenicity. Despite this expansion of regulations concerning preclinical safety testing, nobody doubts that in the last analysis clinical, not preclinical, tests still have to bear the burden of proof in the final assessment of a drug's safety. Clinical studies, however, have their own disadvantages. For instance, side effects with low frequencies of occurrence are unlikely to be discovered in these kinds of tests. The result is that, instead of being governed by the rules of a clear and generally accepted methodology, regulation authorities have to base their assessments of the safety of new pharmaceutical products on a heterogeneous set of test results. It is this fact that led Wade and others to speak of regulatory activity as an art, rather than as a science (Wade 1977). It is also this fact that gives room to the diversity of evaluations of contemporary drug regulation practices one finds in different corners of the social system of medical care.

The question that obviously has to be answered is, Why hasn't it been possible to agree on procedures for testing safety, whereas a consensus on a methodology for efficacy was achieved years ago? It is worth considering a number of possible explanations.

In the first place, one might think of the difference as stemming from differences in the *nature* of the judgments that are involved: Assessments of efficacy might be held to rest on factual judgments as opposed to the value-laden criteria involved in evaluations of safety. A closer look, however, reveals inadequacies in this argument. Assessments of both safety and efficacy constitute complex mixtures of factual and value-laden judgments. No drug is safe or effective for the entire pertinent population. Hence a (value-laden) decision has to be made, for example, whether the relative ineffectiveness of the drug for patients of a certain age, gender, or lifestyle has consequences for the assessment of the drug's overall efficacy. Likewise, safety judgments rest on value-laden decisions. That the nature of the judgments involved in the assessment of safety and efficacy is indeed the same can also be inferred from the fact that they can be curiously interchanged. To take an extreme case, estrogen-progesterone combinations (the contraceptive pill) are described by the Spanish Pharmacopoeia as effective in regulating menstrual cycles, but as having the serious side effect of preventing pregnancy (Veatch 1981).

For a second possible answer to the question of why there is agreement on standards of efficacy but no consensus on standards of safety, one might want to turn to the *logic* of the situation in establishing both judgments. At first, this presents a more promising case, for it may be observed that, although the establishment of efficacy is a question of testing for a few well-delineated effects (application of the drug), the establishment of safety is a matter of testing for a potentially infinite and in practice large number of effects. No wonder then that problems of efficacy and safety are discussed along different lines. On closer inspection, however, this argument turns out to be defective. To see this, note, first, that both judgments are in principle fallible. For safety this is obvious, but it also holds for the assessment of efficacy: Further research may reveal hitherto hidden *ceteris paribus* clauses (for example, related to long-term use) that may imply the revision of a former positive judgment. Hence, when we compare the logical situation of safety and efficacy testing, we are essentially comparing the degree of testability (Popper 1968, ch. 6) of statements about the safety and efficacy of a drug. Second, it may now be observed that, if agreement could be reached on the relevant background knowledge, there would indeed be less doubt that, generally, the degree of testability of an efficacy judgment concerning a few well-defined effects would be much higher than the degree of a many-variable judgment such as safety. It then follows that, if a positive efficacy statement has passed the relevant tests, it will indeed be accepted much more readily than its safety counterpart. As an explanation, however, the argument is not acceptable because it simply begs the question. The fact to be explained is not just that, where agreement on relevant tests and background knowledge is established, consensus about efficacy comes much more easily than consensus on safety; rather, it is that for safety one continues to quarrel about standards, whereas for efficacy one does not. In other words, what is at stake is the proviso on which the explanation is founded. Only when prior agreement on background knowledge exists is it possible to use general methodological criteria (such as degree of testability) to explain the behavior of scientists in disputes.

That we are dealing, for efficacy, with empirical issues rather than with logical subtleties can be inferred from the case of West Germany, where lack of prior agreement about background knowledge eventually resulted in a long-running and continuing disagreement on suitable standards of efficacy of drugs. When, under pressure from the European Economic Community's harmonization policy, drug regulation was legislated in West Germany, as late as 1978 considerable debate preceded the legislation (Burkhardt and Kienle 1980; "The new German drug law," 1980). In this debate advocates of longstanding traditions of medical thought (homeopathy, phytotherapy, and anthroposophical treatment) vehemently opposed such general regulations. The discussions that occurred showed all the characteristics of Kuhnian incommensurability. In these debates efficacy, not safety, was the main issue, with homeopathy-, phytotherapy-, and anthroposophy-oriented physicians arguing that efficacy of drugs could be defined only within a particular medical tradition. Central positions in the proposed law, such as the role of controlled clinical trials as proof of efficacy, excluding other forms of proof, for example, those derived from decision theory, were vehemently criticized. The attack came from the juridical side (Fincke 1977) and from the scientific side (Kienle 1974; Burkhardt and Kienle 1978, 1980) and elaborated on criticisms of controlled clinical trials known within regular medicine but considered to be not so serious as to undermine the agreed-on standards of efficacy. It demonstrates the power of the homeopathic and anthroposophical medical traditions in Germany; in the end three different regulatory committees were set up, each using its own interpretation of efficacy and each surveilling the production and use of medicines in its own segment of the German social system of medical care.

The German case suggests that, rather than science defining how trust in medicines is to be achieved, it is the other way around: The existing structure of the social system of medical care determines the way in which medical technology becomes assessed scientifically. Consequently, instead of looking for further logical and internal scientific explanations of the differences found over standards of efficacy and safety, sociological explanations deserve serious consideration. The fact is that thinking about efficacy and thinking about safety have developed in different nodes of the network that constitutes the social system of medical care. Although in most countries ideas about efficacy developed mainly within the confines of the medical profession, thinking about safety was also influenced by social groups outside the profession. When in the early 1960s drug regulation agencies emerged in a number of countries as complexity-reducing nodes, efficacy became a matter typically discussed against a common background, whereas safety had to be negotiated by a variety of groups with competing interests and diverging backgrounds.

Social Origins of Ideas about Efficacy and Safety of Drugs

To explore in more detail the possibilities of a sociological explanation of the differences between the measure of agreement about standards of safety and efficacy, it may be helpful to review the emergence and expansion of drug regulation in the United States and Western European countries (especially the United Kingdom) with their different traditions of (political) decision making and legislation. The overall pattern of development in these countries is roughly the same; they differ in the stage that they have reached and sometimes, in the mechanisms that brought about the development. We begin by tracing the evolution of ideas about efficacy; a similar description of the development of ideas about safety follows.

In 1962 legislation in the United States by the Food and Drug Administration (FDA) began to include efficacy as a criterion for drug evaluation (Temmin 1980). This event marked a major shift in the status of drug regulation. In 1910 an early attempt to include efficacy had failed; the Supreme Court decided at that time that therapeutic effectiveness was a matter of opinion, not of fact. For a long time evaluation of efficacy thus remained in the hands of individual general physicians, as the FDA was unable to prevent the introduction on the market of drugs with exaggerated or even false effectiveness claims. In 1962, however, opinion had shifted, and therapeutic claims, now conceived of as being provable by scientifically trained experts, had acquired the status of fact. Expert committees were installed by the FDA and staffed with medical specialists. The 1962 law required demonstration of "substantial evidence of efficacy" of drugs, leaving the experts to decide what kind of evidence they would like to see. Eventually, it became general practice to use the open literature and research reports submitted by the drug companies, supplemented with the informal judgment of the panel, that is, by the clinical experience of panel members and their peers.

The American Medical Association, the professional organization of general physicians, criticized the withdrawal of responsibility for a drug's efficacy from the hands of the general physician, that is, from the doctor who prescribed the drug. In 1962, however, it was already too late to make this argument. In 1959 senate hearings on the drug industry had been started in order to seek procedures for protecting consumers against high prices of products, which, as one senator put it, were distinguished by the fact that "those who order do not buy, and those who buy do not order." An outcome of these hearings was the proposal of legislation to increase FDA surveillance over drug manufacturing and new drug introduction. Impressions of clinical practice were no longer perceived as sufficient to discover the truth about a drug's effectiveness. During the hearings a succession of notable physicians had stressed the need for a truly systematic evaluation. The individual general physicians' judgments, which for decades had been conceived of as a satisfactory basis for the assessment of the efficacy of medicines, had become obsolete. In a social system of medical care that increasingly had come to rely on industrially produced drugs, a government-related institution took its place in assessing efficacy of drugs.

A more effective protest against the 1962 law and its subsequent implementation by the FDA came from the pharmaceutical industry. In a series of legal proceedings industries objected to the delegation of decisions to small expert committees and to the operationalization of efficacy claims that had evolved. With an eye on products that were introduced on the market before 1962 and that now had to be reviewed under the new law, pharmaceutical companies wanted general medical practice, and hence the market, to have more voice in decisions about approval. Eventually, in 1970, the FDA won the dispute, but in its course it had changed its policy. Clinical experience was now ruled out for its informality and lack of explicit objectives and methods (and hence for its vulnerability in court). Double-blind testing became the required research methodology. Proceeding along this line, the FDA succeeded in overcoming the political pressure to give medical practice a voice in drug approval, and it reached agreement with industry by providing strict and clear methodological directions for proving effectiveness. Within the political framework of the FDA, a specialized part of the medical profession, in particular clinical pharmacologists, gained a central position.

In the United Kingdom developments eventually led to the same results, although through quite a different route (Dunlop 1972). Until 1968, regulation of drugs in the United Kingdom remained a completely voluntary matter between industry and the Committee on the Safety of Drugs. The committee did not formally approve a drug's introduction on the market but confined itself to the collection and dissemination of information. For this purpose it administered a file specifying adverse reactions of drugs, and it encouraged practicing physicians to report their experiences with new products. The committee focused on safety of drugs; efficacy was considered only in relation to safety matters. Close relations continued to exist between British industry and professional groups. (See Gillespie et al. (1979) for a comparable situation in UK toxicological assessment of pesticides.) In the United Kingdom

most of the leading physicians, pharmacists, veterinarians and pharmaceutical industrialists are familiar with each other and often on friendly terms and the political atmosphere and the competitive commercial pressures are perhaps less fierce [than in the US]. . . . Thus, much of the contact with the applicants and the requests for clarification and amplification took place in robust but usually good humoured encounters over the telephone or in informal meetings rather than in official communications duplicated for the record. (Dunlop 1972, p. 307)

In 1968, in the United Kingdom regulation of drugs also became the subject of legislation. Actual practices, however, changed only gradually. Efficacy became one of the law's criteria, but, in contrast to the safety and chemical-pharmaceutical quality of drugs, its assessment was not made by the newly installed Safety of Medicines Committee; rather, it was delegated to an expert committee that included representatives of medical practitioners. In the late 1970s standards were raised and, in accordance with US policy, most Western European countries now include double-blind testing.

Except for the many informal contacts with industry that seem to be typical for the United Kingdom, development of drug regulation policy in other Western European countries, such as the Netherlands, Switzerland, France, Sweden, and Denmark, followed the UK pattern and took place within the confines of regulatory committees or agencies, without open debates or conflicts with either medical practitioners or industry. This difference in the pattern of development of drug regulation policy between the United States and Europe is also found in other areas (see Brickman (1984) for toxic chemicals regulation).

In contrast to drug efficacy, there has been no long tradition of professional thinking about safety of drugs. For a long time problems of safety were almost exclusively related to the quality of the manufacture of drugs and therefore mainly depended on the skills and accuracy of individual pharmacists. Only after the rise of drugs as industrial products after World War II did potent drugs with potentially serious adverse effects become available on a massive scale. Another reason for the absence of a tradition of professional thinking about safety is related to the fact that incidences of serious adverse reactions often show up in low frequencies or only after a considerable time lag; therefore the incidence is not visible in an individual general physician's practice. Such practices are simply not the proper locus to study safety of drugs.

In the United States legislation of safety of drugs began comparatively early, viz. in the 1930s (Temmin 1980). A 1938 law enabled the FDA to restrict the availability of potentially toxic drugs to prescription only. In most other countries, however, legislation on safety of drugs began only after the so-called thalidomide affair. This tragedy occurred in 1962, when phocomelia—a condition in which children are born with missing limbs—began to appear with alarming frequency in West Germany and other European countries and the drug thalidomide was identified as the source.

Induced by a disaster, the safety of drugs became an international public issue. Nonprofessional groups, such as the Nader Health Research Group in the United States and consumer organizations in Western Europe, participated in discussions and have continued to do so. A tendency to stress absolute standards of safety emerged. It is generally believed that a tragedy such as the thalidomide affair should never happen again.

Of course, the absolute standards of safety had to be implemented by regulatory agencies and had to be encoded in legal terms. Everywhere this turned out to be a Sisyphean task. The volume of animal studies that is required has continually increased over the years, as have the number of species and the number of studies required to trace long-term effects. After the thalidomide affair teratological studies became routinely required, but soon new themes emerged: carcinogenicity, mutagenicity, and, more recently, drug metabolites. The importance attributed to these themes, however, varied over time and became subject to the influence of various nonprofessional groups.

Last but not least, further knowledge in these areas tended to widen, rather than to narrow, the scope of controversies. Toxicologists have continually devised new tests, the stimulus almost invariably being previously identified hazards in relation to other drugs in humans. New analytical techniques also tend to generate new problems, rather than to stabilize standards of testing. The net effect of this development has been that safety evaluation of new therapeutic agents has become more diverse, more expensive, and more time consuming, because it is necessary for the manufacturer to test for all possibilities. Moreover, standards of safety vary considerably among countries, causing troubles for pharmaceutical industries operating on international markets. For industry, safety is an issue still beyond its control, whereas a conclusion that some chemical agent is not an effective drug on certain criteria, in principle, still leaves room for attempts to establish its efficacy for other indicators, and, in fact, quite a few innovations in drug research have come about this way. An inference that the compound does not meet safety standards (for example, a carcinogenicity test) establishes an irrevocable fact.

Not surprisingly, there have been a few deliberate attempts to propose standards of safety for drugs that would be acceptable to the international community of clinical pharmacologists. Thus far, however, they have failed to be accepted. A public statement by one group of leading European pharmacologists (Sestre Levante Workshop, 1977) identified the major obstacle to agreement: Present policy on safety is "dominated by fear of accident." The present "unrealistic" standards will disappear, the statement continues, only when the public faces the fact that "medicines are not and cannot be absolutely safe." It needs to become "aware that treatment with medicines always carries some risk" and that the pursuit of absolute safety regardless of other considerations is achieving more harm than good. In the view of these industrial and academic pharmacologists, "practices however cherished or habitual, which cannot be justified scientifically, should be abandoned. The possibility that any subsequent mishap or tragedy may wrongly be blamed on changed practice must be faced and accepted as an inevitable risk of therapy." The closure of the debate among professionals on the safety of drugs therefore appears to depend on the adoption by the general public and its representatives in those debates of a "professional view," that is, its decision to drop absolute safety as a prerequisite for the acceptance of drugs.

Our historical and sociological observations on the origins of the ideas about efficacy and safety of drugs may thus help to explain the emergence of asymmetries in the agreement achieved with regard to standards of efficacy and safety. Standards of efficacy developed on the basis of a tradition of thought that had evolved, successively, in practices of general physicians, in expert committees, and in the new discipline of clinical pharmacology. When for juridical reasons it became necessary to formulate explicit standards, the FDA was able to do so on the grounds provided by the implicit consensus that meanwhile had grown inside the medical profession. With a lag of a few years, European countries followed this path. Standards of safety, however, evolved along different lines. Because the need for such standards was induced by a disaster, the countries could not develop policies exclusively within the confines of established professional thinking but had to agree with outside bodies, such as consumer organizations representing public attitudes. Not being prepared to accept that the risks of such tragedies as the thalidomide affair were inevitably related to the benefits of modern potent drugs, the public has demanded absolute standards of safety. This poses severe problems, particularly for industry. But, on the other hand, public attitudes cannot yet be neglected.

Drug regulation agencies were originally planned as institutions staffed with scientific experts on which consumers could rely for their protection against possibly dangerous industrially produced drugs. Regulation has now reached a stage at which, in order to come to terms with industry, agencies may be tempted to effect public attitudes. Laymen, who for their trust in medicines have come to rely increasingly on expert knowledge, thus now run the risk of adopting uncritically professional, "realistic," attitudes, that is, attitudes adapted to the reality of pharmaceutical industries.

Conclusion

The stabilization of technological artifacts is bound up with their adoption by relevant social groups as an acceptable solution to their problems (Pinch and Bijker, this volume). For complex technologies such groups, which may include institutions, organizations, and unorganized individuals, may be conceived of as being dispersed over social networks. The evolution of technology takes place in such networks, which by the same processes also evolve. Although specific ideas and artifacts may originate from the hands of individual inventors or firms, technical innovations are thus culturally and socially produced. For a technical solution of a problem to become part of technology, it has to become embedded in existing social networks. This involves complex processes of social management of trust. People must agree on the translation of their troubles into more or less well-delineated problems, and a proposed solution must be accepted as workable and satisfactory by its potential users and must be incorporated into actual practice in the social networks that make up the world in which the technology evolved. In many cases acceptance of solutions is mediated by anonymous processes, for example, by processes related to market mechanisms or by processes that rest on unquestioned cultural traditions. In some cases, however, particularly in complex networks, trust may also become institutionalized, and consciously planned assessments of products may become part of the processes of social acceptance of technical artifacts. Drug regulation is an obvious case in point.

Only a few decades ago the decision to produce and accept chemicals as drugs lay completely in the hands of individual pharmacists and doctors. If one wanted to study the evolution of medical technology in those days, the general physician's consulting room was a good bet as the "consumption junction" (Schwartz Cowan, this volume), where choices between competing technologies are made. But with the growth of the complexity of the social system of medical care and the related introduction of industrially produced rapid action and extremely potent drugs, the decision concerning which chemicals *count* as acceptable drugs for a particular disease has been withdrawn from the hands of individual doctors to become the concern of centralized agencies. Given their position in the social system of medical care, as complexity-reducing-nodes, such agencies are supposed to assess pharmaceutical products in a neutral, rational, and objective way, thereby serving, in principle, the interests of all participants in the social network, consumers no less than producers. The rhetoric of the standard empiricist view of science presents itself as particularly suited for this purpose. Scientific rationality promises to provide an Archimedean point outside the social world, from which the acceptability, and hence survival, of technological artifacts can be decided.

In contrast to what is suggested by the standard view of science, however, a consensus on the acceptability of artifacts such as drugs does not emerge from an application of formal rational rules. Actually, as we have demonstrated, it is quite the reverse. Where a consensus has been achieved already, either implicitly or explicitly, it is possible to agree on formal rules, for example, when for juridical reasons the necessity to formulate them comes to the fore. The history of thought about efficacy of drugs presents a clear example. In areas such as safety of drugs in which a consensus is harder to achieve, formulation of rules tends to lead to disputes about rules rather than to agreement between rival perspectives and interests. Scientific procedures and certified knowledge do not emanate from a social vacuum; they, too, are outcomes of social processes of acceptance. In fact, they are often affected by the same social forces that determine the development of the social networks in which the technical artifacts that they are supposed to evaluate "objectively" evolved. In the evolution of technological artifacts there are no Archimedean points outside the social world.

The Consumption Junction: A Proposal for Research Strategies in the Sociology of Technology

Ruth Schwartz Cowan

The sociology of technology, if it is ever to justify its existence as a subdiscipline, should take as its proper domain of study those aspects of social change in which artifacts are implicated. The processes by which one artifact supplants another (technological change) or by which an artifact reorganizes social structures (technological determinism) or by which an artifact diffuses through society (technological diffusion) are fundamentally sociological in character, subsets, as it were, of the larger topic of social change. Properly constituted historical investigations can be used (and indeed must be used) as the raw material for studies of social change, but many historians of technology have had cause to wonder what "properly constituted" could possibly mean in this particular context.

In their contribution to this volume Trevor Pinch and Wiebe Bijker have provided some useful guidance on this matter. A properly constituted history of science, they remind us, should be impervious to the question of whether or not the ideas being examined historically are true or false by current standards. Similarly, they argue, a properly constituted history of technology should consider artifacts that were "failures" on the same par with artifacts that were "successes." The task of such a historical investigation is not to glorify the successes but to understand why some artifacts succeed and others fail. To fulfill this obligation, historians of technology must be careful, as they have not often been, to track down all the possible technological solutions that have been offered to a given social problem, and then they must be equally careful to examine those solutions in the context of the time period in which the choices were being made. Today's "mistake" may have been yesterday's "rational choice."

Pinch and Bijker go one or two steps further, however. They assert that historical case studies, which can provide new methods for studying technology and new theories for the sociology of technology, must also be alert to the existence of what they call relevant social groups, groups that influence the creation, the demand for, the production, the diffusion, the acceptance, or the opposition to new technologies. In the history of science, they remind us, this prescription is relatively easy to fulfill because science is generated by and fills the needs of relatively circumscribed social groups; no more than a handful of European males were, for example, responsible for the destruction of the Ptolemaic universe and the construction of the Copernican. Four or five centuries later the same could still be said about those who created, confirmed, and affirmed the neo-Darwinian synthesis or the theory of plate tectonics, to cite other examples. In the history of technology the situation is different—and potentially infinitely confusing—because a large number of relevant social groups are involved in the success or failure of any given artifact (ranging, for example, from the small group of craftsmen or engineers who may be responsible for innovation to the somewhat larger group of managers who may make decisions about the innovation, to the even larger group of production experts who must turn the innovation into an artifact, to the even larger group of people who must distribute, market, and sell it, and then to the potentially even larger group of people who will consume it). Any one of those groups, or individuals acting within the context of their group identity or (worse) combinations of those groups or (even worse yet) some other group not yet enumerated (such as purchasing agents for governments), may be responsible for the success or failure of any given artifact. How is the historian of technology who wishes to become "properly constituted" going to cope with such an infinitely expandable universe of relevant social groups? Pinch and Bijker have given us a prescription but precious few suggestions about how it might be filled.

In my own work I have found it possible to begin filling the prescription by focusing on the actual or potential consumer of an artifact and imagining that consumer as a person embedded in a network of social relations that limits and controls the technological choices that she or he is capable of making. The concept of network that I utilize is similar to the one defined by John Law and by Michel Callon in their separate contributions to this volume—a temporal association between heterogenous and interacting elements—but I differ from Law and Callon (and also from Hughes, Bodewitz et al., and MacKenzie in their contributions) in my effort not only to place the consumer in the center of the network (at the consumption junction) but also to view the network from the consumer's point of view. Law wants to know what holds networks together; Callon wants to know why some engineers are such effective network builders; Hughes wants to understand how networks come to be built; MacKenzie and

Bodewitz et al. describe for us the ways in which elements of the network interact with each other. These are investigations of networks from the outside in; the investigators are asking questions with which the people embedded in the network may never have been concerned. My enterprise is somewhat different. I focus on the consumption junction, the place and the time at which the consumer makes choices between competing technologies, and try to ascertain how the network may have looked when viewed from the inside out, which elements stood out as being more important, more determinative of choices, than the others, and which paths seemed wise to pursue and which too dangerous to contemplate.

There are many good reasons for focusing on the consumption junction. This, after all, is the interface where technological diffusion occurs, and it is also the place where technologies begin to reorganize social structures. If technological change is goal directed (and who can doubt that it is?), then a positive consumption decision at the consumption junction is in fact the goal, and an investigation that focuses on the consumer is thus heuristically justified. Such a focus brings into relief (perhaps more clearly than other foci can) the variables that have governed the behavior of all those relevant social groups who influence consumers' choices. It also frees the investigator from the bonds of needing to know, through hindsight, what the consumer could not possibly know with certainty, that is, which choices would turn out to be "right" and which "wrong." The historian need evaluate only those elements of the network that caused consumers to act in the particular ways they did, rather than worry about whether in the long run those actions were right or wrong.

The single most worrisome complicating factor in this mode of analysis is, however, the obvious fact that consumers themselves come in many different shapes and sizes; indeed any single given human being can enter the consumption junction under a number of different guises, depending on what it is that is being consumed. The same individual may, for example, have one set of interests when entering the consumption junction as the Director of the Ordnance Department of the US Army and a quite different set when acting as a homeowner in need of a new basement furnace. Once recognized, this complicating factor can add rather than detract from the usefulness of consumer-focused analysis, because it reminds us that we must define consumers in terms of the artifact about which they are making choices (as, for example, "a prospective purchaser of a washing machine") as well as by other appropriate socioeconomic variables (for example, "middle class," or "rural"). Market researchers and advertising agencies are quite accustomed to this form of categorization; sociologists of technology might do well to adopt the practice.

In this chapter I hope to demonstrate the usefulness of consumer-focused analysis first by using it to solve a historical puzzle that I first encountered when studying the history of home heating systems in the United States (a puzzle that previous forms of analysis had failed either to notice or to resolve) and, second, by using it to generate hypotheses about other aspects of that same historical development that warrant further study. I begin with a quick introduction to the history of the artifacts and the technological systems that Americans have been using to keep their food cooked and their bodies warm from the colonial period to the present.

The History of Home Heating and Cooking Systems in the United States

Over the course of the last 250 years Americans made a transition first from open hearths to cast iron stoves and then from cast iron stoves to coal-fired furnaces and then from coal-fired furnaces to gas, oil, or electric forms of central heating.[1] These transitions were more complicated than one might initially suspect because they involved changes in implements as well as changes in the nature of fuel and fuel-delivery systems. In the American colonies and in the United States before 1800, the vast majority of structures, whether domestic or public (such as churches, schools, and inns), were heated by wood, which was burned in open fireplaces; most cooking and baking was done in such a fireplace or its attendant ovens. Sometime between 1810 and 1840 cast iron stoves began to replace open hearths. This transition began in public buildings and (following much the same pattern that the personal computer is following today) subsequently spread to the domestic setting. Owners of older structures began to brick in their hearths and plumb stove pipes directly into their chimneys; new structures came to be built without hearths and mantels or, rather, with only decorative hearths and mantels.

During these years the cooking stove began to be differentiated from the heating stove. The open hearth had served both functions and, following suit, so had the earliest cast iron stoves; heating stoves were adapted to cooking by the addition of enclosed boxes adjacent to the firebox (for baking or for warming) and of potrings on the top surfaces (figure 1). Within a few years, however, the cooking stove and the heating stove began to evolve along separate pathways, the cooking stove becoming larger and taking on a characteristically stepped appearance (hence the

Figure 1
(a) An early form of the enclosed cast iron stove, intended for both cooking and heating, made in Pennsylvania in 1767. (b) The same type of stove depicted in use in a detail from *Birth and Baptismal Certificate of Margaret Munch*, by Carl Munch, 1826 (National Gallery of Art, gift of Edgar William and Bernice Chrysler Garbisch).

name "range") and the heating stove becoming smaller and more compact as the years progressed (figure 2).

Just as this differentiation was occurring, fuels were changing as well. Wood was becoming increasingly expensive, especially in urban areas, and consumers were beginning to substitute coal (which was becoming cheaper as the cost of transporting it fell). Bituminous coal, which is much easier to ignite than anthracite, became easier to acquire as the canals and then the railroads eased access to the areas west of the Appalachians, where it is abundant. On the western frontiers, particularly on the western plains, variant fuels, such as buffalo chips and corn husks, were used when coal was either unavailable or too expensive. The earliest cast iron stoves were sold with convertible grates to allow for the use of either wood or coal, but later in the century this was no longer standard practice, as coal increasingly displaced wood from shore to shore. By the end of the nineteenth century coal was the principal fuel used for household heating, and heating and cooking implements had been differentiated from each other.

Other variant implements and fuels appeared during the second half of the nineteenth century. Since early in the century inflammable gases had been manufactured in most American cities. Most of this gas was used either for lighting or for industrial purposes, but after the Civil War some manufacturers did develop and market domestic boilers that produced hot

Figure 2
The heating stove (a) and the cooking stove (b) have been differentiated (from an advertisement of the 1880s).

water for household consumption, using inflammable gases as the fuel. Petroleum had been discovered in Pennsylvania in 1859, and after the Civil War several entrepreneurs went into the business of refining and transporting it. The most important petroleum product in those years was kerosene (used principally for lighting), but some households purchased suitably constructed stoves in which kerosene could also be used for cooking and for heating. Some urban households (those that were exceedingly wealthy) acquired the advantages of central heating in those years; basement furnaces were installed (coal-burning) as well as peripheral implements (vents and registers, if it was a warm air system; pipes and radiators if it was steam or hot water) for distributing the heat throughout the dwelling.

During the first half of the twentieth century central heating diffused down the economic ladder, although the poorer half of the population did not benefit from its comforts until after World War II. During most of that time coal remained the dominant fuel for heating, although manufactured, and natural gas came to replace it for cooking. As gas supplanted coal for cooking, the form of the kitchen range altered, losing its characteristically stepped appearance (because the gas flame could be regulated and there was no longer any need to move pots and pans from one burner to another in order to vary cooking temperatures), becoming more compact, and acquiring a porcelainized finish and automatic thermostatic regulation of the oven. After World War II gas cooking lost place to electric cooking (especially in those suburban tracts to which gas service had not previously been provided); the electric range took on the features that the gas range had already acquired. In a similar fashion electric, gas, and oil systems (all of them automatic) came to displace coal systems for central heating, the electric systems being the ones with the most radically different morphology because they were not, strictly, central systems at all. By the last quarter of the century most Americans, whether they lived in privately owned homes or in multiple dwellings, whether they were rural, suburban, or urban, indeed whether they were rich or poor, enjoyed the benefits of automatic central heating and relatively clean, more or less automatic cooking; even the substantial increase in energy prices that occurred during the 1970s did not deter more than a tiny (and now decreasing) fraction of the population from the pursuit of this particular benefit of industrialization.

On the surface the transitions that I have just described seem rather straightforward; unfortunately, a number of historical puzzles appear as soon as one delves below the surface. The first of these puzzles concerns the diffusion pattern of the cast iron stove. I first began to understand the usefulness of the consumption junction as a locus of analysis when I tried to unravel this particular puzzle.

The Cast Iron Stove

The cast iron stove achieved popularity in the nineteenth century; by most indicators (newspaper advertisements, patent records, analysis of surviving structures, statistics on production, statistics on the number of firms entering the business), the transition from hearths to stoves can be said to have occurred, at least in the northeastern United States, by the outbreak of the Civil War in 1860 (Goldmann 1982). Yet stoves were not a nineteenth-century invention. Dutch settlers had built brick and tile stoves in New Amsterdam in the seventeenth century; German and Scandinavian settlers had brought cast iron stoves with them in the eighteenth century (Pierce 1951). Indeed, perhaps the single most famous stove of all, Benjamin Franklin's Pennsylvanian Fireplace, was invented and advertised in the 1740s (Franklin 1960 [1740]). The advantages of a stove over a hearth were numerous, and Franklin as well as others clearly recognized them. First, by controlling the passage of air over the burning fuel (usually wood), a stove permitted much greater fuel efficiency—a not insignificant matter for people who were cutting, hauling, and chopping their fuel for themselves or were paying others to do it for them. Second, the stove provided more comfort for the amount of fuel that it used because it could be placed in a central position in a room and heated by convection rather than by radiation. Third, because the fire in the stove was enclosed, a stove, although not without ash, was potentially cleaner than a fireplace. Given all these advantages and the fact that eighteenth-century iron producers, especially in Pennsylvania, were actually making small quantities of stove plates, how can we possibly explain the fact that the vast majority of nineteenth century Americans failed to adopt the stove? Why was the transition from the hearth to the stove delayed until the nineteenth century? What considerations, other than efficiency, comfort, and cleanliness (which we would value highly today), governed the behavior of ordinary householders in the eighteenth century?

Those few historians who have considered the history of the stove have dealt with this question in one of two ways: either by ignoring it or by appealing to the ethnic prejudices of the English segment of the American population. The open fire, it was said, was uniquely and symbolically English, and the foods that could be cooked on an open fire, also uniquely and symbolically English, particularly roast beef, could not be duplicated on the stove (which bakes instead of roasts). "Would our Revolutionary fathers have gone barefooted and bleeding over snows to defend air tight stoves and cooking ranges?" inquired Harriet Beecher Stowe (an ardent defender of English tradition) in 1864. The open fire, she averred, with "its roaring hilarious voice of invitation, its dancing tongue of flames . . . called to them through the snows of that dreadful winter to keep up their courage [and] made their hearts warm and bright with a thousand reflected memories" (Stowe 1864, p. 42).

Given the vast number of ethnic symbols that have fallen by the wayside on the North American continent, it seems a bit strange that this particular symbol remained potent even in the face of the increased comfort and lowered costs that could have been achieved if its power had been dissipated. Indeed, the history of the Franklin stove itself suggests that the ethnic explanation is wanting. The Pennsylvanian Fireplace (figure 3) was designed by Franklin as a compromise between the best aspects of the open hearth and the best aspects of the enclosed stove. It was a cast iron box with an open front constructed in such a way that the combustion gases released when wood was burned in front traveled through a chamber before passing out of the room; thus Franklin's fireplace provided some

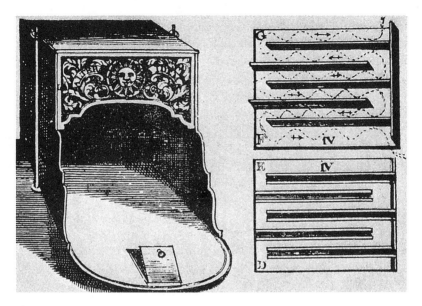

Figure 3
The original form of Benjamin Franklin's cast iron stove, the Pennsylvania Fireplace
(c. 1750).

heat through convection as well as through radiation from the fire itself, the sight of which its inventor called, "in itself a pleasant thing" (Franklin 1960 [1740], p. 432). Thus the Pennsylvanian Fireplace was meant to overcome the ethnic prejudices of the English segment of the population; yet it was a commercial failure. So few were purchased that, when Franklin wanted one sent to him in London, his agent in Philadelphia, Hugh Roberts, had to tell him that the furnace in which it was originally cast had stopped production (Bining 1938, p. 97). Something other than ethnic prejudice must have been at work.

To discover what that "something" might have been, one need only focus on the average American consumer of fireplaces in the mid-eighteenth century, asking both what interests such a consumer might have had and what sort of network might have existed at that time to bring a stove into a home. Figure 4 is an effort to draw a sketch of such a network. It distinguishes urban from rural consumers and reminds us that consumers, located in the household domain, were not producers of this particular commodity. In addition, urban consumers certainly and rural consumers possibly obtained such commodities through intermediaries who were located in the wholesale and the retail domains. A glance at figure 4 easily

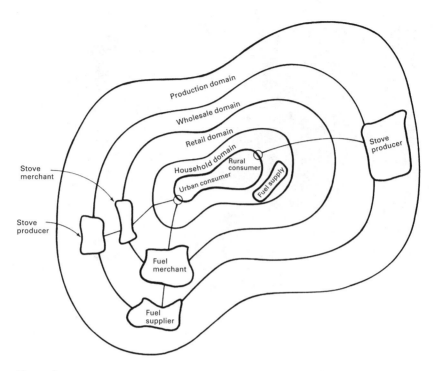

Figure 4
Network sketch of urban and rural consumers considering purchase of a stove c.
1760. The open circle represents the consumption junction.

suggests the possibility that features of the production, wholesale, and
retail domains could easily have affected the willingness of a consumer to
purchase a stove.

In the eighteenth century stoves were produced on iron plantations,
large tracts of land that were the site of the iron mines themselves, the
furnaces that were used for smelting and resmelting, the forges, the wood
lots that supplied the fuel for furnace and forge, the homes of the iron
workers, and the farms that supplied workers and their families with food
(Bining 1938; Paskoff 1983). Stove plates were cast directly from the initial
smelting of iron, from the molten iron, which, if cast into bars, would have
been called pig iron. Of necessity, iron plantations tended to be located far
from centers of settlement, and thus the manufacturer of stove plates was
bound to be located at a considerable distance from the potential purchasers of them. The cost of transportation would have driven up the cost of
the stove, and in the American colonies in the eighteenth century the cost

of transportation was high. Thus all iron goods, and stoves in particular, were expensive when sold at retail. "This stove," one of Franklin's contemporaries remarked about the Pennsylvanian Fireplace, "is an invaluable acquisition to the richer part of the world, but the poor can never enjoy it" (Bining 1938, p. 99).

Safety was another "interest" that, along with cost, may have been on the minds of eighteenth-century consumers when contemplating stoves, for the stoves were widely regarded, to use the language of the day, as "insalubrious." Under the heat generated in the firebox, cracks sometimes developed in the stove plates, and such a crack could (and apparently occasionally did) have fatal consequences for the inhabitants of a household. The continuous draft of air that is needed to feed an open hearth also served to drive the poisonous gases that resulted from combustion directly up the chimney, but a stove had no such draft. Open hearths were, of course, also dangerous, but their dangers were dangers with which people had coped for centuries; the risks of stoves were new and thus potentially more worrisome. In addition, many people believed that the drafts that accompanied open fires were beneficial to health, whereas the airtight stove made a room so dry and so stuffy as to be dangerous. Franklin's stove was intended to correct these defects (because the firebox was open), but it did so only at the cost of creating another defect: The open stove could not be used, as a closed stove could, for cooking. Thus the average eighteenth-century consumer, whether rural or urban, who was interested in purchasing a stove had to make a choice between one that was potentially dangerous as well as exceedingly expensive and one that was clearly not so dangerous but that would increase (rather than decrease, as its inventor claimed) fuel consumption in the household (because the kitchen hearth would still have to be fueled for cooking at the same time that the Pennsylvanian Fireplace was being fueled for heating). In 1796 the American Philosophical Society in Philadelphia organized a contest meant to encourage stove innovation, and one proposal sent to the society summed up the quandary faced by the average consumer at that time:

The close-iron stove was invented to save fuel by casting heat with the least possible loss. But it is far from solving the question in full latitude. By being unattended with free circulation of air it is insalubrious. It is also expensive and admits not of cooking. To remedy some of the inconveniences of the close-stove . . . Franklin invented his open-iron-stove. This by its openness gives a free circulation of air and by its projection into the room . . . causes but a little waste of heat produced by the fuel. But at the same time it increases the consumption of fuel beyond the close stove. It also is expensive and does not admit of cooking. ("An essay on warming rooms," 1796; reprinted in Bining (1938), p. 99)

The innovation that finally resolved these multifaceted problems was not a profound alteration in the design of the stove but rather an alteration in the structure of the industry that manufactured them, an alteration in the production domain. Sometime in the late 1820s a New York coal merchant and stove maker (someone who assembled stoves from plates cast at a furnace and then subsequently sold the stoves at retail), Jordan Mott, began to experiment with casting stove plates himself from pig iron, which he resmelted in a copula furnace using coal as fuel (Bishop 1966 [1868], p. 576). According to contemporary accounts, Mott became the first stove maker who actually made stoves instead of just assembling them. His innovation was important because it profoundly lowered the cost of stoves: A copula furnace was rather small, did not require great capital investment, and could be profitably located in or near an urban area because it burned coal rather than charcoal. The resmelted pig iron, as it turned out, was also less likely to crack under the heat generated in a stove firebox. "Mott's operation gained the attention of iron men and before the close of the year [1835] copula furnaces began to be erected and soon spread over the cities and villages of the Union" (Bishop 1966 [1868], p. 577).

A boom period for the manufacture of cast iron stoves began. Consumers who were emigrating west purchased them because they could be transported disassembled and then placed in operation quickly to provide more efficient heating and cooking, especially in locales that did not have abundant supplies of wood. Consumers who remained in the east began to purchase them and plumb them into their kitchen hearths with increasing frequency as the price of the appliance fell and the price of one fuel (wood) continued to rise. Stoves became even more attractive when, as the result of improvements in rail transport, coal became a lower-priced fuel than wood and bituminous coal became as easy to obtain as anthracite. Fully one-third of all the cast iron products reported in the *Census of Manufactures, 1860* were stoves; and most of those had been made, not on traditional iron plantations but by single-product stove manufacturers. Of such establishments, 220 were reported in the *Census*, which was the first to enumerate stove manufacturing as a branch of business separate from general iron founding (Temin 1964, p. 38). Mott's innovation was thus a part of the gradual process of differentiation in the iron industry itself; a group of businesses was created that specialized in manufacturing one product and serving only one kind of consumer, the householder.

Historians who have examined the history of the stove also have not looked at the history of the industry that was producing them (Giedion 1948; Wright 1964; Keep, n.d.; Pierce 1951), and historians who have examined the history of the industry (Temin 1964; Paskoff 1983) have almost never considered stove production as worthy of more than passing attention (railroad tracks, yes; stoves, no). Thus a perusal of the existing literature would not have led a scholar to the hypothesis that the history of the industry would have had such a significant impact on the history of the implement. I was initially able to see the value of combining these two perspectives by imagining a diagram such as the one illustrated in figure 4. This kind of consumer-focused diagram led me to evaluate the universe of choices that was open first to an eighteenth- and then to a nineteenth-century consumer, in terms of the consumption-production network in which such consumers might have been embedded. Who sold stoves? In 1760? In 1860? At what price? Were there wholesalers? How were stoves transported? How were they used on a daily basis in the home? Only after asking questions of this sort did it occur to me that the reason for the failure of the stove in the eighteenth century (and its subsequent success in the nineteenth) might have had something to do, not with the technical character of the implement, but with its price; only then did I come to realize that what was "better" in technical terms was not necessarily "better" in consumption terms.

Using Networks to Generate Hypotheses

Such an analysis—focused on the consumer, extending its causal reach into other socioeconomic realms, open to various criteria for "betterness"— seems to me to be essential in making sense out of the history not only of stove technology but of all technologies. All technologies have consumers, and all technological development is oriented toward a positive consumption decision, whether the ultimate consumers are located in the consumption domain (as householders are) or in some other domain (wholesale, retail, or production). Different kinds of consumers, making choices about artifacts other than stoves in time periods other than the ones already discussed, would no doubt need to be sketched in quite different ways, but the basic principles on which such a network diagram might be constructed need not alter. One needs to move from the consumer domain to the household domain and to evaluate the ways in which the special social and physical relations of the household might influence consumption choices; then one should move from the household to the retail domain;

then from the retail to the wholesale; and then from the wholesale to the production domain—all the while evaluating the pressures and the interests that may be affecting actors in each domain, understanding, in particular, that the criteria for "betterness" are different in each domain. One also needs to be aware that, although the network diagram may be focused on the individual consumer, consumers of different types can appear in almost any domain. These principles of diagram construction and analysis can be illustrated through two examples, again drawn from the history of home heating and cooking.

The gas cooking range did not diffuse quickly in the United States (Bacon 1942, pp. 75–80). Devices utilizing manufactured inflammable gases as fuel for cooking were on the market during the first half of the nineteenth century, and they became quite popular in England during the second half (Wright 1964; Ravetz 1968), but the gas range (which is, basically, not a terribly much more complicated device than a Bunsen burner) did not begin to supplant the coal cooking stove in the United States until roughly 1920. This was true, apparently, even in cities that were well supplied with manufactured gas. During the early decades of the twentieth century the American gas cooker also underwent several rather radical changes in design (Busch 1983), a sure sign that its market was expanding: Additional burners were provided (the earliest American models had only one or two); then ovens were added; then thermostats; then the surface was porcelainized; and finally the arrangement of burners, ovens, and broiling trays was altered so as to create the now familiar tabletop design. Why was the American consumer in 1918 (say) considerably more interested in purchasing a gas range than her or his forebears in 1870? And why were American consumers in 1870 considerably less interested in a gas cooker than their English cousins? A sketch such as the one illustrated in figure 5 can help at least to develop some hypotheses about how that question can he answered.

Figure 5 reminds us that the availability and the price of fuel may have something to do with the diffusion rates of implements that burn fuel; hence the difference between American and English acceptance of the gas cooker may have something to do with the spread of gas service in English cities and the price of coal in both nations. In addition, the sketch also reminds us that in the early decades of the twentieth century there were two easily distinguishable consumers on the market for cooking devices in the United States. The first type was the traditional individual consumer, located in the household domain, who was purchasing directly for use. The second type was what might be called multiple consumers—landlords

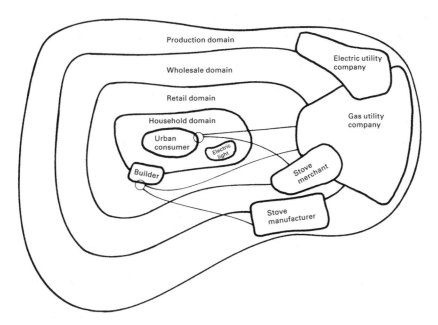

Figure 5
Network sketch for urban consumers considering purchase of a gas range c. 1920.
The open circle represents the consumption junction.

and real estate developers—who were purchasing for tenants and home buyers. My diagram needs, therefore, to include at least two consumption junctions. At the retail level gas stoves were being sold principally by the organizations that manufactured gas, that is, by the gas utility companies, who themselves appear to have been buying directly from the same kind of single product stove manufacturers who were also, simultaneously, producing coal stoves (as well as kerosene and oil cookers). The existence of the gas utility companies at the retail level reminds us, however, that this was precisely the historic period (between 1880 and 1920) in which the American gas companies were losing a considerable part of their market to a new competitor, the electric companies (Hughes 1983). And thereby hangs my hypothesis.

Perhaps the gas cooking range became popular in the United States, eventually displacing the coal stove, in large part because the gas utilities, under pressure to salvage a declining business, developed various techniques to enlarge the market for their product: perhaps by selling the stoves at or below cost to individual consumers, perhaps by entering into wholesale arrangements with landlords and real estate developers, perhaps by creating advantageous rate schedules for householders who purchased gas ranges (and gas water heaters). All these techniques were tried by at least one major utility company, about which a history has been written (F. Collins 1934). The fact that the trade association of gas companies, the American Gas Association, is known to have sponsored developmental work on a thermostat for gas ranges, to have marketed gas ranges under its own name (AGA), and to have developed cooperative advertising schemes with the manufacturers of gas ranges makes this hypothesis even more likely (Busch 1983). At the least this hypothesis suggests a strategic research site at which the puzzle of the delayed diffusion of the American gas range might be solved: not the patent records, as one might initially think, but rather the records of the gas utility companies and the publications that were addressed to them (as well as, perhaps, the publications that served the community of apartment house owners and real estate developers). By itself, figure 5 cannot solve the riddle of the gas cooker, but it can help us to generate plausible hypotheses that can be investigated, as well as plausible hypotheses that, because they deal with the interests of actors who operate in different social domains, are sociologically meaningful.

Figure 6 is another sketch, this one constructed in an effort to understand the shift from coal to gas, oil, and electricity (for home heating) that occurred, in the United States, within two decades after the end of World War II. In 1940, 55 percent of all the residences in the country were heated with coal or coke (77 percent of the central heating systems); by 1950, just five years after the end of hostilities, that figure had dropped to 35 percent, and by 1970 to 2 percent of the total (US Bureau of the Census, 1940, 1950, 1970). If one can rely on the accuracy of the consumer price indexes, the price of coal remained roughly on a par with the price of electricity, gas, and fuel oil during this period. What, then, can explain the flight from coal?

Figure 6 reminds us that there was a housing boom in the two decades immediately following the end of World War II, which means that, during the years in which coal was declining as the fuel of choice for home heating, a significant portion of the consumers who were on the market for home heating systems were multiple consumers—builders and developers. As with the gas range, this suggestion produces a research strategy that involves discovering the nature of the relationship between builders of tract housing and manufacturers of the presumably "more advanced" heating systems. But the sketch also reminds us of something more: Another significant domain must be added to figure 6 because some of the

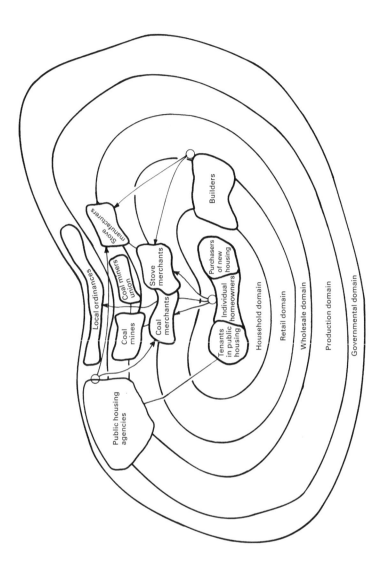

Figure 6

Network sketch for various consumers making decisions about coal-burning furnaces c. 1950. The open circle represents the consumption junction.

consumers on the market for home heating systems in the postwar years were neither individual nor multiple consumers but government agencies involved in the construction of public housing. Such governmental agencies may have had interests similar to those of the builders (the best heating system for the lowest cost), but they almost certainly had other interests as well: Some may have been required by local political conditions to opt for one fuel source rather than another (natural gas, for example, because the gas utility companies may have been politically powerful; or waste steam, for another example, because the municipality already owned electrical generating plants), and some may have been expressly prevented by local ordinance from using coal because of the threat of air pollution (Tarr and Lamperes 1980). In the production domain figure 6 also calls attention to a factor that might otherwise be overlooked: labor relations in coal mining. The Coal Miners Union, then under the leadership of John L. Lewis, was one of the most militant unions in the country, repeatedly calling its members out on strike between 1940 and 1960—and thus repeatedly reminding consumers (of all types) of the possibility of shortages in the supply of coal. Consumer preferences for heating systems that were automatic and clean no doubt also contributed to the rapid disappearance of coal stoves and furnaces in American homes, but figure 6 suggests the not unreasonable hypothesis that there were other factors in the decline of coal that are worthy of investigation.

Conclusion

In the preceding examples I have tried to explain why a focus on the consumer and on the network of relations in which the consumer is embedded can not only produce historical case studies that will be valuable grist for the mill of the sociology of technology but also generate hypotheses worthy of investigation. This form of analysis does not penetrate deeply into the processes of invention, innovation, development, and production (the initial stages in the evolution of a technological system), but it can open up the "black box" of diffusion (the final stage). Most artifacts have different forms (as well as different meanings) at each stage in the process that ends with use, so that an analysis that ignores the diffusion stage does so only at its peril. In any event, a consumer-focused analysis that deals properly with the diffusion stage can also shed important light on invention, innovation, development, and production. This last feature was demonstrated in the case of the cast iron stove; focusing on the consumer revealed that the diffusion of the stove was dependent not so much on the inventions that altered the form of the stove but rather on innovations that altered the pattern of its production.

Consumer-focused analysis also satisfies many of the criteria of social constructivist sociology outlined by Pinch and Bijker. Technological systems that eventually fail (such as the open hearth) are put on an equal footing with those that eventually succeed (such as the cast iron stove) because, when seen from the perspective of those who were faced with making consumption choices between them in any given historical period, the outcome of the historical process (success or failure) becomes irrelevant to the analysis (because the consumer could not possibly have known what the outcome would be). Careful attention to the network in which the consumer is embedded necessitates that attention be paid to various social groups (and their interests) who might not have been otherwise considered (because of their apparent distance from the process), for example, the possible importance of public housing agencies in the postwar flight from coal.

At the least, this form of analysis suggests research strategies with which to broach any given historical topic, strategies that stand a good chance of producing sociologically sophisticated results, if by "sophisticated" we mean moving smoothly and sensibly between several levels of social structure. This is why no one who had attended only to the historical literature on the evolution of the stove could possibly have succeeded in constructing an explanation for its popularity in the United States in the nineteenth century that would be satisfying to an inquisitive sociologist; to be admissible, such an explanation requires attention not just to the history of the artifact itself but also to economic history (the price of fuels), demographic history (growth of cities, western expansion), and industrial history (alterations in iron production).

Yet one more virtue, at least to my mind, of consumer-focused analysis is that it allows room for the one characteristic without which no sociological or historical explanation should be taken seriously: the "unintended consequence." By detailing the network of social relations in which a consumer is embedded, this form of analysis reminds us that different social groups, acting in what they perceive to be their own best interests, can, because they are embedded in a complex network, produce effects that may be quite different, perhaps even diametrically opposed, to what they intended. The striking coal miners, for example, certainly did not intend their actions to undermine the market for coal, but that may have been precisely what they ended up doing in the long run. Every perceptive historian understands that history is laden with unintended consequences—they are as inevitable as death and taxes—but few socio-logical models have been adept at generating explanations for them, especially not in the history and sociology of technology. Network analyses (no matter what they are focused on) have the potential for bridging the gap, often gaping, between historians who like to tell good stories and sociologists who would like to turn those stories into case studies.

Note

1. The material summarized in this section is derived from my book (1983, ch. 3, 4), as well as from the following works: Bacon (1942), Ferguson (1974a), Giedion (1948), Handlin (1980), Keep (1916), Peirce (1951), Strasser (1982), and Wright (1964).

Seeing with Sound: A Study of the Development of Medical Images

Edward Yoxen

This chapter is a contribution to a growing debate on the sociology of technological innovation. It is organized around a case study that describes some of the stages in the origination of a medical technology, the ultrasound scanner. In particular, I discuss differing ways of trying to generate two-dimensional images using high-frequency sound waves in the 1940s and 1950s. Some of these attempts were productive, and some were unsuccessful, in the sense that no fully functional equipment that could perform to the standards of diagnostic accuracy required was produced by the group concerned. However, to write of success and failure in this way only poses the problem to be explained. That is to say, the variation in design objectives, development strategy, institutional and professional background, and tenacity in pursuing the research objectives forces us, first, to ask what it means when one says that a particular technological artifact "works" and, second, to try to explain both success and failure without taking the definition of either as self-evident. As Pinch and Bijker (1984) have argued, one of the lessons of the sociology of scientific knowledge for the sociology of technology is that the selection of technological forms early in the innovation cycle and the "stabilization" of such forms are social phenomena requiring explanation, analogous to the formation of a consensus in science. Questions of inventive success and failure can be made sense of only by reference to the purposes of the people concerned.

In the case of clinical ultrasound we may be tempted to think of doctors and engineers as being the only people whose perceptions of innovative success would be important. If a piece of equipment generates information of demonstrable diagnostic value, then we could say that it "works." However, those asked by doctors to operate the equipment may find it burdensome or demeaning to do so, and their enthusiasm for and evaluation of the technology will differ. Thus in some UK hospitals midwives resist the idea that they should perform ultrasound scans because they consider that it devalues their own clinical skills. Radiographers, on the other hand, tend to be much more enthusiastic about the technology because their status derives from their competence in creating images, and ultrasound adds to the variety of tools available to them. But, although radiographers' acceptance played a part in the diffusion of this technology, there is little evidence of its playing any role in the invention of ultrasound.

Furthermore, the whole question of what the use of ultrasound in obstetrics achieves is now being questioned by some of those on whom it is used, namely, women who have been through programs of prenatal care in the hospital or in other clinical centers. Ultrasound images have considerable psychological power. Combined with the experience of being scanned, they can significantly affect feelings about a pregnancy, usually positively, although sometimes not (Hyde 1986). Doctors rely increasingly on scans during pregnancy, and the questions of what this reliance actually achieves and what comes with it are now being asked with increasing confidence by women's organizations (Association for Improvements in the Maternity Services, 1985).

A major theme in this paper concerns image generation as a process. The role of visual images in science has recently begun to receive more attention as part of a more general concern with conventions of representation and procedures and rhetorical devices for building a consensus (Rudwick 1976; Shapin 1979, 1984; Latour 1983, 1986; Lynch 1985a, b; Yoxen 1986). In this paper I emphasize the search for conventions of representation by means of which three-dimensional forms can be rendered as two-dimensional images. Lynch (1985b) has argued that scientific images are a special kind of construction, arrived at by schematizing, geometrizing, and highlighting salient features from a mass of detail. Nature is *rendered* in ways that accentuate certain features of interest. It is imaged, in a transitive sense, by operations on it. For such images to be relied on as evidence, there must be general agreement as to their value and reliability, and there must exist a set of procedures for generating them. In this case I consider the formation of a medical consensus that ultrasound images have value. Interestingly, Lynch suggests that images tend to become more schematic and more abstractly geometric as representational technique evolves. With ultrasound this is not apparent; rather the emphasis seems to have been on increasing clarity and on improving the depiction of form. Even after fifteen years of development, when the scanning machines could be said to work, there remained considerable skepticism as to the comparative value of these images. What then began was a process of enrollment of doctors, in Callon's sense, developing the view that their clinical judgments could be strengthened by a reliance on images generated through the work of others (Callon, this volume). I do not consider the mechanics of enrollment here, although it is a question fundamental to the evolution of many medical technologies.

Ultrasound in Medical Diagnosis

The physics of ultra-high-frequency sound was discussed by Lord Rayleigh in his textbook on sound (Rayleigh 1877–1878). In the aftermath of the Titanic disaster, a British patent had been taken out by E. G. Richardson on the use of ultrasound to detect icebergs at sea. With the appearance of submarines in naval warfare, attempts were made, particularly by Langevin in France, to use ultrasound to detect them. The problem was to generate sufficient power. By 1917 Langevin had succeeded, establishing the foundations of *so*und *na*vigation and *r*anging (sonar; Hackmann 1985).

In 1928 the Soviet physicist S. Y. Sokolov suggested that ultrasound could be used to reveal discontinuities in metals, for example, a flaw in a welded seam. It was 1940 before the physicist Floyd A. Firestone, at the University of Michigan, developed and patented what he called a super-sonic reflectoscope. This work was done with funds from General Motors and Sperry Rand. The device was actually produced in 1940 and patented in 1942, but it was not publicized until 1944; papers describing it appeared after the war, when it became apparent that Desch and his colleagues in Great Britain had independently developed similar equipment (Firestone 1945).

In the 1920s and 1930s the predominant medical interest in ultrasound was in its possible therapeutic and destructive effects; in other words, the energy rather than the information in the sound wave was used. Two German scientists, Gohr and Wedekind, published a review in the *Klinische Wochenschrift* in 1940 that sums up the general view. They discuss the use of ultrasound in rehabilitative medicine, the production of novel thera-peutic compounds, and the removal of dust from factory air. But they say nothing about imaging, although they do mention that, in studying the effects of ultrasound on tissues, fluids, and animals, they had tried measur-ing the amount of energy transmitted through an animal (Gohr and Wedekind 1940).

But it was the idea of a *pattern* of differentially attenuated ultrasound that occurred to Karl Theo and Friedrich Dussik, Austrian brothers, one a neurologist and the other a physicist, in the late 1930s. In other words, differences in the amount of energy transmitted through an organ could be used to create a pattern that would represent the form of that organ. This led the Dussiks to make the first claim that ultrasound could be used diagnostically.

Karl Theo Dussik was born in 1908 and trained at the University of Vienna. From 1938 to 1940 he was head of the neurology department at the Allgemeine Polyklinik in Vienna. In 1937 he obtained the first pictures of the patterned attenuation of an ultrasound beam passing through the skull at a series of positions. The intention was to discover abnormalities in the shape of the ventricles in the brain without using x-rays and dyes. The energy of the transmitted beam could be registered on a photographic plate, so that as the transmitter was moved across the skull a pattern of dark and light patches was built up on the plate. Although the process took far longer than the scanning to produce a television picture and although the image was not displayed on a cathode-ray screen, there is some similarity here with the technology of television, on which Friedrich Dussik had worked. Their apparatus was rather rudimentary, but it was based on the idea of scanning an object to build up a representation of it. They worked with great difficulty under wartime conditions and in the turmoil of the early postwar period. Karl Dussik published a paper in 1942 and another in the *Wiener Medizinische Wochenschrift* in 1947 (Dussik et al. 1947). In 1949 he established his own private clinic at Bad Ischl.

In the 1947 paper, Karl Dussik also mentioned the possibility of using ultrasound for surgical purposes, by focusing beams from different directions onto tumors in order to destroy them. This idea was later developed in the United States by William J. Fry. Essentially, though, Dussik seems to have regarded what he called hyperphonography as a research and diagnostic technique in neurology that did not require the introduction of air or of radio-opaque dyes into the brain and that was thus likely to be safer than the existing procedure using x-rays.

Mapping the Ventricular Space

The intention was to create a two-dimensional representation of the shape of the fluid-filled ventricles in the brain, because abnormality of this space would suggest abnormal growth of brain tissue and the presence of a tumor. In effect, it was a way of mapping interior surfaces of the brain. The Dussiks were trying to develop a particular kind of representation. It is important to recognize the principles and graphic conventions involved.

To some extent the representational conventions resembled those of radiography. But there were also differences. Rather than registering the outline of a tumor as a shadow, indicating the presence of opaque tissue, the hyperphonogram purported to reveal distortions of the interior surface of a fluid-filled space. Such structural changes were to be deduced from deviations from normal values of the transverse dimensions of that space. This dimension was measured by registering the smaller absorption of energy when the ultrasound passed through the ventricular fluid rather than simply through brain tissue. The energy absorption was detected by the shading of heat-sensitive paper. The darker the patch, the less the attenuation and thus the bigger the ventricle at that point. Shade stood for transverse cross section, and the shaded patches were added together by the eye to form a shape roughly similar to the longitudinal section of the ventricle. Looking at the hyperphonogram one could "see" the ventricle in three dimensions, even though the dark patches in the image had extension only in two. That at least was the theory. The value of the method rested on the accuracy with which one could delineate the shape of the ventricles and compare them with what was believed to be the norm. An example is shown in figure 1.

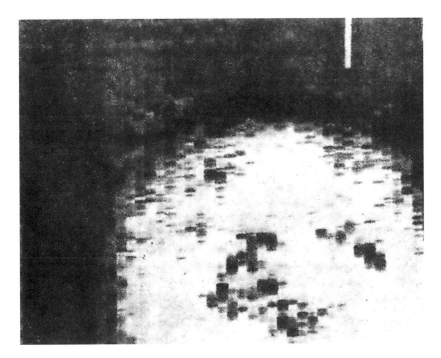

Figure 1
A hyperphonogram of a normal cranium. From Dussik et al. (1947).

Uptake of Dussik's Work

Karl Dussik's work came to the attention of the American physicist, Richard Bolt, who was director of the Acoustics Laboratory at MIT from 1946 to 1957. In those years Bolt built up the laboratory into a major center of research in acoustics, much of it financed by the US Navy. By 1951 there were thirty-one active projects under way. A breakdown of funding from 1945 to 1952 shows that some $1,757,000 had been received over this period, of which about 83 percent had been supplied by the navy, about 8 percent by the air force, and the remaining 9 percent by some nine different sources. The number of faculty members was two or three from 1945 to 1950, expanding to ten in 1951. There were eleven research assistants in 1946, twenty-two in 1951. The Acoustics Laboratory was then a rapidly growing center, with a strong interest in underwater sound and architectural acoustics. The medical acoustics project was just one of a number of projects, although it is the one with which Bolt was involved.

In April 1949 Bolt, his colleague Leo Beranek, and their collaborator at the Massachusetts General Hospital, the brain surgeon H. Thomas Ballantine, wrote to J. R. Killian, the president of MIT, seeking funds for their proposed research.

A group of officers from Headquarters European Command, USA, recently visited Dr. Dussik's hospital in Bad Ischl, Germany. Their report has been received and indicates that Dr. Dussik, a man of forty years of age, was on the staff of the University of Vienna until 1938 when he was displaced by the Nazis. He is a neurologist, at present on the staff of one of the Tyrolean universities and in charge of the Austrian State Neurological Institute. He developed his ultrasonic apparatus with the aid of his brother, who is a physicist. Dr. Dussik told the American medical officers that he fully realised his apparatus was crude, having been made in a local shop under rather adverse circumstances, but maintained that the principle is sound although the machine needed considerable refinement. (Bolt 1949)

They go on:

During the past year, another investigation into the possible use of ultrasound has been underway at the Naval Medical Research Institute, Bethesda, Maryland. This development has been undertaken by Dr. George Ludwig primarily as an aid to surgery of gall bladder disease. Briefly this project was an attempt to produce an instrument which by utilisation of apparatus quite similar to that employed by industry would give evidence as to the presence and location of gall stones. Dr. Ludwig has, with the aid of General Precision Laboratories, Pleasantville, New York, developed such an ultrasonic probe. He has been successful in localizing gall stones in dogs and this localization has been precise in a three-dimensional manner. He has also assembled basic data relating to the behaviour of high-frequency sound waves in human tissue. Dr. Ludwig will be on the staff of the Massachusetts General Hospital, July 1 1949. (Bolt 1949)

In October 1949 Bolt and Ballantine visited Dussik's clinic. Despite some doubts about his research strategy, they endorsed his work but concluded that it was unlikely to be refined as it needed to be, as Dussik himself acknowledged, under the conditions in Austria. Bolt and Ballantine were accompanied on this trip by a young German engineer, Theodore Hueter, whom they recruited to work at MIT.

By the beginning of 1950 the project was underway. One of the first tasks was to investigate the pain threshold and, using a continuous EEG recording, to determine whether ultrasound appeared to affect the functioning of the brain. At the power levels Bolt and Ballantine planned to use, there was no pain and no detectable change in EEG. At higher levels they were able to cause histological damage to the nerves of cats.

Testing the Representational Strategy

The course of the MIT—Massachusetts General work on ultrasound can be followed in some detail from the annual reports of the president of MIT, the quarterly reports of the Acoustic Laboratory, and the various publications that appeared. In the spring of 1950 a scanner was produced that more or less resembled that used by the Dussiks; it was used to examine fixed brain sections and two living subjects, one a member of the research team, the other a neurological patient with a brain tumor, whose heads were surrounded by a water bath. Almost immediately it became apparent that variations in skull thickness would be a problem because the acoustic signal was attenuated in passing through bone.

Nonetheless, papers were given at professional meetings that summer, and a paper appeared in *Science* in November 1950 (Ballantine et al. 1950). Also, Ludwig published an important paper on the velocity of ultrasound in various kinds of tissue and their "acoustic impedance" in the *Journal of the Acoustical Society of America* (Ludwig 1950).

The problem with the variation of bone thickness endured, and various ingenious attempts were made to get around it by varying the gain, comparing signals at different frequencies, and so on. One would imagine that highly trained electronic engineers used to processing noisy signals to extract information from them would not have been too daunted by this. At the beginning of 1951 the Massachusetts collaborators reported work on sound velocity in tumor tissues.

A value for sound velocity in this type [meningioma] of 1.54×10^5 cm/sec was obtained. This value is practically identical with the sound velocities in normal tissues reported by Ludwig. Since there is little change in the densities of various tissues, no change in sound impedance between normal brain tissue and this type of tumor tissue can be expected and no reflections would occur. (*Quarterly Progress Report of the Acoustic Laboratory*, 1951, p. 24)

This is an interesting passage because it suggests that they were thinking about specular reflections from tissue boundaries as a means of diagnosing tumors. As I describe later, this possibility was being studied by two other groups in the United States at around this time, and one question that suggests itself is why the group at MIT did not switch from transmission studies to reflection work when the former began to prove difficult. After all, radar, sonar, and flaw detection all work using echoes. Perhaps the experts on sound velocity in biological materials (Hueter and Ludwig) "knew" that reflection would not work with brain tumors? Another explanation might be their commitment to working on studies of the head, where to this day it has proved almost impossible to get useful information from the immense complexity of reflected ultrasound signals. Whatever the reason, the fact that they stuck to their original research strategy is something to be explained.

Over the next two years attempts were made to use the scanner in diagnosis at the Massachusetts General Hospital, but the results were largely inconclusive. In the April–June 1951 issue of the *Quarterly Report*, they published their decision to discontinue the research, concluding that they were unable to distinguish the information in the transmitted signal from the "noise" introduced by medically irrelevant variations in skull thickness.

When compared with x-rays, diagnostic ultrasound in neurology had failed its clinical test. Bolt and Ballantine also compared ultrasonograms created from the skull of a living patient and from an empty cranium immersed in a water bath. In this comparison, essentially the same picture was obtained, suggesting the pattern produced was better understood as a representation of the variations in bone thickness over the skull than as a representation of the form of the brain. The ultrasound methodology was then tested in two separate comparisons, and in each a reference image was used to set standards of reliability. But it was the reliability of the way of generating an image that was being put to the test, not the representational conventions as such.

Hueter and Ballantine went on to work on the surgical applications of higher-power ultrasound, but there are clear signs that the impetus had gone out of the medical acoustics initiative at MIT. Hueter left MIT in 1956 to work for the Submarine Signalling Division of the Raytheon Corporation, before moving on to work for Honeywell, of which he became a vice-president. That they should have abandoned their work to apply ultrasound in neurology is not surprising, but it does need fuller explanation. At first sight it is more puzzling that they did not consider other areas of medicine to develop diagnostic applications, as two other groups in the United States in the early 1950s had published papers indicating that ultrasound scanners could detect tumors in other parts of the body. Perhaps they felt restricted to neurological patients?

Wild and A-Mode Scanning

The origins of diagnostic ultrasound lie in neurology, where it proved difficult to use. While the work still continued at MIT, other researchers explored different ways of using ultrasound more in line with its engineering and naval uses, which employed reflection. This much more successful approach called forth another set of graphic conventions for displaying an image.

One research group was based in Minneapolis and centered on Dr. John Wild in the Department of Surgery at the University of Minnesota Medical School. He had qualified as a doctor in Great Britain in 1942 and took up this post in America in 1946. He began using an ultrasound generator that had been employed to train naval pilots to recognize the kinds of radar echoes that they would see flying over the islands of Japan. The generator worked at 15 MHz, an order of magnitude higher than the frequencies employed in flaw detection. This improved the resolution of ultrasound images but reduced the penetration through tissue. The ultrasound was transmitted in short pulses rather than continuously, as was the case with Dussik's equipment.

Wild's initial apparatus was simple in design. It consisted of a crystal mounted inside a water-filled chamber, at one end of which was a rubber membrane. The tissue was not scanned as such; the acoustic beam was sent into it, and the echoes from within the tissue were picked up and displayed on an oscilloscope screen. This gave an indication of the nature of the tissue from the kind of echoes that were returned. Wild compared his early method to a needle biopsy. Originally he had been interested in measuring the thickness of the intestinal wall, in the belief that an increase in thickness would represent cancerous tissue.

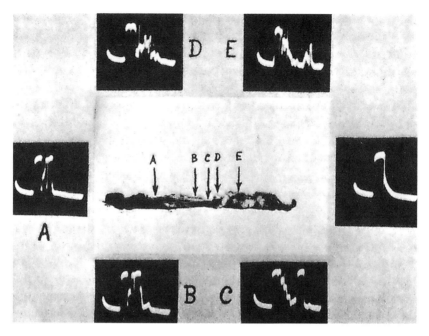

Figure 2
Results of experiments with a strip of human stomach containing a carcinomatous ulcer. The letters indicate the region of the specimen (shown in the center) being tested. The tracing to the right is of a control rubber membrane, from Wild (1950).

Wild's first paper, which describes experiments on tissue samples from dogs and human patients, shows photographs of the images obtained. One is shown in figure 2. Again there is a convention at work here. Vertical amplitude of the trace stands for the strength of the signal, and horizontal displacement across the screen is a measure of the time to detect the echo. The baseline interval is then a measure of the thickness for tissues of known density or a measure of density for tissues of constant thickness. Wild believed that tumors could be picked up as abnormally dense or abnormally thick regions. But he also noticed that tumor tissue was abnormally reflective, and he believed that this could be used in the early detection of tumors.

Whereas the graphic conventions underlying Dussik's approach bore some relation to those in a television camera—building up a two-dimensional image using points or patches of light arranged in lines—Wild's conventions were like those in metallurgy, in which changes in one dimension through a sample are rendered as a two-dimensional image.

This kind of analysis came to be known as A-mode. Although A-mode imaging continues to have a specific use in fetal cephalometry—and was also used in the 1950s in studies that sought to pick up reflections from the midline of the brain in order to see if it had been displaced by injury—it has given way to other more complex forms of representation. Its advantage is that it can be used to make accurate positional measurements, such as those involved in calculating the diameter of the fetal skull at its widest point, an indicator of fetal maturity. But at the same time, because measurement is restricted to one dimension, the technique can give no direct representation of form. In 1951 Wild and his collaborator, Donald Neal, an aeronautical engineer, published a paper in *The Lancet* describing work with postmortem material to detect brain tumors and on the diagnosis of breast tumors in living patients.

Howry and the Idea of Sector Scanning

Someone else who took up the diagnostic use of ultrasound was Dr. Douglass Howry. In 1948, while an intern at the Veterans' Administration Hospital in Denver, Colorado, Howry began a literature review and some theoretical studies on how ultrasound might be used. His first experiments were done in 1949, somewhat after Ludwig's work on gallstones at the Naval Research Institute in Bethesda. Howry used a reflectoscope of the kind that Firestone had invented. Although Howry and his collaborator, Rod Bliss, an electronics engineer, found that they could get echoes from tissue interfaces and foreign bodies embedded in tissues, the results were too erratic to be useful. Thus they built an oscillator that would work at higher frequencies and a system for scanning a tissue specimen by moving the acoustic beam through an arc. They called this equipment a Somascope. In other words the beam was swept through a narrow angle, creating a two-dimensional pattern of echoes. By timing the echoes electronically, they could build up a representation of a slice through an insonated object. The first results were obtained in 1950.

In their first paper, published in 1952, they showed pictures of the oscilloscope patterns created using a series of objects immersed in water: first, a water-filled condom containing a glass rod; then a normal gall bladder and one with gallstones; a slice of liver with a match, a nail, and a plastic rod stuck in it; and, finally, an arm held in a water bath. They claimed that the reflections from the radius, the ulna, the extensor tendons, the muscle-fat junction and the skin surface could be seen on their "somagram." The pictures they published were taken in the spring and summer of 1951.

More particularly they claimed that their equipment was capable of revealing details of soft tissue of a kind that was completely unavailable from an x-ray. They felt confident enough to circulate their results widely and to send them to the National Research Council for review in August and September 1951 (Howry and Bliss 1952).

Howry's apparatus created a rather different kind of image from those generated by the MIT group and by Wild and Neal in Minneapolis. In effect it is A-scanning, repeated many times by sweeping the ultrasound beam through an arc. The geometry of graphical space is polar rather than Cartesian. What appears on the screen is a series of bright spots, each of which registers a reflection from a discontinuity. As the beam moves through the arc, these points on the screen merge into a pattern that appears as if it were a slice through the insonated object. The partial rotation of the beam and the electronic recording of the echoes as spots of light thus "renders," in Lynch's sense, the internal two-dimensional structure of an organ or a limb or a test object in a given plane. The resulting image is certainly not artifactual. It registers features, like the fat-muscle interface, that really exist. Yet it picks out only those features that reflect ultrasound.

An example is shown in figure 3, along with a drawing of a transverse section through the mid-forearm showing the two bones of the arm, various tendons, membranes, and arteries, the subcutaneous fat, and the skin surface, which serves as a reference image. In Howry's 1952 article the two images were juxtaposed, and the reader was tacitly invited to compare the position of bright patches on the ultrasound image with elements in the anatomical drawing. Anomalous bright patches are also identified as reference reflectors held behind the arm. Furthermore, the lack of detail in the lower half of the image is explained as the result of absorption of the ultrasound by the bones of the arm. Within the image there are areas in acoustic shadow.

With these directions the image in figure 3 is now fairly easy to read. However, if one tries to compare it with the exterior form of an object, then it is in fact much harder to work out what one would be seeing at any point on a slice through that object. There is both a cognitive and a perceptual barrier to be surmounted. One needs to know what should be where, and one needs to be able to visualize how any given slice will appear. Thus the ultrasound images of normal and calcified gall bladders in the same article, for which the reference image is an exterior view of each object, are much harder to decode. Moreover, this is precisely the perceptual transformation that has to be made with obstetric ultrasound scans.

TRANSVERSE SECTION THROUGH MID-FOREARM

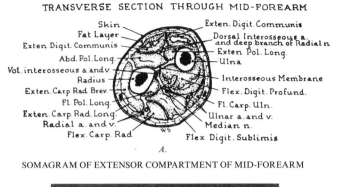

SOMAGRAM OF EXTENSOR COMPARTMENT OF MID-FOREARM

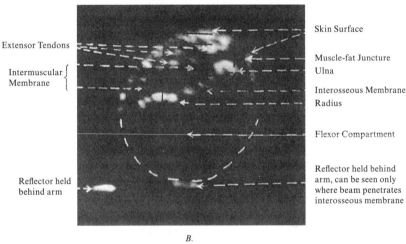

B.

Figure 3
A sample somagram From Howry and Bliss (1952).

This point about ease of visualization was made in a report for the US Atomic Energy Commission, published in 1955, although the contract work on which it was based was done in 1952 through 1953.

Probably the most difficult aspect of the problem is to interpret the data yielded by the echo-ranging system. The familiar "A-scope" presentation, which is a plot of reflected sound amplitude versus distance along the sound beam, is quite seriously limited in its information. It is necessary to integrate or scan a whole series of probe positions in order to consolidate the data so that any analysis can be performed on it. The reflection patterns are so complex that only a comparative analysis against a normal pattern can be interpreted. The problem is so difficult because we are not used to mentally interpreting data which shows not only the exterior details, but all the internal configurations of an opaque solid presented in three dimensions.

An analogy would be to observe a series of cross-sectional x-ray planigrams of an opaque object oriented such that the original spatial references are retained. (USAEC, 1955, p. 23)

The report goes on to consider various other display systems that would be perceptually easier to deal with but more technically complex to create, including Howry's sector scanning method. This section of the report concludes:

In conclusion therefore it must be realised that there is no prospect of adapting this tool to detecting intracranial lesions, for even if the skull did not have high absorption properties of its own, the distances involved would make the application foolish. The use of ultrasonics in this manner or anywhere on the body would have to be limited to very shallow penetrations. Such an instrument must have an intregrating scanning system and reduced aperture transducers. (USAEC, 1955, p. 45)

No author is given for the report. Its unenthusiastic attitude toward the neurological work is said to have inhibited research in the United States in other areas as well. Interestingly, the first section deals with imaging methods based on isotopic labeling of blood proteins. These are favorably reviewed, and work at the Department of Surgery at the University of Minnesota, although not by Wild, is used to illustrate what is possible.

Both Howry and Wild began to refine their equipment and to extend its versatility. Both groups eventually obtained funding from the NIH. For the Denver group the geometric complexity of the scanning was increased to remove some of the artifacts from the image and to improve its definition. One of the problems they discovered was that significant reflection occurs only from an interface when the beam crosses it at right angles or close to it. Rotating the transducer around the object being examined was one way of tackling that problem. But to make this possible, subjects were immersed in water while the scanner ran on a circular track around them. One version of their equipment was made from a B-29 gun turret, filled with water in which the subject sat with a lead weight around the waist to prevent him or her from floating to the surface. This arrangement could be used to scan the liver or other internal organs. It gave remarkably clear pictures but was obviously unsuitable for sick patients. The next development was a scanner with a semicircular water bath, the rotating transducer head being immersed in the water with the patient sitting against a window in the flat surface of the bath, out of the water. This kind of equipment gave clinically useful images, and the impetus in Denver to develop new forms lessened through the 1950s, even though the equipment was rather bulky and not at all portable (Holmes 1980).

By 1955, then, the experience with diagnostic ultrasound was mixed. Some people had found it hard to use. By concentrating on soft tissues and internal organs not surrounded by bone, some researchers had found that the complexity of echoes could be simplified by modifying the equipment to complexify the scanning. The resulting transformations of the internal structures of insolated objects were then somewhat simpler to comprehend. But this perceptual assistance was bought at the price of mechancial or electronic complexity. But usable systems could be engineered and results obtained that were at least promising, if one chose to spend the time building the equipment to generate them and had the confidence to decode them.

This is not an exhaustive account of all the work in ultrasound in the late 1940s and early 1950s. More detail is available from other secondary sources (Wells 1978; Hill 1973; Holmes 1980; Oakley 1984). I have deliberately ignored other work in the United States, France, Sweden, and the United Kingdom for the sake of brevity. I am not able to consider the significance of work in cardiology, ophthalmology, and physical medicine for the development of imaging technologies. But it should already be clear that there was a diversity of approaches to the technical and representational problems of using sound to visualize internal organs. Different groups pursued different strategies and explored the utility of different graphic conventions, even though a common aim was improved diagnosis. What seemed an acceptable engineering solution was somewhat variable, although in each case the basic challenge to be faced was the validation of the resulting image through some sort of visual comparison. For the Denver group, by the mid 1950s they believed that they had what one could call in the idiom of the new sociology of technology a "stable" artifact that generated usable images and was not thought to need radical design modifications. The quality and reliability of the images was seen as a major problem, however, and attention eventually turned to the question of how to standardize the equipment with test objects (Holmes 1967). In Minneapolis the apparatus was much more compact, but there remained considerable skepticism as to whether it was capable of the kind of diagnostic sensitivity that Wild claimed for it. Ultimately this difference of judgment led to the termination of Wild's research funding and massive litigation between him and the agency managing his research.

As far as I can discover, although electronics and engineering firms were involved with this research in various ways, funding some work, lending equipment, and following the results, there were no actual plans to develop and market diagnostic as opposed to therapeutic equipment in the mid-

1950s. By the end of the decade General Precision had an interest in ultrasound equipment in ophthalmology, and Smith Kline marketed in 1964 an echocardiograph based on work done outside the United States ten years before. By the late 1950s the Glasgow firm of Kelvin-Hughes, who made weld-flaw detection equipment, was investing in the development of a diagnostic scanner. Their Diasonograph was put on the market in the early 1960s. Aspects of the development of their product are described in what follows, but against the background of difficulties in application their judgment that this technology was worth a significant investment over a number of years is quite remarkable. A convincing explanation of it requires more research. What is interesting about it is that Kelvin-Hughes's interest came in two stages: initial low-level assistance and then a much more substantial commitment to development of a product. The move from the first to the second stage was accompanied by a major design change to create a new mode of representation, and the commitment to market a scanner went ahead despite a low level of interest from doctors.

Ultrasound in Obstetrics: Early Interpretative Labor

The next set of developments arose through the work of Ian Donald, Regius Professor of Midwifery in Glasgow from 1954 to 1976. Donald had qualified as a doctor before the war and had served in the Royal Air Force from 1942 to 1946. He met John Wild in London in the early 1950s and was stimulated to try to use ultrasound himself. When he obtained his chair in 1954, his opportunities outside the rather conservative ethos of medicine in London and his power as a Regius Professor in Scotland increased and he began to consider how he could realize his plans. He had a patient whose husband was a director of the engineering firm of Babcock and Wilcox, and through this contact he obtained an ultrasound flaw detector. This led him to the manufacturer Kelvin-Hughes, and Donald's first experiments were done on their premises in 1955, using the equipment on an ovarian cyst, fibroids, and a large piece of steak. Their initial results, recorded by the company artist, were promising. As Wild had found, different tissues could be distinguished ultrasonically. The production of recognizable images in the hospital turned out to be much more difficult, and it was another three years before Donald, his junior colleague, John MacVicar, and an engineer, Tom Brown, published a paper on their work in *The Lancet* (Donald et al. 1958).

Several themes stand out in Donald's recent accounts of this development, of which there have been several (Donald 1969, 1974a, 1974b, 1976,

1980). First, the problems of producing clinically useful images in the 1950s were considerable, and many hours were spent in modifying the equipment, changing the procedures, and trying new approaches. One could not see this as a marginal activity occasionally taken up nor as yet a routine task that could be entrusted to a subordinate. Equally, the facts that funds were made available by the Scottish Hospitals Endowment Trust and that Brown was seconded from Kelvin-Hughes indicate that ultrasound was also taken seriously by others as something that would pay off. When Smiths Industries withdrew, having bought up Kelvin-Hughes, the National Research Development Corporation also took an interest. Second, moving into the clinical setting was enormously helpful, although it exposed women and their babies to a highly experimental research technology. Certainly Donald saw their early rejection of postmortem material and laboratory experimentation as decisive (Donald 1969).

Third, this work was undertaken knowing that others had published promising results. Even if some of these earlier claims looked rather dubious, nonetheless it must have been reassuring to know that the technique could be made to work. Donald maintains that the principal stimulus for him came from Wild and that initially he was unaware of most of the early work. After making contact with Kelvin-Hughes, he went to visit Professor Mayneord at the Royal Marsden Hospital in London; Mayneord was using some of Kelvin-Hughes's equipment to extend Dussik's work. When Mayneord gave up, Donald was lent that equipment, through the intervention of Dr. Donald Gordon, a radiologist interested in ultrasound himself (Gordon 1959). There were therefore several individuals in Great Britain able to give Donald some initial practical support.

Gradually the group in Glasgow increased their skill at generating and interpreting A-mode images. Interestingly one of the crucial cases, as it came to seem afterward, involved a disagreement between MacVicar and Donald over what the image meant. Clinical and visual judgments were inconsistent. Subsequent surgery showed that MacVicar was in fact correct and that the prior diagnosis of a tumor, which Donald had been asked to confirm, was wrong.

Compound Contact Scanning: The Most Complex Graphic Space

By 1957 the Glasgow group came to the conclusion that more complex forms of representation were necessary. They set about designing equipment in which the probe was in contact with the surface of the abdomen while being moved across its surface. Their early A-mode work had used a

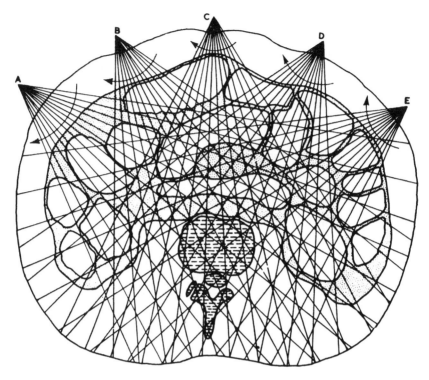

Figure 4
Diagram illustrating the principle of compound sector scanning. Here A through E
are the origins of typical sector scans. From Donald and Brown (1961), who repro-
duced it from *Medical Electronics*.

small water column to "couple" the ultrasound transducer to the patient's
body, which had been difficult to use in practice and produced many wet
beds. The problem was somehow to allow movement across the abdomen—
or longitudinally down it—and to use the sector-scanning technique
employed by Douglass Howry without immersing the patient in water. The
solution was ingenious and created an exceedingly complex graphic space,
which is shown in figure 4. This figure is taken from an article for electronics engineers (Brown 1960).

The resulting oscilloscope image is thus the integration of a whole series
of sector scans, which are themselves summations of axial probes into the
body. By 1958 Donald and his colleagues felt that they had enough material to publish a paper in *The Lancet*. This paper describes 275 scans on 100
female patients at the Western Infirmary in Glasgow. They show scans of

the bladder, the femur, ovarian cysts of various kinds, a normal abdomen (MacVicar's in fact), an abdomen distended by ascites, the fetal skull, showing reflections from the midline of the brain, hydramnios, and two twins in the breech position. They then cover a range of obstetric complications and related clinical conditions. The original purpose of the paper was the differential diagnosis of abdominal distension and the exclusion of malignant disease. The concentration on pregnancy came only as the work progressed, and interest in fetal abnormality and rates of growth, with which obstetric ultrasound is now often associated, came even later.

They concluded:

Our experience of 78 cases in which diagnosis was quickly verified by laparotomy and subsequent histology indicates that ultrasonic diagnosis is still very crude, and that preoperative diagnosis of histological structure is still far off, although such a possibility in the future is an exciting prospect. The fact that recordable echoes can be obtained at all has both surprised and encouraged us, but our findings are still of more academic interest than practical importance, and we do not feel that our clinical judgement should be influenced by our ultrasonic findings. Our most spectacular results have been obtained in dealing with fluid-filled cavities, which certainly show up well; but it is only fair to point out that the illustrations shown herewith are among the very best that we have so far been able to produce out of about 450. They do however encourage great efforts to refine our technique. (Donald et al. 1958)

Great efforts were needed, for it was three years before Donald and his co-workers published another paper. Although two papers on obstetric and gynecological ultrasound in 1959 by other workers and one in 1960 had appeared, it was five years before others began to publish papers replicating Donald's work (Taylor et al. 1964). Certainly one cannot say that Donald's was a technique that others immediately sought to copy, even though there was considerable interest in the *Lancet* paper. Comparative data from the *World Bibliography of Ultrasound,* which is supposed to be an exhaustive compilation of all papers on all aspects of the subject published this century, shows that publications on the use of ultrasound in obstetrics lagged those relating to its use in other fields, such as cardiology and ophthalmology, although obstetric publications are now vastly more numerous.

Conclusions

What implications does this case study have for the sociology of technological innovation? Is it consistent with a social constructivist view of technology? I believe it is, in the sense that the process of development described here is one in which different strategies were pursued with mixed results; at various stages in the work of each group, significant appraisals of overall process took place, and the decisions to continue as before, to diversify, or to abandon the work were made by reference to a whole set of factors. One could not say that some intrinsic "best solution" guided the success of particular groups or that some set of intrinsic difficulties can explain the failure of those who pulled out. It is, of course, the case that those working in neurology had to cope with a complexity of reflections and other problems that did not exist in studies of other parts of the body. But that is not in itself a sufficient explanation of the particular moves that they made.

It would also be difficult, and I believe rather perverse, to see this technology as emerging through a unilinear process of development from some initial scientific breakthrough. Although one could make much more of developments in applied science, particularly those in electronics in the late 1930s and early 1940s that produced fast switching circuits that allowed transducers to be switched from sending to receiving rapidly, it is apparent even from the restricted data that I have presented here that each such technical breakthrough was mediated through all kinds of different applications. In other words, this case conforms to the model of overlapping and multilinear development, discussed by Bijker (this volume), with inputs from applied science at various stages and not just at the origin. Given the time frame that I have chosen, from the late 1930s to the late 1950s, this study is not like those of large technological systems discussed in this volume by Hughes, Constant, and MacKenzie. My focus has been much more on invention and early development to the prototype stage, rather than on the modification of elements within a mature system, in which control of the commercial or political environment may be of crucial importance.

This case study limits itself to the inventive activity of various medical men, assisted by applied scientists and engineers, around the time of the Second World War. It comes, then, after the scientization of medicine in the early years of the twentieth century, within what Reiser has called "the reign of technology" (Reiser 1978). All the doctors here clearly believed it appropriate and worthwhile to devote time to producing a diagnostic technology. Their full motivation I am not able to describe, but it is striking in the case of all but Dussik that the experience of the war and acquaintance with radar or sonar precipitated their interest.

One could certainly link this story to other analogical and practical borrowings from wartime work with radar that led to new research techniques, technologies, and models. The examples include Huxley and Hodgkin (nerve conduction), Porter (flash photolysis), and Ryle and Lovell (radio astronomy). My point is that it was not just a technical familiarity with electronics that was helpful; it was the idea of propagating a signal or a transient perturbation that could reveal characteristics of the system being analyzed that was of heuristic value. For ultrasound the ideas of reflection and image composition seem to have been as important as an appreciation of what the electronic equipment could and could not do.

This leads to consideration of the attributes common to the researchers involved. Bijker has proposed the notion of a technological frame in order to characterize commonality of approach or perception, without the more usual reference to specific disciplines or research traditions (Bijker, this volume). He also writes of the degree of inclusion within the frame as a determining variable in explanations of inventive strategy and performance. I myself am not persuaded that these terms are sufficiently precisely defined to move the debate forward, but the phenomenon to which they are directed is an interesting one, namely, the influence of professional socialization or technical facility on imagination and inventiveness. If one speaks of a shared commitment to the application of ultrasound in diagnostics as a technological frame, then it is apparent that in this case study the individuals concerned drew on their instititional and intellectual resources in markedly different ways. The groups that failed were those that stuck rigidly to an initial strategy: in Dussik's case with limited financial and technical resources, in the case of the MIT group with massive amounts of money and technical experience and support. The groups that were more successful tried a wide range of different approaches and built a range of different devices; they did not just modify one initial piece of equipment. If we treat inclusion as a synonym for rigidity and thus abandon its use as an explanatory concept, the more highly included individuals seem to have lacked the flexible puzzle-solving strategy of the more successful centers.

This is really a question of how different people defined the technical constraints to the use of ultrasound, given the goal of eventual clinical application. The events at MIT suggest that in technological development, as in more fundamental studies, there is no such thing as a crucial experiment, in the sense of a result that instantly terminates a line of research.

Thus the difficulties with variations in skull thickness were recognized by them, if not by Dussik, early on, and they tried a variety of ways of getting around them. There were other problems that they might have pursued, such as studies of reflections from the midline of the brain, or studies of other organs. They did not do so because they took a particular result as indicative of more general problems with the technology, a view not shared by researchers elsewhere. This study supports the view that technical judgments are rarely, if ever, dictated by experimental evidence alone and that what has to be explained is how the interpretative flexibility inherent in research is actually used.

Finally one could group the various devices produced in Minneapolis, Denver, Cambridge, Glasgow, and Bad Ischl and elsewhere in Sweden, France, Germany, and Japan, into some kind of evolutionary tree. If one chooses to construct a tree, there is a kind of trajectory or pattern of relation and descent from bulky lab equipment closely modeled on or built with industrial flaw detectors with awkward systems for "coupling" the transducer to the patient to the machines with solid-state circuitry and much more complex projective geometry. One can find closure of inventive cycles and stabilization of design thinking around particular forms. Clearly, these are phenomena within the innovation process that need to be explained, and, as MacKenzie's work shows, the variable that comes to be decisive is only one of several possible (MacKenzie, this volume). I am not going to comment on this for the case of ultrasound, except to say that my hypothesis is that ultrasound technology stabilized when doctors came to the view that the generation of images on which they could rely for making clinical decisions could be safely entrusted to others, such as radiographers and other subordinate personnel.

I began with a reference to the contemporary debates on the value of ultrasound and the way it is used in the hospital environment. I believe that such discussions have much to gain from taking account of the fact that scanning now is work. The job of the person performing the scan has a history. The tasks involved have been designed, negotiated, and defined in relation to the work of others and depend on the exercise of specific skills. Who has these skills and how they are valued by others has changed through time. Thus the experience of having an ultrasound scan depends on how various individuals are able to work, how they are intended to work, and how their constantly shifting relations with doctors are managed. Explanations of the stability of technologies must take account of the social relations of work as one aspect and, in the case of ultrasound, of the reliability of the images produced.

Acknowledgments

I am grateful to Beverly Hyde, John Churcher, Charles Weiner, Chris Lawrence, Trevor Pinch, and Wiebe Bijker for help of various kinds in the production of this chapter and also to the Archives Staff of the MIT Library and the University of Michigan. I am also grateful to the Royal Society and the Wellcome Trust for support of the research on which this chapter is based.

IV Technology and Beyond

Introduction

Much of the argument of this book has dealt in one way or another with the relationship between technology and society. Rather than studying the social impact of technology, the authors have been more concerned with showing how technology itself can be understood as a social product, or at least as possessing a social dimension. This entails a radical shift in how we conceive of technology and the innovation process more generally. The social aspects of technology do not start when a technological process or product is taken up by the wider society; rather they are always present.

This new image of technology can itself be expected to have an impact, particularly in the domain of technology policy (Wynne 1983). Understanding of the social (and for that matter the political and the economic) dimensions of technology may be crucial in understanding the success and failure of technologies in different contexts. In restricting the social, political, and economic dimensions of technology to the area of application, policymakers may have been misled. Clearly, the consequences of different models of technology for policy are an area that deserves further attention.

Apart from its possible implications for technology policy, the new understanding of technology that is evolving can potentially feed back into the development of technology itself. This is one of the issues that is considered by the last two papers in this collection. They both take as their topic artificial intelligence (AI), a field that promises "to found one of the key technologies of the later part of the twentieth century" (Collins, this volume). AI researchers, in attempting to provide an account of the working of knowledge-based systems, can be seen to be engaged in a similar task to that undertaken by historians and sociologists of technology. And that is the second issue at stake in this part: how the results of analyzing AI with the newly developed tools of the history and sociology of technology feed back into that research program and into the philosophical questions behind the new approaches to technology study presented in this book. The concern of AI is, of course, not merely to understand the basis of technical knowledge but to put that understanding to use by building "intelligent" machines that will be capable of carrying out human activities. Both Steve Woolgar and H. M. Collins take up the challenge of AI in their contributions, but in interestingly different ways.

Steve Woolgar extends the argument of an earlier paper on the divide between humans and machines (Woolgar 1985) and shows how this boundary is socially mediated. The debate over what counts as an intelligent machine and over the likely limitations of the AI research program can be seen to be a continuation of the long-running controversy over the uniqueness of mankind. In this controversy properties that define human capabilities as unique are, according to Woolgar, shifted such that, whatever tasks machines carry out, those tasks become defined as *not* part of mankind's uniqueness. The interpretative flexibility of the technology is demonstrated by this debate.

Woolgar takes up Callon's metaphor that technology, because it also involves sociology, can be used as a tool for sociological analysis. He proceeds to show how the methods of sociologists in their critique of AI and cognitivism construct a particular view of the human/machine relationship, which, in the end, imposes severe limitations on their critique. Woolgar takes the basis of the AI research program to be "cognitivism," namely, AI's attempts to explain behavior by reference to cognitive or mental states that can then be codified within some algorithm. The core of much of his argument is that any attempt to replace cognitivism by a systematic sociological understanding of human behavior makes little difference to the AI task because any such sociological understanding involves the delineation of human actions, thus rendering such actions capable of codification and hence of incorporation within the AI program.

Collins offers a model of knowledge transfer based on his sociological studies of how a particular technical task (building a TEA laser) is accomplished. Collins argues that the "enculturational model" of knowledge, in which knowledge is equated with culture and hence has an irretrievably social element, provides a better understanding of laser building than the "algorithmic model" in which all the technical actions to be followed can be completely specified by a set of algorithms capable of transfer to a digital computer. The tacit knowledge component of technical tasks in particular cannot be completely described by algorithms, and hence the building of expert systems or of knowledge-based systems in AI cannot proceed by the algorithmic model alone. Collins shows the different layers of knowledge embedded in technical tasks and draws attention to the possibilities whereby different types of knowledge can be rendered into forms suitable for AI purposes.

The message of Collins's paper is that the goal of AI of building knowledge-based systems designed to replace human skills can be accomplished only by taking into account the social nature of technical skills. In short, a complete sociological understanding of technology is necessary before AI researchers can write expert systems for the types of technical tasks encountered in technology.

Woolgar seems to be suggesting that the technology of AI itself (broadly conceived) is commensurate with or even part of certain sociological methodologies. Thus, by analyzing sociology, Woolgar can claim, at the same time, to be analyzing technology. Here we are back at the beginning of this book, where Callon argued that studying technology is, at the same time, studying sociology. Collins's paper is perhaps the most full-blooded application of the sociology of technology to the technology itself found in this book. Indeed, his sociological study was awarded a prize by AI researchers as a technical contribution to their field. Clearly, a separation between the sociology and the technology has no meaning for researchers developing the new technology. They know the lessons taught by Edison. Thus these two papers bring us back to our starting point: Woolgar by applying the "seamless web" *adagium* on a methodological level and Collins by demonstrating this metaphor as a participant, so to speak, in the engineers' as well as the sociologists' game. The circle is closed.

Reconstructing Man and Machine: A Note on Sociological Critiques of Cognitivism

Steve Woolgar

The social study of technology is currently undergoing an expansion and a transformation. One impetus for this development is the application of many of the ideas and approaches of the sociology of scientific knowledge to the study of technology. Thus we now find the same post-Kuhznian reevaluation of preconceptions about technology as that which occurred previously with respect to science. This reevaluation produces the following kinds of argument. Distinctions between the technical (scientific) and the social must be broken down. Social analysis should attend to the content of technology (scientific knowledge). Technological (scientific) growth can no longer be thought of as a linear accumulation of artifacts (facts), each extrapolated from an existing corpus of technological achievement (scientific knowledge). Technology, like science, involves process as well as product. In short, both scientific facts and technological artifacts are to be understood as social constructs.

Pinch and Bijker (1984, this volume) articulate these parallels between science and technology with specific reference to the work of the "empirical programme of relativism." They highlight the interpretative flexibility associated with the design and interpretation of technological artifacts. There is no unique way of designing (or interpreting) technology. Designs and interpretations vary across time and among different social groups. When competing views and ideas come into conflict, the upshot of the ensuing controversy is determined by various social contingencies. Pinch and Bijker's main concern is to map the passage of controversy and the formation of consensus to document the social processes whereby technological artifacts come into being and are accepted.

The approach adopted in this chapter is somewhat different. My aim is not so much to specify mechanisms of closure and consensus formation but to examine how technology provides an important tangible focus for continuing debates on the fundamental qualities of man. The particular example of technology discussed in this chapter is artificial intelligence (AI). My strategy is to focus on one side of the AI controversy, namely, recent sociological critiques of "cognitivism."[1] These critiques take issue with the model of human action presupposed by cognitivism. They also challenge the assumption that the achievements of AI have any bearing on our understanding of human action. The dispute between cognitivism and its sociological critics thus entails a struggle between alternative conceptions of man. However, an examination of sociological critiques of cognitivism also reveals the crucial role of these sociologists' commitment to particular *methods* for constructing the nature of man. I argue that the sociological commitment to certain procedures for representing the character of human action and behavior crucially underpins preconceptions about what man is. Despite the vigor of the sociological challenge, I suggest that the sociologists' commitment to particular modes of representation ultimately imposes severe limitations on the likely success of their attacks on cognitivism. First, we need to look more closely at the idea of technology as the focus for continuing debates about the essential qualities of "humanness."

Technology as the Mirror of Man

Discussions about technology—its capacity, what it can and cannot do, what it should and should *not do*—are the reverse side of the coin to debates on the capacity, ability, and moral entitlements of humans. Attempts to determine the characteristics of machines are simultaneously claims about the characteristics of nonmachines. Clearly, as Pinch and Bijker suggest (1984), technology lends itself to radically different interpretations. But the central idea here is that differences in interpretations of technology both express and give rise to competing preconceptions about the essential qualities of humans. In discussing and debating new technology, protagonists are reconstructing and redefining the concepts of man and machine and the similarity and difference between them. As well as providing a tangible focus for continuing debates about the uniqueness of man, technology may also act as a catalyst for changing conceptions of the nature of man. Assessments of the character and success of new technology can both reify existing assumptions and provide powerful images for further attempts to establish (construct) "the character of man."[2]

This approach has some affinities with Hughes (this volume) when he refers to "institutional structures that nurture and mirror the characteristics of the technical core of the (technological) system." If we allow that these institutional structures embody competing notions about human uniqueness, human intellectual capacities, and so on, it is perhaps not surprising that AI is so controversial. The exact technical characteristics of this particular technology are much disputed precisely because institutionalized notions about the character of man are at stake. The present approach also has some resonance with Callon's argument (this volume) that technology is a "tool for sociological analysis." A close examination of discussions of the character of technology—what it is, what it can and cannot do—shows that discussants themselves perform "societal analyses." In the present case "analyses" performed by discussants of AI constitute claims about man's basic character, human abilities and capacities. By following protagonists in the AI controversy, I follow the construction and reconstruction of different models of man's basic attributes.[3]

The controversy over AI is the latest chapter in the long-running debate about the uniqueness of man.[4] Throughout the history of human fascination with artificial devices, the substance of these debates has varied according to the perceived capacity of the technologies under discussion. For example, when technology predominantly was composed of prosthetic devices (functional additions to the mechanical abilities of the human body), it could always be argued that humans are unique by virtue of their intellectual faculties; no prosthetic device could emulate a human's ability to reason, know, and understand. The work of AI, however, attempts to develop a technology that emulates action and performance previously accredited to unique human intellectual abilities. Consequently, the advent of computers, and of AI in particular, has raised questions about the uniqueness of man in a slightly different form. For example, in some discussions, emotion is now invoked as the category of attributes that testify to man's uniqueness, just as intellect was invoked when the debate focused on prosthetic technologies.[5]

Cognitivism and the Possibility of Machine Intelligence

Cognitivism is the general doctrine that behavior can be explained by reference to cognitive or mental states. Cognitive science (theory) is the specific form of cognitivism that seeks to develop explanations of conduct in terms borrowed from computer science. Cognitive science thus deploys a model of human action that, although long entrenched in the Western human sciences, is now being reified in the form of new computational artifacts (Suchman 1985, p. 1). It claims both to support and to be reinforced by the effort of AI to design and build artificially intelligent machines.

A main aim of AI is the design and construction of machines to perform tasks assumed to be associated with some cognitive (sometimes intellectual) ability. This is not, however, a uniform position within AI. Some AI researchers flatly declare their disinterest in "cognitive abilities." For them the machine's performance of tasks is the sole technical goal of their research, independent of whether or not such task performance would require intelligence in a human. Nonetheless, both practitioners within AI and spokesmen on its behalf (philosophers, marketing entrepreneurs) explicitly see the raison d'être of AI as the attempt to design machine activity that mimics what they construe to be cognitive behavior. This latter position is referred to as the strong version of AI (Searle 1980).

The various diverse research efforts subsumed under the AI rubric make use of a wide variety of mental predicates. These include such behavior as reasoning, thinking, knowing, believing, deciding, seeing, learning, understanding, and problem solving. To the extent that the claims of (and on behalf of) AI articulate the cognitive character of machine performance, they depend on the basic proposition that these kinds of behavior can be explained in terms of cognition. In claiming the cognitive character of behavior, in describing a certain behavior as, for example, intelligent, one effectively construes a task as requiring intelligence for its performance. The performance of the task is thus presumed to derive from some cognitive state or ability. This is the basis for the computational theory of mind, which lies at the heart of cognitive science. Changes in cognitive states are said to be effected by various computational procedures, and the behavior (actions, the performance of tasks) is to be understood as the outcome of these computational procedures. Of course, the designation of actions as a "behavior" that can be explained in this way is itself contentious. As we will see, the relationship among terms such as "performance," "action," and "behavior" is at the heart of the dispute between cognitivism and its sociological critics.[6]

Critiques of Cognitivism

Coulter's (1983) discussion is by far the most comprehensive sociological critique of the foundations of cognitive theory yet available. Coulter argues that an overreaction to earlier behaviorist tendencies to reduce human conduct to habit (the construal of conduct as a series of trained responses to environmental inputs) has led cognitive theorists such as Fodor (1975) to oversimplify the meanings of relevant predicates of human conduct and to ignore the "complex logical grammar" (Coulter 1983, p. 9) of their use in language. For example, Coulter shows that, whereas a whole variety of actions is in practice formulated as "deciding," Fodor assumes that "deciding" is invariantly the upshot of a process of deliberation about a decision. Similarly, "understanding" is to be treated as a wholly inner occurrence even though, Coulter points out, there is no way ever to ratify claims to have understood independently specific instances of situated conduct. Instead of examining the formulation and attribution of activities in situated conduct, Fodor postulates an internalist conception of the meaning of mental predicates. Coulter thus complains that Fodor's theoretical program depends (at least in part) on a unilateral redescription and reformulation of ordinary concepts and predicates, both of conduct and of the putatively "mental" domain. More generally, Coulter's point is that, in order to secure a scientific basis for their investigation, writers such as Fodor do violence to the way in which cognitive concepts, such as deciding, understanding, and seeing, have currency in ordinary language.

Coulter similarly takes issue with the program for reducing human behavior to certain physical events. For example, Putnam (1960) postulates an equivalence between human behavior and such organic occurrences and emissions as the motions of bodies and the production of sound waves. But Coulter argues that human behavior comprises activities that are not reducible to phenomena explicable by physical or neurophysiological science. For example, the description of an arm movement as greeting a friend (or wafting away a fly or signaling someone to stop) is a description of personal conduct; the person is the originating agent, not the body. Conversely, descriptions of bodily behavior (for example, "his body was producing too much adrenalin") denote an absence of personal or human agency. According to Coulter, the mistake made by Fodor and Putnam is to confuse involuntary organic episodes with intentional human behavior and to treat the involuntary episodes as causally explicable of intentional behavior.

Although Coulter argues against psychophysiological determinism,[7] he does not challenge the physiological basis of human behavior: "No one wants to deny the obvious fact that our central nervous system processes have something to do with our speaking" (Coulter 1983, p. l5). For Coulter such processes merely provide the *possibility* of human (interpretative) behavior. They do not explain it. In Fodor's scheme (1975) the central nervous system (CNS) is responsible for translation between physiological processes and human behavior correlates. Thus the CNS undertakes encoding and decoding of acoustic waveforms and performs computational analyses of sentence structure in order to arrive at the meaning of the message received by the listener. Coulter notes that this scheme hinges on a unilateral redescription of what speakers hear in terms of waveforms. But having denuded the socially meaningful character of the activity—"stating," "claiming," "denying," and so on are all reduced to "producing sound waves of such and such a frequency"—we must then postulate some mechanism for "adding back the meaning." This mechanism, in Fodor's view, lurks in the unconscious mind. It is by these maneuvers, says Coulter, that Fodor establishes the object for investigation: "In order to *invent* a phenomenon for elucidation, we have a mythical process ... turned into a quite superfluous *central nervous system* function" (Coulter 1983, p. 18, emphasis in original). Coulter's position is that the CNS has no more than an enabling function: "My nervous system enables me to speak, to say what I do, but it is *I*, not my brain, that does these things" (1983, p. 26).

What Seems Cognitive Is Actually Social

Central to this critique of cognitivisim is a difference over the appropriate application of the categories "social" and "cognitive." For Coulter what is at stake is the most appropriate way of conceiving of conduct. His central complaint is that too much has been appropriated by cognitive theory, that the realm of the social *is* being sold short.

Coulter makes this point by effecting a kind of deconstruction of the ways in which cognitive theory tries to establish its scientific character. Coulter thus argues that cognitive theory claims a set of activities as cognitive by stripping mental predicates of their praxeological character. It posits a physical/neurophysiological mechanism whereby human behavior is the upshot of inner computational processes. Cognitive theory thus attempts to constitute itself as a science by initially denying dependence on the very phenomenon that it claims is relevant to its work, namely, human conduct. The activities it seeks to topicalize as cognitive are stripped of their social character in order to posit the existence of inner processes for study; but the connection between these inner processes and the behavior to be explained is then obscure and has to be arbitrarily reasserted. Having established a science of the central nervous system, rather than of the person, Coulter says that the relevance of this science's findings for the activities of the person are unclear. Thus, although it is the brain that supposedly governs computational procedures for message decoding, the ultimate claim of cognitive theory is to be able to explain how the person understands. Coulter argues that cognitive theory fails to bridge the enormous gap between "the brain's control of message decoding" and "a person's understanding."

It is important to note, however, that Coulter's critique itself depends on a distinction that is foundational to cognitive theory. In accepting as unproblematic the notion that human behavior has a physiological basis, Coulter implicitly endorses the distinction between physical and social phenomena and questions only the nature of the connection between them. Thus for Coulter some activities *are* involuntary organic episodes (and hence the province of physiological investigation), whereas others *are* intentional human actions (and hence the province of social science). But in concurring with this division of labor, Coulter underplays the accomplished character of the distinction. One is reminded here of Ryle's (1949), and Geertz's (1973, ch. 1), example of winking. What is entailed by the rapid movement of a single eyelid (whether it is indeed winking or an involuntary twitch or an attempt to mimic a wink) is the upshot of localized interpretation in a specific context. Whether or not the movement is involuntary (a "nervous tic," as we say) is itself a practical accomplishment. By accepting this as a given for purposes of his sociological analysis, Coulter concedes the construction that human behavior has a physiological basis. He selects certain cognitivist constructions for critical analysis but overlooks the constructed character of other classifications (for example, "involuntary organic episodes").[8]

The important point is that Coulter's critique deals only partially with the problem of describing behavior. He argues against the way in which cognitive theory appropriates actions as "cognitive" and is more concerned with redesignating them as social than with questioning the basis for making designations in the first place. This is important, as we will see, because it undervalues the role of description.

The Language of Behavior[9]

Of course, what counts as behavior, and what distinguishes human from other kinds of (machine or animal) behavior, is crucial to the dispute over cognitivism. As already mentioned, behavior can refer to and is implicated by a wide range of other descriptive terms: action, conduct, performance, and so on.[10] It is not surprising that parties to the debate each use these terms in quite different ways. For example, action for the neo-Wittgensteinian sociologist is not what the cognitive scientist conceives of as action. Clearly, this is more than just an argument about the causes of certain well-observed phenomena. Parties to the dispute are offering quite different definitions of the phenomena to be explained.[11]

Descriptive terms pertaining to behavior are "loaded" by virtue of the expectations of their typical usage. For example, it seems straightforward to refer to a machine's performance and only slightly less natural to speak of its activity. But some awkwardness sometimes arises when we refer to a machine's action or behavior. The oddity of applying mental predicates to machine performance is equally revealing of the prejudices built into our use of these phrases: "The machine thinks . . ."; "the machine reasoned that . . ."; and so on. This kind of awkwardness is sometimes exacerbated by certain pronominal usage. For example, "machines *who* think" (McCorduck 1979).

In general terms, what we experience as awkwardnesses can be understood as violations of our expectations about the "correct" order of performance, the moral order of abilities, and entitlements. Our everyday linguistic usage reflects the institutionalization of these expectations. By the same token, debates occasioned by the advent of AI can be understood as attempts to renegotiate this accepted moral order. The successful institutionalization of the concept of, say, "thinking machines" might be reflected in the future "normal" usage of many phrases that now appear strange.

The situation is more complicated, however, than this general account might lead us to believe, for one need not reflect much to see that there is no simple one-to-one correspondence between categories of action and particular mental predicates. There are many occasions of use when potential awkwardness fails to materialize. Thus we are all familiar with cars that "refuse" to start, with "temperamental" television sets, and with space shuttles that (*pre-Challenger*) are "behaving perfectly." On these occasions predicates often connoting features of human intentionality are unproblematically applied to nonhuman objects. Conversely, there are many occasions when humans unproblematically appropriate descriptions of their behavior sometimes used to connote the absence of "humanness." For example, Turkle (1984) relates how depictions of a person's "machine-like" behavior can confer great esteem within the community of AI hackers.

The variation in the sense and application of mental predicates can be further illustrated with a specific example (taken from Woolgar (1985)). Consider an imaginary device that detects the onset of advertisements, commercial messages, and other nuisance interruptions during television programs.[12] This device would permit the television to be turned off (or at least muted) during these interruptions. Two quite different reactions to the operation of the device are possible. We might be completely satisfied with its efficient execution of the desired task; we might speak of the device as "knowing" when to spare us from the misery of "messages from our sponsor." Alternatively, we might be disappointed to discover that the device works "merely" by detecting some change in the electronic signal at the onset of commercial breaks; its operation, in this view, might be said to be "entirely mechanical" and not "really" intelligent. We might say it is "unable to determine changes in story line," that it "fails to see" that the commercial is substantively different from the interrupted program. Thus, although on one level we could be perfectly happy with its "intelligent operation," we could also argue that the device is "not really intelligent."

As has been argued elsewhere (Woolgar 1985), the interpretative flexibility of mental predicates provides an important dynamic for the AI research enterprise. In the given example the view of the device as mere mechanism redefines and thus reserves the attribute of "intelligence," for some future assessment of performance. The way is thereby cleared for further research into devices that are "really intelligent," where this is (temporarily) equated with the capacity to analyze story lines, content, tenor of presentation, and so on. Similarly, we can see from this example that the assessment of intelligence in terms of mechanism projects a further ambiguous assessment of performance, namely, "the capacity to analyze story lines" and so on. More generally, tests of what counts as intelligence appear to build in a facility for redefining intelligence. Instead of bringing research to a close, a "successful" manifestation of intelligence occasions the redefinition of what, after all, is to count as intelligence. In the field of expert systems research, for example, the "success" of any expert system ironically guarantees its own failure, in the sense that "real expertise" then becomes the topic for future exploration.

The Science-Technology Relationship

For my purposes the important consequence of the interpretative flexibility of behavior-related descriptions is an ambiguity about the relationship between the premises of cognitive science and the perceived achievements of AI.

I have already mentioned that some sociological critiques of cognitivism deny that the achievements of AI have any bearing on our understanding of human action. This echoes the position of AI practitioners who hold that machine performance is a technical goal with no bearing on the question of whether or not such performance would require intelligence in a human; the soft version of the AI program is consistent with the sociological challenge to the claim that cognitivism supports and is reinforced by the achievements of AI.

In effect, the argument between the soft and the strong versions of AI is an argument about the relationship between science and technology.[13] On some occasions supporters of the strong AI program appear to advance a symmetric view of the relationship between science (cognitive theory) and technology (AI). They point out both that theories of mind in cognitive science are explicitly based on computational metaphors *and* that the results and findings of cognitive science increasingly inform the work of AI. When they suggest, however, that the success of AI is a vindication of the cognitive theory, on which the design and construction of intelligent machines is based, they appear to lean toward a hierarchical view: the creative discovery of facts of nature by science (cognitive theory) and their routine application by technology (AI).

By contrast, critics such as Coulter hold that there is no necessary connection between cognitive theory and AI. For Coulter cognitive theory depends on a false equivalence between the achievements of AI and the operation of cognitive processes; arguments for establishing a phenomenon (the mind) for scientific investigation are spurious, and the notion of "cognitive processes operating at the unconscious level" is fallacious. There is no reasonable basis to suppose that such phenomena exist, according to Coulter, except through the illegitimate bludgeoning and reformulation of ordinary concepts of human behavior and action. Coulter says that the denial of "unconscious cognitive processes" as an appropriate topic of investigation is not meant to imply a criticism of the technical achievements of AI; it is simply that the assessment of such achievements in terms of their relevance for mysterious (and artificially construed) cognitive processes is inappropriate. "I do not think that [AI] need be assessed *at all* in connection with psycho-physiological meta-theory, any more than progress in advanced cartography needs to be assessed in terms of the need for theorising about children's (or non-technical adults') map-using capacities" (Coulter 1983, p. 25, emphasis in original).

Coulter seeks to deny *any* relationship between cognitive science and AI. The difficulty with this is that both cognitive science and AI deploy mental predicates in their descriptions of processes and performance. Given the flexibility of language use associated with mental predicates, it seems difficult to sustain the case that there is *no* association between descriptions of machine performance and metaphors of cognitive process.

My earlier example of the "intelligent" advert suppressor can help elucidate this point.[14] One implication of the different responses to the device is that we might be willing to attribute intelligence on the basis of its effect (performance) alone but unwilling to do so on the basis of the way it operates (mechanism). This distinction mirrors the difference between those (strong AI) researchers committed to the simulation of cognitive mechanisms and those (soft AI) researchers who believe that the prime task is to mimic human behavior (performance) by whatever means. The example shows, however, that *criteria of performance and mechanism are not unproblematically distinct.* The mental predicates characteristic of AI discourse enjoy an interpretative flexibility; a description of a machine's performance may well involve an assessment of the means by which it performs. Similarly (and *contra* Coulter), descriptions of progress in cartography may involve an assessment of the means by which map reading is done. This suggests that an important topic for investigation is the way in which descriptions of machine activity are construed as descriptions of, say, performance rather than mechanism.

The Determination of Behavior by Rules and Plans

Suchman's critique of cognitivism (1985) takes issue with the conception of "purposeful action," which underlies the design of certain machines. In particular, she argues that artifacts based on the cognitivist conception of action embody a crucial confusion between the ideas of *plans* and of *situated action.* She analyzes transcripts of videotaped interaction between users and a large photocopier controlled by a computer-based "expert help system."

This expert help system embodies a conception of human action shared by designers, the behavioral sciences and our commonsense. Briefly, the conception is that the significance of what people say and do is best understood as the reflection of their underlying plans. Applied to communication, this view holds that the coherence of action is individual, and is given in advance, and identifies the problem for conversants as the transmission and recognition of their respective plans. (Suchman 1985, p. 4).

Suchman suggests that the characteristics of human-machine communication revealed by her study demonstrate the failings of the view that purposeful action is planned action. The cognitivist conception of action admits to a vagueness in plans but regards this as a fault to be remedied. When action does not follow from prescribed plans, the cognitivist reaction is that further precision is required in the plan's specification. By contrast, Suchman endorses the position that plans are *essentially* vague. In other words, plans (like rules, instructions, and so on) have an *inherent* indeterminancy that guarantees the impossibility of exhaustively specifying their meaning.

Suchman argues that plans should be understood, not as determinants of behavior located in the actor's head but

> as formulations of antecedent conditions and consequences of action, which account for action in a plausible way. . . . As ways of talking about action, plans *per se* neither determine the actual course of stimulated action, nor adequately reconstruct it. (Suchman 1985, p. 4).

Similarily, Coulter claims that cognitive theory's account of language ability needlessly overintellectualizes the mundane achievements of language performance in order to posit the existence of inner linguistic processes, as if a child's capacity to add numbers was evidence of her comprehending the principle of (say) commutativity. As Coulter points out, the reliance of computational theories of the mind on rule-guided mechanisms embodies a confusion between the notion of behavior being "regarded as in accord with a rule" and behavior being "guided by a rule." For example, we do not need to know the principles of Newtonian motion to be able to catch a ball. Indeed, one might claim that a defining characteristic of interpretative interaction is that any form of conduct can be accounted for by (that is, said to be in accord with) different and sometimes contradictory rules. Rules are resources for the "telling" of behavior, post hoc rationalizations of conduct, and not principles that determine that conduct.

The Codifiability of Behavior

We have seen that critics of cognitivism take issue with the construal of behavior as "cognitive," rather than as "social," that some of them deny the association between cognitive theory and AI, and that they question the view of behavior as rule (or plan) determined. It appears, however, that the alternative approach recommended by these critics espouses an axiom of description and explanation that is itself fundamental to cognitivism.

Recent sociological critiques of cognitivist explanations of action have much in common with the well-known "anti-computer" writings of authors such as Weizenbaum (1976) and Dreyfus (1965, 1979). However, a distinguishing feature of the recent work is the explicit avowal of a neo-Wittgensteinian alternative to the cognitivist explanation of action. I wish to suggest that this particular invocation of Wittgenstein actually jeopardizes the success of their critique.

The thrust of Suchman's (1985) argument is that cognitivism confuses plans with situated action and that this is detrimental to our understanding of practical action. Cognitivism has failed because of its unwillingness to engage the complexity of situated conduct. It regards this complexity as a mere technicality, a source of vagueness to be overcome by more precise articulation of computational procedures. For Suchman the task for a science of practical action is to produce descriptions of this complexity. She says that we have to examine actors' use of particular occasions to provide for their actions' developing purpose and intelligibility. In short, we have to "*strengthen our characterizations* of what interaction is" (Suchman 1985, p. 123, emphasis added).

The danger is that this way of expressing the argument against cognitive theory's abstracted view of action can be read as embodying an implicit hope for another kind of scientific approach to the phenomena, a "more realistic" view of how interaction "actually" goes on. Is the appeal for stronger characterization intended to invoke some alternative criterion of accuracy of description? In exactly what sense might the sociologist's descriptions be superior to those of cognitivism? Without careful attention to this issue there is a danger that anticognitivism will merely reproduce the assumption of codifiability that characterizes the cognitivist position itself.

Similarily, in arguing that cognitive theory should take more account of the "mental," as this has currency in ordinary discourse, Coulter exhibits his basic commitment to the possibility of a scientific approach to the study of the properties of mind (Coulter 1983, p. 3). Coulter argues that cognitive theory is unsound and requires reorientation by way of input from studies of ordinary language usage. As I have already discussed, Coulter is clearly right to insist on greater attention to the use of key terms associated with cognition. But to regard examination of the use of mental concepts as a way of improving a science of the mind surely concedes too much to established (but unexplicated) notions of what should count as properly scientific. When Coulter says that the postulation of cognitive processes operating at the unconscious level depends on *illegitimate* reformulation of ordinary concepts of human (social) behavior and action (Coulter 1983, p. 25), he implies that there are other legitimate criteria for making a science out of such concepts.

The difficulty here is that, as Coulter himself indicates, the establishment of scientific concepts depends on their having been successfully "desocialized." In other words, an essential feature of any display of "being scientific" is that the concepts being treated are not just those of the commonsense world. Coulter notes how Dennett (1978) addresses this point by proposing that a mature cognitive science should simply avoid the contamination of (most of) our ordinary, vernacular mental predicates. Everyday attributions in mentalistic or intentional language are regarded by Dennett as something of a mess. Dennett thus views "beliefs," "pain," "understanding," and so on as essentially defective theoretical constructs because they do not unproblematically refer to definable states. Coulter argues that this application of scientific standards of theory construction is "quite arbitrary," that Dennett has "no good arguments" for treating such concepts as theoretical concepts. By these standards, suggests Coulter, are not all concepts something of a mess?

We see here how Dennett's invocation of theoretical criteria is crucial to the endeavor of "arguing science" (that is, seeking to establish a scientific basis for a theory). From Dennett's point of view it simply would not do to base a science on concepts that are regarded as incoherent, messy, and ill-defined. From Coulter's point of view this evaluation of concepts is simply inappropriate. Yet it is not clear how Coulter's alternative program will "scientifically" accommodate the social character of commonsense concepts.

Although these critics of cognitivism often refer to themselves as neo-Wittgensteinian, it is clear that their implicit endorsement of a scientific study of behavior (practical action) depends on their ignoring an important element of Wittgenstein's argument. Wittgenstein raises the possibility of the indeterminacy not just of rules and meanings but of *all* "explanations" of behavior. Because Wittgenstein's argument specifically addressed examples of mentalistic accounts of action, however, recent sociological interpreters have seized on Wittgenstein as an attack on mentalism (cognitivism). They have overlooked (or conveniently ignored) the reading that it is an attack on explanations of behavior per se.

Bloor (1983) exemplifies this use of Wittgenstein's later writings in his attempt to legitimize a "social theory of knowledge" that advances sociological explanations of conduct previously explained by cognitivism. He admits, with admirable candor, that he is "going against certain of Wittgenstein's stated preferences, his chosen method and perhaps his deepest prejudices" (Bloor 1983, p. 5). But Bloor evidently fails to consider the important consequences of using Wittgenstein in this way. "A sociological reading points up the ways in which Wittgenstein could have taken his work forward. It makes us wonder why he stopped where he did" (Bloor

1983, P. 4). But if Wittgenstein is to be understood as denunciating the search for causes and the construction of explanatory theories, we have a good reason for stopping and not taking Bloor's particular route "forward." Despite this, critics of cognitivism have used Wittgenstein to legitimize arguments that, although behavior is *not* reducible to psychological and cognitive explanans, it *can* be reduced to sociological explanans. These sociologists' commitment to the explanatory enterprise is thus in danger of merely substituting a social for a cognitive mechanism. Instead of facing the critical issue of whether or not human behavior can be codified (formalized, reduced to a series of rules such as instructions or to an algorithm) *at all,* critics of cognitivism have pursued the narrower path of debating which kind of codification is the more appropriate.

To the extent that they purport to be explanations, sociological characterizations of behavior implicitly claim codifiability for their descriptions. We might thus envisage such results to be a potential input to an expert system (Collins, this volume). Does their utility in computer programs provide one possible criterion of the strength of these characterizations? None of the expert systems researchers to whom I talked were willing to treat the upshot of sociological deliberation as results in this sense.[15] For example, in discussing the problem of natural language understanding, expert systems researchers indicated an awareness of conversational analysts' work on what counts as "understanding" in ordinary conversation. Their general opinion, however, was that such work is more useful as a heuristic, a way of suggesting things to think about, than as a source of findings that could be incorporated into a language-understanding program.

Conclusion

In this chapter I have examined sociological critiques of cognitivism in line with the general idea that discussions about technology embody fundamental preconceptions about the character of mankind. Discussion and controversy over the characteristics and capabilities of intelligent machines is usefully understood as part of a process of reconstructing the essential qualities of human behavior. It turns out, however, that these sociologists' version of the essential characteristics of humanness is closely tied to a specific belief in ways of representing, describing, and explaining behavior.

Recent sociological critiques of cognitivism pose a strong challenge to theories that conceive of human behavior in terms of its determination by mental states. Although this is to be welcomed, considerable caution should be exercised before adopting the recommended alternative, for these sociologists' alternative vision of human behavior depends heavily on the same ideology of representation that sustains cognitivism. Sociological critiques of cognitivism display the same affinities for producing definitive descriptions of behavior, albeit under a social rather than a cognitive rubric. Critics of cognitivism fail to realize that the premises for producing definitive descriptions of human behavior are at issue here. The battle over whether these descriptions should use the language of mentalism or of sociology may be less important than their common dependence on representative ideology.

What is to become of these sociological critiques of cognitivism? The dilemma is clear. On the one hand, sociological critiques pitched at too general a level are unlikely to have any impact on AI. For example, humanist reassertions of the voluntaristic basis of human behavior are unlikely to affect cognitivism's continued use of a deterministic model. On the other hand, criticisms of cognitivism that are successful at the level of detail will entail alternative (sociological) descriptions of behavior, which are themselves amenable to codification and hence programming. The sociological commitment to the revelation of social regularities is thus in danger of merely supplying new facets of "intelligence" for further axiomatization.

The fundamental problem is that these sociologist-critics are themselves committed to an ideology of representation that places paramount importance on the ability to make descriptions stand on behalf of "real world" entities hidden from immediate experience and observation. The basis of their argument is a commitment to the documentary method of interpretation, but the object of their critical attention (cognitivism) is a realm of argument that is the most highly developed form of this commitment. This perhaps provides a clue as to why the odds are stacked heavily against these critics of the new technology. So long as they continue to be uncritical about their own use of representation, they will continue to reproduce an impoverished version of what they are actually fighting against.

Notes

The construction of this argument has greatly benefited from helpful comments made by Brian Bloomfield, Jeff Coulter, Trevor Pinch, Lucy Suchman, Anna Wynne, and Steve Yearley.

1. Sociologists have been relatively slow to appreciate the availability and significance of AI as a social phenomenon. In part, this is because of the prevalence of a limited conception of the social compared with pre-Kuhnian notions of the social in the sociology of science (Woolgar 1985). Notable exceptions are the works by Coulter (1979, 1983), Fleck (1982, 1984), Turkle (1982, 1984), De Mey (1982), Suchman (1985), and Gilbert and Heath (1985). See also Collins (this volume).

2. Feuerbach has been credited with noting that humans create technology in their own image. The point here is not simply that the nature of technology is driven by conceptions of the human state but that in the development of (and discussion about) technology, prevailing conceptions of the human state and technological capacity are renegotiated.

3. This approach also has the advantage of reminding us how certain analysts' problems (for example, the relationship between science and technology; see note 13) are themselves lively issues for protagonists in disputes over technology.

4. A full history of this debate would include consideration of attempts to compare and contrast humans and animals—Darwinism, the ape language controversy, and so on—as well as comparisons between humans and machines.

5. The strategy of invoking a "reserve" category of attributes has in general been relatively unsuccessful in averting fears about the erosion of man's uniqueness by the creation of machines. Attempts to establish *alternative* criteria of uniqueness have usually failed to overcome deeply held preconceptions about the special character of man. Certainly this kind of argument did not prevent considerable alarm over the ever-increasing capacity of devices to mimic human mechanical abilities (see, for example, Butler's 1872 (1970) marvelous parody) nor did it prevent speculation and considerable controversy about the fundamentally mechanical character of all human action (see, for example, the discussions of La Mettrie's 1784 work by Needham (1928), Rignano (1926), Rosenfield (1941), and Vartanian (1960)).

6. It would have been nice to find a neutral term to designate that thing which is variously labeled "performance," "action," and "behavior" by different protagonists. My own usage unfortunately implies a degree of asymmetry in my analysis; see for example, the preceding sentence, in which I talk about the designation of actions as "behavior." On the other hand, the idea of a truly neutral term (my candidate for which is "that thing") misleadingly implies the possibility of producing a (pure) description that is free of all connotations.

7. Psychophysiological determinism is the notion contained in computational theories of the mind that human behavior is invariably the outcome of prior calculation or reasoning whether or not the person is aware of it. Neuronal discharges and synaptic transmissions comprise the computation of descriptions of the behavior in which a person is engaged, and the brain governs these computational procedures by recourse to a set of "wired in" rules. In language performance this accords with

the familiar claim that the child's acquisition of natural language is tantamount to the internalization of propositional rules (grammars). On Fodor's account this is internalization directly into the nervous system.

8. It is probably futile to suppose that any account can simultaneously challenge the constructed character of all its presuppositions. In this paragraph, for example, I have assumed the givenness of "movement." The point, however, is that inattention to this phenomenon merely reproduces the rhetoric of representation that lies at the heart of cognitivism.

9. This section could equally well be entitled the "The Language of Action" or "The Language of Performance," and so on. Readers are invited to assess the difference in tenor of the ensuing argument.

10. I use these terms interchangeably in this paper (see note 6).

11. Collins's 1975 study of replication in physics similarily suggests that disputes about correct experimental method are tantamount to disputes about the nature of the phenomena being experimented on; negotiations about the adequacy of the experimental method were negotiations about the nature of gravity waves. Here, "negotiation" extends beyond the adequacy of experiment. The very conceptualization of behavior (see note 6) represents a claim about its essential character.

12. Since first writing this, I have been told that such a device indeed exists. It has been developed in the United States by a well-known electronics company in order to enable editing of videotaped TV programs. But it is not being marketed, so the story goes, because of enormous pressure from the TV networks.

13. Some writers have recently argued that it is more realistic to think of science and technology as enjoying a *symmetric* rather than a *hierarchical* relationship. Viewed as a hierarchical relationship, science and technology are interacting cultures, both of which generate products (facts and artifacts) from their own bodies of lore and competence and neither of which is necessarily beholden to the aims and achievements of the other (Barnes 1982a; Barnes and Edge 1982). Viewed as a hierarchical relationship, technology is definitely the poorer relation; technology merely deduces the implications of science. Viewed as a symmetric relationship, feedback is possible in either or both directions. Studies such as Gibbons and Johnson's discussion of semiconductor research (1970) provide examples in which the relationship between science and technology is such that both technology and physical theory are simultaneously advanced.

14. This paragraph is based on a suggestion made by Lucy Suchman.

15. This comment is based on interviews with expert systems researchers at MIT and Stanford University. For a contrary view, see contributions to Gilbert and Heath (1985).

Expert Systems and the Science of Knowledge

H. M. Collins

Artificial intelligence (AI) and its offshoots are of more than normal interest to historians and sociologists of technology. It looks as though AI will be the foundation of one of the key technologies of the later part of the twentieth century, and this is justification enough for a theoretically informed documentation of its development. In addition, however, AI has "internal" relationships to technology studies. Because "knowledgeable machines" can be seen as storage and transfer media for technological knowledge, the way they work must interest those concerned with the nature of technological culture or the transfer of technology. Historians and sociologists, at least those who practice their craft in order to investigate the way in which knowledge originates, changes, spreads, and grows, should be anxious to examine what can be learned by using AI as an instrument—a kind of experimental probe—for exploring the nature of knowledge itself.

There is another internal relationship. As much as history of technology and sociology of knowledge can learn by looking to AI, they in turn can shed light on the possibilities, limitations, and fruitful routes for development of knowledgeable machines. Self-imposed evaluative neutrality is less easy for the analyst to maintain in the case of AI than in other areas of science because both AI and recent science studies share a topic: knowledge. This means that the findings of the whole research program of modern science studies, as well as specific examinations of AI, have implications for what AI researchers are trying to do, whereas for most science and technology the visitor from science studies can be an onlooker, in this case he or she is also an expert, a knowledge specialist. Thus my findings must contribute not just to the documentation but also to the *form* of the developing relationship between human and machine. This, then, is not a reflexive paper; it is not even a relativist enterprise as regards the subject of AI; it is an attempt to build up the internal relationships of AI and the science of knowledge.[1]

Algorithmic and Enculturational Models of Knowledge

To understand the significance of studies of science and technology for AI, I give a schematic description of the existing debate on whether or not artificial intelligence is possible (for a historical sketch of this debate, see Turkle (1984)). In simplified terms, the key to the opposing views are preferences for one side or the other of a dichotomy—what I call the algorithmic/enculturational dichotomy. I describe the debate in these terms because the dichotomy underlies a number of other debates that are more familiar to social scientists and because the dichotomy is used later for an analysis of the components of knowledge.

For the social scientist the dichotomy is first encountered in Max Weber's distinction (1949) between explanations that are "causally adequate" and explanations that are adequate at the level of "meaning." Weber's unsuccessful struggle to reconcile these approaches to the study of social life has left its legacy in the competition in sociology between *verstehende* methods (also underpinned by the phenomenological school) and aggregative statistical approaches. Methods of research, such as use questionnaires and other techniques for gathering "information" about societies, rest, one might argue, on the assumption that interesting and useful knowledge can be transmitted to members of society and the researcher alike in discrete bits capable of being written down, classified, and counted. This might be called the algorithmical model of sociological research. On the other hand, the idea of "understanding" implies that the means of transferring whatever is to be transferred among members of society and from members to the social scientist is face-to-face interaction. When this interaction is deep and extended, it is called participant observation, or, more exactly, participant comprehension (Collins 1984). This method goes with a culture-, or skill-like model of knowledge in which socialization, not verbalization, is thought of as the way that knowedge is transmitted.

One might say that the dichotomy also underpins widespread if oversimplified ideas about the difference between science and technology. Thus it has been suggested that scientific knowledge is knowledge about "closed systems," whereas the rest of life takes place in open systems (Bhaskar 1975). Although the laboratory is by no means the closed system that it has been taken to be, as many recent studies have shown (for example, the papers in Collins 1981a), the distinction between open and closed systems is useful because it does inform our everyday thinking. We do generally think of scientific theories and apparatuses as acting in the predictable fashion that pertains to closed systems. If we no longer think of the whole universe as clockwork, we do still think of well-understood and controllable subsystems within it in this way. To use a more contemporary analogy, we think of scientific theories and the corresponding apparatuses as existing in a closed world like that of a digital computer program; here, any residual unpredictability is a matter of complexity or impenetrability rather than fundamental indeterminacy or unforeseen environmental disturbances. On the other hand, the design of crafted artifacts is design for open system use and is not an exact science; in addition to formal theories, principles of design rest on traditional ways of doing things—on artisans' skills and on bodies of tacit knowledge that cannot be fully articulated (Polanyi 1967; Collins 1974).

There is also a correspondence between this dichotomy and the traditional perception of the appropriate mode of communication of knowledge among scientists and among technicians or craftsmen. Scientific knowledge, it is often thought, can be properly transmitted through information channels, such as scientific journals, for what there is to be known can be contained in a set of discrete bits of information and logical instructions. Again, this is the algorithmical model of communication. The transfer of craft knowledge, by contrast, is a matter of acquiring skill. An apprenticeship, or at least a period of interpersonal interaction, is thought to be the necessary prelude to the transfer of skill-related knowledge. This process rests on the enculturational model.[2]

Artificial Intelligence

The traditional approach to AI, which aims to create a model of thinking from a relatively limited number of principles, can be said to reflect a pattern of thinking characterized by the algorithmic model of knowledge, learning, and communication. It works from principles, just as Euclidean geometry works from axioms; hence it is often called "top-down" AI. To proponents of this approach the difficulties that have been encountered in making machines act like humans seems to be a matter of time and complexity. The human brain is thought of as an immensely complex and, perhaps rather untidy, computer. Understanding the brain and building a computer model of it are rather similar problems. The current capabilities of computers reveal their long-term potential. In playing chess, in helping translators, in recognizing a subset of spoken words and a few patterns, computers are already acting intelligently even if the level of their ability is low. That they are quite good at chess and such makes even the claims about low ability less than straightforward (Turkle 1984).

On the other hand, the enculturational model often informs those who believe that computers will never be able to act like humans, or, if "never" is too big a word, that what has been done so far does not make the route to genuine artificial intelligence any more foreseeable. For example, the computer's ability to play good chess, however anthropomorphic it makes us feel, is taken to be a matter of its number-crunching superiority (that is, its ability to evaluate many moves ahead). Although any chess-playing computer must narrow its search by using heuristics (rules of thumbs) and although these heuristics have been developed by studying human play (and are therefore inductive, or "bottom-up"), the computer knows far too little to be able to succeed by calculating as few moves as the human player, and therefore it beats humans only because it can look much further ahead within its relatively unskilled (machinelike!) strategy. From this viewpoint the chess computer's ability looks after some analysis little more remarkable than the ability of a pocket calculator to out-calculate its owner.[3] In any case, chess takes place with a fixed number of pieces with fixed moves on a board with a fixed design. Although there is a vast number of positions possible, a game of chess is an archetypical closed system in which meaning and context are not salient problems. Of far greater significance from the cultural perspective is the fact that computers are almost total failures when it comes to open system tasks that humans find trivial, such as reading handwriting and understanding natural speech. This seems to have much to do with our grasp of social context.[4]

Optimists who believe in the algorithmic approach think that, although the AI programs of the future may be complex beyond our understanding, the underlying logic will nevertheless be the familiar logic of mathematics and (the canonical version of) science; methods will be developed for programming meaning and context that will be no different, in principle, from what we have now. The other side of the dichotomy is well represented by the phenomenologist Hubert Dreyfus in his book *What Computers Can't Do* (1979). Dreyfus writes:

Intelligence requires understanding, and understanding requires giving a computer the background of common sense that adult human beings have by virtue of having bodies, interacting skillfully with the material world, and being trained into a culture. (p. 3)

What I "know" about the cultural practices which enables me to recognise and act in specific situations has been gradually acquired through training in which no one ever did or could . . . on pain of regress, make explicit what was being learned. (p. 60)

From this point of view the project of reproducing human abilities by modeling them theoretically is doomed because humans themselves seem to have abilities that transcend any conceivable representations. As Dreyfus says, to succeed, AI researchers would have to solve problems that have baffled philosophers for thousands of years.

Expert Systems

I have talked about AI and its offshoots, but top-down AI is of little interest as a tool in the study of knowledge because it is too ambitious and unrealistic.[5] This is being increasingly realized even in the AI community, where it has been found that the only promising route forward is not more analysis of human action but giving the machines more *facts* about their environment and more rules of thumb for solving complex problems. This trend has reached a new high with the development of the bottom-up approach to AI, known as expert systems or intelligent knowledge-based systems (IKBS). This idea seems, in principle, much more likely to lead to knowledge-impregnated machines that will be not only commercially useful but also valuable as instruments for exploring technological culture.

The idea of an expert system is that, instead of trying to create intelligence by building models de nova (so that each time the machine "uses its mind" it makes a fresh deduction), a body of knowledge is extracted from an expert and stored, and the machine simply refers to this store for what would be the expert's answer to a problem. There is little of the deductive, model-building approach to this; it is essentially pragmatic. The expert systems designer modestly accepts that we do not know much about what we know; he or she simply tries to reproduce it in the computer's memory so that it can he retrieved in a way that will help the nonexpert.[6]

It is claimed that "knowledge engineers," as the expert systems designers call themselves, have written systems that can replace experts in some areas. Medical diagnosis is an area in which much work has been done, and it has been claimed that success rates of more than 90 percent correct diagnoses have been obtained in narrow specializations. There is also a celebrated case in which a system is supposed to have made a major mineralogical discovery. In this paper, however, I do not have any existing system or systems in mind; it is not entirely clear how well current systems work once the rhetoric has been stripped away. (See Dreyfus and Dreyfus (1985) for a recent skeptical assessment.)[7]

However primitive existing systems are, they still precipitate interesting questions. For example: What sorts of things can experts learn from a machine and what can they not? What can and what cannot be stripped from the human mind and encoded? And, of course, What is their potential as replacements for experts? This last question is of equal moment to those concerned with the nature of human knowledge and those interested in the impact of new technologies on the place of work. In the *New Scientist* (September 26, 1983, issue) a small article is entitled "Apprentices Lose Out to Computers" (p. 34). The article claims that expert systems are being used to capture human expertise and that apprentices are being replaced. The article concludes that sound assessment of the utility of expert systems in such a role is required, a conclusion with which I cannot but agree.

Components of Expert Knowledge

To start to answer the questions we need to demystify culture to some extent—to get it out of its black box without trivializing it into information. We can start by thinking of expert systems as links in chains of communication. We can think of them as intermediaries between expert providers of knowledge and end-users; thus

Expert → Expert System → End-user.   (*)

Although a knowledge engineer often intervenes between expert and system, we can think, fundamentally, of a two-step process, as represented in (*).

The naive interpretation of (*) is "left to right," following the arrows; the expert inserts knowledge into the system, and the system then passes it to the end-user. For example, medical experts put their knowledge into an expert system, which can then be consulted by less knowledgeable end-users; the end-users must provide some information, such as details of patients' symptoms, pulse, and blood pressure, but it is the system that uses its knowledge to provide a diagnosis. The simple left-to-right interpretation, however, ignores key philosophical and sociological ideas having to do with the contribution of the recipient in a chain of communication. Unless, minimally, the end-user can interpret the symbols that the expert system uses to give its advice, nothing useful can be transferred (Oldman and Ducker 1984; Suchman, personal communication, 1984; Collins et al. 1986). When this principle is extended to its limit, of course, the end-user becomes as much an expert as the original expert at the left-hand end of the chain, and then the system has been used as no more than a store for bits of information; all the interpretation and expertise has been put in at the right-hand end, and the mystery of the machine as a repository of culture is dissolved. The crucial questions, then, are questions about the extent to which an expert system could in principle store knowledge of a sort that would be useful to an *unskilled* (unenculturated) end-user, for example, the extent to which it could act as a master to an apprentice. This is also the extent to which it could *replace* skilled persons in the process of enculturation, as the article in *New Scientist* promises.

Ignoring a priori principles and starting from the encounters that scientists and expert systems builders actually have with human knowledge, we can say that what must be transferred when a non-expert is being turned into an expert is exhausted by the following four categories: (1) facts and formal rules, (2) heuristics, (3) manual and perceptual skills, and (4) cultural skills. I use the term "facts" in a commonsensical way, but the term "heuristics" is taken from the expert systems literature. This literature reveals a burgeoning range of knowledge typologies (for example, see Welbank (1983)), but their almost universal characteristic is stress on the dividing line between factual information and experts' informal rules of thumb, or "heuristics." Because heuristics are rarely written down and because the knowledge engineer has to extract them from the expert in a painstaking manner, they are seen as crucially different from ready codified facts and are usually taken by themselves to exhaust the informal aspects of what is known. From the point of view of the algorithmic/enculturational dichotomy, however, heuristics are closely related to facts and formal rules because, like them, they can be articulated. The crucial dividing line that existing dichotomies ignore is the dividing line between explicable rules and facts and the nonexplicable tacit component of knowledge. In the fourfold classification that follows, the first two categories appeal, as it were, to the algorithmic way of thinking, whereas the second two appeal to the enculturational way.

Facts and Formal Rules

Facts and formal rules include facts of the sort that are readily transferred in written form to those in the same cultural universe, for example, the melting points of specific chemicals, the range of sizes and shapes of the electrodes of a laser, or the type of bacilli that are associated with diphtheria.[8] The sorts of rule I have in mind in this category include Ohm's law and Boyle's law and calculus and so forth. Again, these make sense only within certain cultural contexts.

Heuristics

Heuristics are composed of explicable rules of thumb, standard practices, etc. They are not normally found in written sources because the scientific literature has developed in such a way as to exclude all but the most schematic information. There is nothing in principle that stops heuristics from being written, however. Heuristics are found, for example, in sports coaching manuals and, sometimes, in Ph.D. theses. Knowledge engineers are interested in uncovering the heuristics that experts use so that they can be encoded in expert systems.

Examples of heuristics are: "Always start the melt cooling from well above the putative melting point." "The tolerances in the design of TEA-laser electrodes are such that the exact shape of the electrodes is unlikely to be responsible for laser failure." "If a throat swab contains rod-shaped bacilli and it is suspected that the patient has diphtheria, then the rods are almost certainly diphtheria bacilli."

Manual and Perceptual Skills

Manual and perceptual skills include such things as the ability to recognize the presence of shapes that suggest the existence of crystals within an otherwise undifferentiated lump of solid material, the ability to distinguish the different sounds that can be made within a prototype TEA-laser cavity, and the ability to take a throat swab and see relevant shapes under the microscope. Of course, because recognizing things is a theory-laden process, it is not easy to distinguish these abilities from cultural skills.

Cultural Skills

Michael Polanyi's idea of tacit knowledge provides the link between manual skills and what I call cultural skills. Polanyi's well-known example of tacit knowledge is the skill associated with bicycle riding. The formal dynamics of balance on a bicycle do not comprise the rules of riding. A rider may know nothing of centers of gravity and gyroscopic forces yet can still ride, whereas the most expert bicycle engineer may not be able to do so. The rider knows how to ride but without being able to say how; indeed, too much self-conscious attention tends to be counterproductive. The same tacit qualities apply equally to nonphysical cultural skills, such as recognizing significant objects in a "noisy" environment.

Cultural skills include the abilities required to understand and use facts, rules, and heuristics. By this I mean the ability to read and comprehend information and advice, but not only this. Cultural skill, as I intend it, includes that which enables, say, an English or American person but not, say, an English-speaking Chinese native to continue the sequence 2, 4, 6, 8 appropriately as "10, 12, 14, 16" on one occasion and appropriately as "Who do we appreciate" on a different sort of occasion. The way we manage this kind of thing, as Dreyfus says, cannot be made explicit except on pain of regress.

"Socialization" might be a better description than cultural skills, but the exact nuance is related to phenomenological ideas such as "taken for granted reality" (Schutz 1962, 1964) or "natural attitude" or Wittgensteinian terms such as "form of life" (Wittgenstein 1953; Winch 1958; Bloor 1983). The notion that there is a hidden component to scientific, intellectual, and theoretical abilities is hard to grasp; thus I use the term "cultural skill" to stress the relationship with manual and perceptual skill, with respect to which we readily accept that there is an inexplicable component.

Examples of Knowledge and Its Transformations

The four components of knowledge described are not stable. Sometimes, it would appear, a piece of knowledge found in one category at one time is found to belong to a different category at another time. Perhaps the best known movement of this sort is when knowledge moves from the heuristic box to the fact/rule box. This is one of several things that happens as science develops and the rules of thumb of building and engineering become theorized—moving from being the craftsman's private skill to being a piece of publicly available science. Of course it must be borne in mind that studies of science have shown how much craft knowledge is still required to operate the instruments of supposedly publicly available science and how potentially open to reversal is this private to public movement (Collins and Pinch 1982). There are always two ways of describing this kind of change in the status of knowledge. One way is the language of scientific discovery, technical development, and theorization, and the other way is in the language of modern science studies. Thus the process could be described in terms of the demodalizing of scientific descriptions (Latour and Woolgar 1979) that result in local, private events coming to be seen as merely singular representations of universal publicly available scientific phenomena (Collins 1985). Apart from a few asides, in this chapter I use the language of scientific development, although the usually assumed epistemological implications of this language are by no means a necessary consequence.

Sometimes, pieces of knowledge move in the direction opposite to that described previously. Perhaps a rule of thumb is used, or invented, because the underlying theory is too complex for practical use. For example, in the case of semiconductor crystal growing, figures for the exact pressure reached within ampuls used to contain the melt could, in principle, have been discovered—it is well within the compass of modern science—but rules of thumb based on the vapor pressure of the individual elements within the melt were combined with "folk knowledge" of the strength of typical ampuls in the practice of the art (see Collins et al. 1986).

Again, in the laser case, which I observed closely, the builder, Dr. Robert Harrison, began with some exact, "theoretically derived" figures for the tolerances allowable in the clearance between the trigger wires and the close and parallel bottom electrode. The tolerances, according to his information, were so demanding that careful measurements would be required. Later, however, he found that a quick glance at the configuration, to make certain that the tubes that held the trigger wires were roughly "flat," was sufficient. In both these examples a piece of knowledge either potentially or actually found in category 1 turned into a heuristic, something belonging in category 2.

It might be useful to represent these movements of knowledge diagrammatically (figure 1). The two moves discussed so far look like A and B in the diagram.

Another kind of transformation takes place when machinery and instruments are used to transform a manual and perceptual skill into something much more akin to a fact (D in figure 1). For example, measurement of the temperature in a steel manufacturing process may change from a skilled judgment of color to a matter of reading a dial. This is still a perceptual skill but one so widely distributed that its skill aspect is invisible unless subject to deliberate attention. In the crystal growing case, verification of the existence of satisfactory crystals within the solidified melt, traditionally determined by skilled observation, could be "automated" by using electron microprobe analysis. (This technique can determine the ratios of elements in local areas of the frozen mass; if the ratios correspond exactly to the constitution of the desired crystal, then crystalline material must almost certainly be present at that point.) In the laser case, in which Harrison became adept at working out what was happening in the discharge cavity by listening for pings or cracks, we can imagine a machine being made to analyze the spectrum of the sounds and produce an automatic diagnosis.

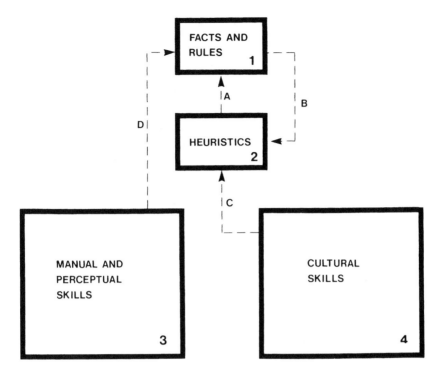

Figure 1
Transformations of knowledge.

Such transformations of sensory skill into more articulated categories is vital to the prospects of automation and robotics. Once more it is important to be aware of the extent to which practical craftsmanship has not always proved to be easily replaceable by automated measuring devices.[9]

When it comes to the prospects of AI, IKBS, and the big problems in the science of knowledge, the crucial shift is from the category of cultural skills into one of the explicable categories. One might say that top-down AI depends on shifting such knowledge all the way into the facts and rules category (routes C plus A in figure 1), whereas expert systems depend only on shifting it as far as the heuristics category (route C alone). That is why expert systems are so interesting to the social scientist; their success depends not on discovering the deductive rules of human action but only on rendering the contents and mechanisms of culture explicit—precisely the goal of interpretative sociology and history.

Moving Knowledge from the Tacit to the Explicit Category

Continuing with the pragmatic, empirical approach, I now look at an example of a piece of knowledge that moved from the cultural skills category into the facts and rules category in the 1970s. This is an example drawn from the development of the TEA laser (Collins 1974, 1985; Collins and Harrison 1975).

When Harrison, the laser scientist I worked with, first set about building a TEA laser, he knew that the leads from the top capacitors to the top electrodes had to be short. It turns out that he did not interpret this knowledge properly, and consequently his laser had top leads that were short as he saw it but too long to allow the device to work. Eventually, he found that in order to make the leads' linear dimension count as "short" in TEA laser building society, the top capacitors had to be mounted in an inverted position in a strong and complex steel frame so that their terminals could be near the top electrode. Still later, as this aspect of the design of the laser came to be seen as a piece of *electronics*, quantitatively theorized it came to be understood that the *inductance* of these top leads was the crucial variable and that the leads must be 8 inches or shorter in length.

Knowledge about lead length can be seen as actually or potentially belonging to three different categories at different times and places. In the first instance Harrison knew the verbal formulation "keep the top leads short" but did not have the tacit knowledge required to interpret the meaning of "short." Later, he acquired knowledge about the meaning of "short," that is, "short" did not mean "less than 1 mile" (note that we all know enough about the culture of lasers and electricity to see the funny side of that) nor did it mean "less than 1 yard." It meant "substantially less than 1 foot." Within different contexts all these are possible interpretations of "short," but only the last one is correct in laser society. Eventually, Harrison learned the meaning of "short," where TEA lasers are concerned; we can describe his knowledge of the length of the top leads as belonging to the heuristic category—category 2.

We can also imagine a prior stage (which, as a matter of fact, Harrison did not pass through) in which apprentice laser builders, in copying their masters' design, invert the top capacitors in a strong iron frame as a matter of tradition and thereby have leads of the appropriate length as a matter of course. Such apprentices need know nothing of lead length; it need not even be a concept within their laser building world; nevertheless, there is a sense in which those who could reproduce their masters' skills satisfactorily could be described as having sufficient knowledge of lead length to build a laser. One would have to say that it was part of their tacit knowledge of laser building. Note that what they knew of lead length would be completely impossible to explicate—firmly in category 4 (with aspects of category 3). Later, when lead length became articulated as "invert your capacitors so as to make sure that your top lead is short enough," it would have moved to category 2.

At the still later stage when the matter of lead length was passed into the vocabulary and culture of electronics, the knowledge became formalized and theorized. At this final stage, when it was known that the leads must be shorter than 8 inches in length, the knowledge belonged to the facts and rules category, moving from category 2 to category 1.

These shifts of category are complex; they involve shifts of context, individual enculturation, and changes in the "cultural ambience" as much as changes in the way the knowledge is formulated. Harrison's learning to interpret "short" and so attain the heuristic involved enculturation on his part; he had to internalize the laser builder's context for the interpretation of "short." The shift from heuristic to formal rule also involved a shift of context, into electronics. What is more, we can consider the "8-inch rule" to be only in category 1 because of the wide distribution of measuring skills within our cultural ambience. If the laser was to be built in a more "primitive society," the rule would have to be considered to be lower down the diagram in figure 1 at best.

To see what is going on, we must remember that the notion of technical and scientific progress is not the only way of describing the sorts of change that we have looked at. One might say that it is important to remember the relativistic aspect of skill. Thus we say that reading a dial or measuring the length of a lead requires "less skill" than was needed before (meaning either that measuring skills have become much more widely distributed or that modern "meters" as opposed to older instruments make use of skills that are already much more widely distributed) because we do not think of widely distributed skills as skills at all. For example, typewriting and car driving were once esoteric skills but are no longer such. This has little to do with changes in what has to be done in order to typewrite or drive a car and much to do with the wide distribution of these abilities. Ironically, one of the outcomes of research on AI is the discovery that some widespread abilities, such as speaking, writing, and interacting socially, are difficult to explicate and that the things we think of as the preserve of clever people, such as rapid calculation or playing good chess, are relatively easy to encapsulate in formulas. This is why we cannot think about the development of knowledgeable machines without thinking about how they are going to be used, the way they will fit into the cultural ambience. A machine that can interact quite acceptably in one social setting may be unacceptable in another.

Now try to imagine Harrison making his laser with the help of an expert system. If the system had been designed at the imaginary apprentice-master stage, there would be nothing in it about lead length, however persistent the knowledge engineer who built it, because knowledge about lead length was not consciously known and therefore could not be explicated. There is no special reason to suppose that inverted capacitors would come to figure in the system's heuristic rules because they might well be thought of by everybody as no more than a traditional part of the design, a decorative conceit. Most lasers built by properly trained apprentices would still work, but the crucial importance of lead length would remain unrecognized. Such an expert system would work only where there was already a substantial tradition of laser building in the community. If, however, the system was built at a later stage, say around Harrison's starting point, then it would have included an instruction about keeping the top leads short, but this instruction would still only be usable by someone who was expert enough to know how to interpret the rule in the context of laser building. Cultural competence is required, but the amount is less than would be required for success in the previous instance. Later, when electronics was discovered to be the relevant context, the rule was made less ambiguous (for this specific design of laser). The instruction "keep the leads less than 8 inches in length" enabled many more users with less specialized skills to succeed. To put this in another way, the rule depends on far more widely distributed cultural competence; the meaning of "8 inches" is unambiguous for most natives of Western societies. What we call basic technical education is enculturation of this sort.

Cultural Skills as Unexplicated Rules

In the last section a piece of knowledge was seen to move from an unexplicated to an explicable category of knowledge. At the end of the series of real and imagined transformations, the knowledge was firmly in the facts and rules category, which is to say that it could be understood by reference to widely distributed cultural competence. Thus it could be used by what we might as well call unskilled end-users, bearing in mind that the unskilled end-user is still culturally competent enough to understand many things, such as handwriting and natural speech, which are well beyond the reach of any current computer.[10]

The model I have developed so far helps us to understand two things: the immediate potential and the limitations of expert systems. By using the model, we can see that the success of today's expert systems depends on more than the brilliance of computer builders and programmers and the perserverance and ingenuity of knowledge engineers as they elicit knowledge from experts. It depends on the nature of the skill that is being encoded into the system (if this is already highly structured and theorized, then few problems of interpretation will arise), and it depends on advances in understanding within the science *and* on the skills of the end-user *and* on changes in the ambient culture such that relevant cultural skills become more widely distributed. But the model also presents a paradox.

On the one hand, it seems impossible to explicate cultural abilities because, as the phenomenologists stress, this leads to an infinite regress; each explanation begs a further explanation until the foundation of shared culture, which provides meaning, is reached. Without shared culture, the explanations go on forever.[11] On the other hand, we do share aspects of culture, and therefore we can learn, and, as we have seen, the amount that we can explicate seems to grow in extent. What is more, the explication of certain aspects of our cultural competence is a common experience. When we interact with those who are close to us in cultural terms, we make do with few explicit remarks, and these convey meaning because so much is shared at the outset. But, as cultural and contextual distance increases between communicators, the potential ambiguity of more and more messages becomes realized, and more needs to be made explicit and *can* be made explicit. Social scientists make use of this fact when they study cultural dislocations; at such times native actors are forced to articulate aspects of their taken-for-granted lives.[12] Thus, although cultural skill, like manual and perceptual skill, has components that cannot be made explicit, some things can be articulated if the necessity is forced on us; culture cannot be as impenetrable as all that.

To cope with these theoretical tensions and with the potential successes of expert systems, we need a working model of culture. The model I propose takes cultural skills to comprise, for all intents and purposes, an indefinite, open-ended set of rules of action. This model can account for what we have seen taking place as knowledge transforms itself from one type to another. In the model explicable rules can convey knowledge only because they rest on the indefinitely large foundation of taken-for-granted and shared reality. Although the whole of this taken-for-granted reality can never be explicated, any specific and limited part of it can. We have seen bits of culture become articulated but only by virtue of the way they rest on cultural competence of users and interpreters.[13]

In the model categories 3 and 4—manual and cultural skills—comprise the infinite bottom part of the iceberg of knowledge, unstateable but understood and usable by human beings. Categories 1 and 2 are the visible tip of the iceberg, a tip that grows bigger day by day, accounting for the improving interactive competence of machines but without sensibly diminishing what lies beneath and therefore without implying that current progress means that machines are a model of man. Some of the tip even falls back into the foundation from time to time and some of the tip becomes moribund as portions of our cultural skills that support it die out with changes in the cultural ambience.[14]

The research program suggested by this model, the research program of the science of knowledge, is the way in which the various transformations of knowledge from category to category take place. These changes, as I have said, depend on transformations of the cultural ambience as much as anything else, and changes in the cultural ambience are certain to be enmeshed in wider social and political affairs. It should be clear that limited aspects of this program are already the subject of modern sociology of scientific knowledge.[15]

But the interaction between the knowledge sciences and AI is, as I suggested in the introduction of this paper, a two-way process. The model also suggests fruitful ways of looking at the development of AI machines, such as expert systems. For example, a hypothesis that is immediately suggested is that those whose professional competence rests on stores of codified information, such as solicitors and medical specialists, will be more easily replaced than those whose expertise is less tidily organized, such as barristers and general practitioners. It also suggests that, although expert systems may soon become useful aids to those who are already skilled, the replacement of skill is a prospect involving many more orders of difficulty. See Collins et al. (1986) for an apparently successful attempt to draw out these ideas for the expert systems technical community.[16]

Summary and Conclusion

In this chapter I have suggested that historians and sociologists of technology can use expert systems as experimental probes into the nature of knowledge by looking at knowledge engineers' attempts to render culture visible. I have also argued, and illustrated, that the way we can learn about the potential of these machines is by looking at them through the perspective of philosophy and modern sociology of knowledge. To be teleological about it, we might say that this is one of the purposes of sociology of knowledge and its related disciplines.[17]

I have argued that a machine cannot be understood aside from its end-user and the cultural ambience in which it works. The role of the end-user is to insert that part of the iceberg of cultural knowledge that cannot be programmed. Progress in IKBS will come with the articulation of more of the iceberg's base, and this will allow systems to be used by less and less skilled end-users. But the role of the end-user's skill will never disappear, for when I talk of unskilled end-users, I mean only end-users whose skills are widely distributed. The speed and extent to which the need for esoteric skills in the end-user is reduced is a matter for more detailed research. Only when experts and end-users try to use systems to communicate about subjects with which they have little in cultural common does the major problem of AI, the explicit representation of indefinitely ramifying commonsense human knowledge, begin to present itself. In the meantime, the problems facing the knowledge engineer are much more tractable and have pragmatic solutions that rely on the end-user's capabilities.

Finally, it is worth noting that the major problems—encoding cultural knowledge and transmitting it to the nonenculturated—are already within the purview of the sociologist and anthropologist. The point of participant observation (comprehension) is to transfer cultural abilities to an investigator who is initially an outsider. This is what sociologists are used to. The problem of representing these abilities in some less culturally specific manner is exactly the problem of the successful participant who then tries to transfer the new insights to the nonexpert professional readership of the research journal. The extent to which what is tacit and cultural can be made explicit is part and parcel of the same problem. Participant observation—that "softest" of social science methodologies—may oddly enough be of direct relevance to the new breed of knowledge engineers.

Notes

The empirical aspect of this research was supported by the Economic and Social Research Council under grants HR 3453/1 and CO8250002.

I am grateful to Ed Constant, Bryn Jones, Arie Rip, Trevor Pinch, Richard Velleman, Jay Gershuny, Andrew Kerslake, and an anonymous referee for encouragement and helpful comments.

1. This is not to say that interesting work cannot be done on AI from a symmetric and relativist viewpoint. For example, Woolgar (this volume) examines the way that

ascriptions of intelligence are granted or withheld from machines without thinking about whether or not such decisions are reasonable. Nevertheless, only a puritanical reflexivity could hide the significance of such a study for the evaluation of the products of AI research.

In an interesting paper Fuhrman and Oehler (1986) discuss the meaning of reflexivity. They conclude that a paper can be rendered reflexive merely by claiming that it is such, for example, by putting in a sentence along the lines of "This applies to me, too." This suggests that too much concern about reflexivity may be misplaced.

2. An important feature of modern sociology of scientific knowledge, which has its philosophical roots in the later work of Wittgenstein (for example, 1953; Collins 1974, 1975; Bloor 1983), is that it shows that a great deal of scientific work, including what might be thought of as purely mental or theoretical activity, is best thought of as a skill-like activity. This goes along with the redescription of laboratory activity as open-system-like.

3. Having been consistently beaten by a small chess computer, I am under no illusion about the transparency of the machinelike aspects of intelligent machines.

4. Our comprehension of speech and so forth is only *nearly* unproblematical. Ethnomethodology points to the amount of "repair work" we have to do even in normal interaction. Nevertheless, it is self-evident that my comprehension of English speech needs more to explain it than methods of repair when it is compared with my comprehension of Chinese speech. I cannot find the way to express this within the enthnomethodogical vocabulary.

5. The top-down version has been much used by psychologists who feel that computer models offer a test for theories of the working of the brain. See, for example, Boden (1981).

6. For an uncritical account of expert systems, see Feigenbaum and McCorduck (1984). For a definition, see Hayward (1984).

Traditional top-down AI has begun to look like IKBS as it is realized that an AI program requires a large amount of background knowledge about the world in order to begin to make the same sort of sense of things as humans. Top-down AI and IKBS are actually opposite poles of a continuum of trade-offs between theory and knowledge. Top-down AI stresses theories of knowledge; expert systems stress discrete bits of knowledge. Both approaches use some of each, however.

7. This is why I refer only to the *idea* of expert systems.

8. I select these illustrations of types of knowledge for the following reasons: The chemical mixtures example refers to work I am currently doing that looks at the problems of encoding the tacit skills involved in semiconductor crystal growing into an expert system (see Collins et al. 1986). The TEA laser example refers to earlier work of my own on transfer of tacit knowledge and is discussed at greater length

in a later section of this chapter. The diphtheria example refers to L. Fleck's *Genesis and Development of a Scientific Fact* (1979 [1935]).

9. See Jones (1984). See also Latour and Woolgar (1979), who tend to treat apparatuses in rather too reified a manner.

10. Using a realist, "progressive" descriptive language in this chapter is doing no more than using the existing conventions of current work in the sociology of knowledge. In such work realist language is used to describe aspects of knowledge that are not under relativist scrutiny. An alternative treatment of, say, the 8-inch laser building rule is, of course, possible and can be found in my other work on the TEA laser.

11. Elsewhere (Collins 1985) I have described the game Awkward Student, which is a device for bringing out the nature and limits of the explicability of simple rules.

12. For example, areas of naturally occurring controversy in science and other cultural activities have been used as fruitful sites for sociological research. Garfinkel's early experiments (1967) can also be interpreted in this way; deliberately disturbing social settings forces actors to be self-conscious about the way in which regular order is maintained.

13. The idea of knowledge as composed of a large number of rules is, of course, appealing to the knowledge engineer. For example, Hayes-Roth (1985) suggests that "a commercially practical system may require as few as 50 rules," that "expertise in a profession requires about 10,000 rules," and that "the limits of human expertise are at about 100,000 rules." In the current state of the expert system art, 100,000 is virtually as distant as the effectively infinite number that I suggest comprises human competence. My guess would be that, as machines become more powerful and if the 100,000-rule machine becomes a possibility, then either human expertise will cease to be seen as a collection of rules or the number of rules that is claimed as the human limit will increase dramatically so that it remains well over the horizon.

Something that does not follow from my version of the model is the inevitable success of the strong AI research program, nor even speedy improvements. If the strong program requires an indefinite and open-ended collection of rules to represent human culture, then complete success, whatever that means, is no more attainable than travel at the speed of light. Speedy improvement *may* not be possible either. If the number of new rules that is needed for each "unit increase" in cultural competence grows exponentially and at a rate faster than improvements in computer power, then "small" advances will become harder and harder. To put this more in terms of modern AI experience, moving from domain-specific applications to more general abilities is a much greater problem than developing competence in many specific domains. In other words, a human is more than the sum of the expertises belonging to the many domains in which he or she operates.

14. This idea seems to be at variance with Popper's notion (1972) of a third world of objective knowledge. Popper seems to think that this third world exists, in librar-

ies and so forth, in such a form that, even if human culture were destroyed, the knowledge would survive. It is knowledge independent of human beings. But my argument shows that such sets of symbols would lose their meaning coextensively with the destruction of human culture. The bottom of the iceberg can survive only by being transmitted through the generations and only if there is continuity of social life. Thus printed symbols would lose all significance if human culture were destroyed. The existence of a supposed third world then, is organically related to the existence and continuity of human culture.

When aspects of that culture die (for example, when craft skills are lost), they cannot be recaptured from written accounts without being "reinvented" in practice. A written account may guide the reinvention process, but it can do so only because there is enough continuity of culture to allow the written words to have meaning. That part of the bottom of the iceberg that supports the esoteric skill must be developed anew.

15. I think it would be correct to say that sociology of knowledge has concentrated mostly on route A in figure 1, but see Gooding (1986) for another way of looking at attempts to move knowledge from categories 3 and 4 to categories 1 and 2. (In Gooding's article Michael Faraday's scientific and experimental skills are the topic.)

16. Interestingly, Hayes-Roth (1985) suggests that a useful rule for expert system builders is "Seek problems that experts can solve via telephone communications." Given the analysis here, one can see why this is such an apt rule: It ensures that the solution to the problem is verbalizable.

It may be of interest of researchers in the knowledge sciences that the paper by Collins et al. (1986) found a ready audience in the technical community. In fact, the paper shared the prize for technical merit at the 1985 expert systems annual conference of the British Computer Society, Expert Systems 85. I hope this will encourage knowledge scientists to try to speak more frequently to the technical audience.

17. Actually, this "use" for technology studies is quite similar to the work on automation being done by Jones (1984), Jones and Wood (1985), and Manwaring and Wood (1985). For example, these researchers have looked at the limits and significance of deskilling in the workplace as attempts are made to replace skilled machinists with numerically controlled and more comprehensively computer-controlled machine tools. Jones and Wood, in particular, have used ideas similar to those discussed here to argue that the possibilities for the replacement of machinists' skills are limited.

References

Arieti, S. 1976. *Creativity: The Magic Synthesis*. New York: Basic Books.

Arkin, W. M. 1984. "Sleight of hand with Trident II." *Bulletin of the Atomic Scientists*, December, 5–6.

Armacost, M. H. 1969. *The Politics of Weapons Innovation: The Thor-Jupiter Controversy*. New York: Columbia University Press.

Aron, R. 1968. *La Révolution introuvable*. Paris: Fayard.

Association for Improvements in the Maternity Services. 1985. "A commentary on the *Report of the RCOG Working Party on Routine Ultrasound Examination in Pregnancy*." London: AIMS. Available from 163 Liverpool Road, London N1 0RF.

Bacon, E. M. 1942. "The growth of household conveniences in the United States, 1860–1900." Ph.D. thesis, Radcliffe College.

Baekeland, L. H. 1907a. "Condensation product of phenol and formaldehyde and method of making the same." US Patent 942,700. Filed on December 4, 1907.

Baekeland, L. H. 1907b. "Method of making products of phenol and formaldehyde." US Patent 942,699. Filed on July 13, 1907.

Baekeland, L. H. 1909a. "Bakelite, a new composition of matter: Its synthesis constitution, and uses." *Scientific American*, supplement, 68: 322–323, 342–343.

Baekeland, L. H. 1909b. "Bakelit und Resit? Erwiderung auf Dr. H. Lebachs Veröffentlichung über Resinit." *Zeitschrift für Angewandte Chemie* 41: 2006–2007.

Baekeland, L. H. 1909c. "On soluble, fusible, resinous condensation products of phenols and formaldehyde." *Journal of Industrial and Engineering Chemistry* 1: 345–349.

Baekeland, L. H. 1909d. "The synthesis, constitution, and use of Bakelite." *Journal of Industrial and Engineering Chemistry* 1: 149–161.

Baekeland, L. H. 1912. "Phenol-formaldehyde condensation products." *Journal of Industrial and Engineering Chemistry* 4: 737–743.

Baekeland, L. H. 1916. "Practical life as a complement to university education. Address of acceptance of the Perkin Medal." *Metallurgical and Chemical Engineering* 14: 151–158.

Baeyer, A. 1872. "Uber die Verbindungen der Aldehyde mit den Phenolen." *Berichte der Deutschen Chemischen Gesellschaft* 5: 25–26, 280–282.

Bailes, K. 1976. "Technology and legitimacy: Soviet aviation and Stalinism in the 1930's." *Technology and Culture* 17: 55–81.

Ballantine, H. T., Bolt, R. H., Hueter, T. F., and Ludwig, G. D. 1950. "On the detection of intracranial pathology by ultrasound." *Science* 112: 525–528.

Barnes, B. 1974. *Scientific Knowledge and Sociological Theory*. London: Routledge and Kegan Paul.

Barnes, B. 1982a. "The science-technology relationship: A model and a query." *Social Studies of Science* 12: 166–172.

Barnes, B. 1982b. *T. S. Kuhn and Social Science*. London: Macmillan.

Barnes, B. 1984. Review of *Networks of Power, Social Studies of Science* 14: 309–314.

Barnes, B., and Edge, D., eds. 1982. *Science in Context*. Milton Keynes: Open University Press.

Beard, E. 1976. *Developing the ICBM: A Study in Bureaucratic Politics*. New York: Columbia University Press.

Beaujouan, G. 1966. "Science livresque et art nautique au XVe siecle," in *Actes du cinquieme colloque international d'histoire maritime, Les aspects internationaux de la decouverte oceanique aux XVe et XVIe siecles*, M. Mollat and P. Adam, eds. Paris: SEVPEN, 61–85.

Becker, H. A., and Porter, A. L., eds. 1986. *Impact Assessment Today*. Van Arkel: Utrecht.

Bell, F. G. 1969. "Schuler's principle and inertial navigation." *Annals of the New York Academy of Sciences* 147: 493–513.

Beniger, J. R. 1984. *The Control Revolution*. Cambridge, Mass.: Belknap Press.

Beranek, L., Bolt, R., and Ballantine, H. T. 1949. Research proposal accompanying a letter to J. R. Killian, April 11, 1949. J. R. Killian Correspondence, MIT Archives, Collection AC4, Box 72, Folder 4.

Berman, R., and Baker, J. 1982. *Soviet Strategic Forces: Requirements and Responses*. Washington, D.C.: Brookings Institution.

von Bertalanffy, L. 1968. *General Systems Theory: Foundations, Development, Applications*. New York: Braziller.

Bhaskar, R. 1975. *A Realist Theory of Science*. Leeds: Leeds Books.

Bijker, W. E. 1984. "Collectifs technologiques et styles technologiques: Eléments pour un modèle explicatif de la construction sociale des artefacts techniques," in *Travailleur collectif et relations science-production*, J. H. Jacot, ed. Paris: Editions du CNRS, 113–120.

Bijker, W. E., and Pinch, T. J. 1983. "La construction sociale de faits et d'artefacts: Impératifs stratégiques et méthodologiques pour une approche unifiée de l'étude des sciences et de la technique." Paper presented to L'atelier de recherche (III) sur les problèmes stratégiques et méthodologiques en milieu scientifique et technique. Paris, March.

Bijker, W. E., Bonig, J., and van Oost, E. C. J. 1984. "The social construction of technological artefacts." Paper presented to the EASST Conference. A shorter version of this paper is published in *Zeitschrift für Wissenschaftsforschung*, special issue 2, 3: 39–52 (1984).

Bining, A. C. 1938. *Pennsylvania Iron Manufacture in the Eighteenth Century*. Harrisburg, Penn.: Pennsylvania Historical Commission, vol. 4.

Bishop, J. L. 1966 [1868]. *A history of American Manufactures from 1608–1860*, third edition. New York: A. M. Kelly, vol. 2.

Bloor, D. 1973. "Wittgenstein and Manheim on the sociology of mathematics." *Studies in the History and Philosophy of Science* 4: 173–191.

Bloor, D. 1976. *Knowledge and Social Imagery*. London: Routledge and Kegan Paul.

Bloor, D. 1983. *Wittgenstein: A Social Theory of Knowledge*. London: Macmillan.

Boden, M. 1981. *Minds and Mechanisms: Philosophical, Psychological and Computational Models*. Brighton: Harvester.

Bottler, M. 1919. *Uber Herstellung und Eigenschaften von Kunstharzen und deren Verwendung in der Lack- und Firnisindustrie und zu elektrotechnischen und industriellen Zwecken*. Munich: Lehmanns.

Bottler, M. 1924. *Harze und Harzinustrie*. Leipzig: Max Janecke.

Bourdieu, P. 1979. *La distinction*. Paris: Editions de Minuit.

Bourdieu, P., and Darbel, A. (n.d.) *L'Amour de l'art*. Paris: Editions de Minuit.

Bourdieu, P., and Passeron, J. C. 1970. *La reproduction*. Paris: Editions de Minuit.

Boxer, C. R. 1953. "The Portuguese in the East, 1500–1800," in *Portugal and Brazil: An Introduction*, H. V. Livermore, ed. Oxford: Clarendon Press, 185–247.

Brickman, R. 1984. "Science and politics in toxic chemical regulations: US and European contrasts." *Science, Technology and Human Values* 9: 107–111.

Brown, T. G. 1960. "Direct contact ultrasonic scanning techniques for the visualisation of abdominal masses," in *Medical Electronics*, C. N. Smyth, ed. London: Iliffe, 358–366.

Bunge, M. 1966. "Technology as applied science." *Technology and Culture* 7: 329–347.

Burke, A. (n.d.) "Memorandum for Rear Admiral Clark (Op-51) and Rear Admiral Raborn (Office of the Secretary of the Navy), 2 December 1955," in *The Fleet Ballistic Missile System*, Lockheed Missiles and Space Company, Inc. Sunnyvale, Calif.: Lockheed, 6–7.

Burkhardt, R., and Kienle, G. 1978. "Controlled clinical trials and medical ethics." *Lancet* 3: 1356–1359.

Burkhardt, R., and Kienle, G. 1980. "Controlled clinical trials and drug regulation." *Controlled Clinical Trials* 1: 151–164.

Busch, J. 1983. "Cooking competition: Technology on the domestic market in the 1930's." *Technology and Culture* 24: 220–245.

Butler, S. 1970 [1872]. *Erewhon*. Harmondsworth: Penguin.

Callon, M. 1980a. "The state and technical innovation: A cast study of the electrical vehicle in France." *Research Policy* 9: 358–376.

Callon, M. 1980b. "Struggles and negotiations to define what is problematic and what is not: The sociology of translation," in *The Social Process of Scientific Investigation*, K. Knorr, R. Krohn, and R. Whitley, eds. Dordrecht and Boston: vol. 4, 197–219.

Callon, M. 1981a. "Boites noires et operations de traduction." *Economie et Humanisme* 262: 53–59.

Callon, M. 1981b. "Pour une sociologie des controverses technologiques." *Fundamenta Scientiae* 2: 381–399.

Callon, M. 1986. "Some elements of a sociology of translation: Domestication of the scallops and the fishermen of St. Brieuc Bay," in *Power, Action, and Belief: A New Sociology of Knowledge?* J. Law, ed. London: Routledge and Kegan Paul, 196–233.

Callon, M., and Latour, B. 1981. "Unscrewing the big leviathan: How actors macrostructure reality and how sociologists help them to do so," in *Toward an Integration of Micro and Macro Sociologies*, K. Knorr-Cetina and A. V. Cicourel, eds. London: Routledge and Kegan, 277–303.

Callon, M., and Latour, B. 1986. "Comment suivre les innovations; clefs pour l'analyse sociotechnique." *Prospective et Santé* 36: 13–25.

Callon, M., and Law, J. 1982. "On interests and their transformation: Enrollment and counter-enrollment." *Social Studies of Science* 12: 615–625.

Campbell, D. T. 1974. "Evolutionary epistemology," in *The Philosophy of Karl Popper, The Library of Living Philosophers*, P. A. Schlipp, ed. La Salle, Ill.: Open Court, vol. 14-I, 413–463.

Campbell, B. 1983. "The specialization of experts and institutions in the English patent court." Department of Sociology, Mount Allison University, Sackville, New Brunswick. Unpublished.

Carlson, W. B. 1983. "Elihu Thomson: Man of many facets." *IEEE Spectrum*, October, 72–75.

Caro, H. 1892. "Ueber die Entwickelung der Theerfarben-Industrie." *Berichte der Deutschen Chemischen Gesellschaft* 25: 955–1105.

Castoriadis, C. 1968. "Technique," in *Encyclopedia Universalis*. Paris: Encyclopedia Universalis, vol. 15, 803–809.

Caunter, C. F. 1955. *The History and Development of Cycles (as Illustrated by the Collection of Cycles in the Science Museum); Historical Survey*. London: HMSO.

Caunter, C. F. 1957. *The History and Development of Light Cars*. London: HMSO.

Chandler, A. D., Jr. 1966. *Strategy and Structure*. Garden City, N.Y.: Doubleday.

Chandler, A. D., Jr. 1977. *The Visible Hand: The Managerial Revolution in American Business*. Cambridge, Mass.: Belknap Press.

Chandler, C. F. 1916. "Presentation address of the Perkin Medal to L. H. Baekeland." *Journal of Metallurgical and Chemical Engineering* 14: 148–151.

Chaunu, P. 1979. *European Expansion in the Later Middle Ages*, K. Bertram, trans. Amsterdam: North-Holland.

Cipolla, C. M. 1965. *Guns and Sails in the Early Phase of European Expansion, 1400–1700*. London: Collins.

Collins, F. L. 1934. *Consolidated Gas Company of New York: A History*. New York.

Collins, H. M. 1974. "The TEA set: Tacit knowledge and scientific networks." *Science Studies* 4: 165–186. Reprinted in *Science in Context*, B. Barnes and D. Edge, eds., Milton Keynes: Open University Press, 1982.

Collins, H. M. 1975. "The seven sexes: A study in the sociology of a phenomenon, or the replication of experiments in physics." *Sociology* 9: 205–224.

Collins, H. M., ed. 1981a. "Knowledge and controversy." *Social Studies of Science* 11: 3–158.

Collins, H. M. 1981b. "The place of the core-set in modern science: Social contingency with methodological propriety in science." *History of Science* 19: 6–19.

Collins, H. M. 1981c. "Son of seven sexes: The social construction of a physical phenomenon." *Social Studies of Science* 11: 33–62.

Collins, H. M. 1981d. "Stages in the empirical programme of relativism." *Social Studies of Science* 11: 3–10.

Collins, H. M. ed. 1982. *Sociology of Scientific Knowledge: A Source Book*. Bath: Bath University Press.

Collins, H. M. 1983a. "An empirical relativist programme in the sociology of scientific knowledge," in *Science Observed: Perspectives on the Social Study of Science*, K. D. Knorr-Cetina and M. J. Mulkay, eds. Beverly Hills: Sage, 85–113.

Collins, H. M. 1983b. "Scientific knowledge and science policy: Some foreseeable implications." *EASST Newsletter*, November, 2: 5–8.

Collins, H. M. 1983c. "The sociology of scientific knowledge: Studies of contemporary science." *Annual Review of Sociology* 9: 265–285.

Collins, H. M. 1984. "Concepts and practice of participatory fieldwork," in *Social Researching*, C. Bell and H. Roberts, eds. London: Routledge and Kegan Paul, 54–64.

Collins, H. M. 1985. *Changing Order: Replication and Induction in Scientific Practice*. Beverly Hills: Sage.

Collins, H. M., and Harrison, R. 1975. "Building a TEA laser: The caprices of communication." *Social Studies of Science* 5: 441–445.

Collins, H. M., and Pinch, T. J. 1979. "The construction of the paranormal: Nothing unscientific is happening," in *On the Margins of Science: The Social Construction of Rejected Knowledge*, R. Wallis, ed. Keele: University of Keele, 237–270.

Collins, H. M., and Pinch, T. J. 1982. *Frames of Meaning: The Social Construction of Extraordinary Science*. London: Routledge and Kegan Paul.

Collins, H. M., Green, R. H., and Draper, R. C. 1986. "Where's the expertise: Expert systems as medium of knowledge transfer," in *Expert Systems 85*, M. J. Merry, ed. Cambridge: Cambridge University Press, 323–334.

Constant, F. W., II. 1978. "On the diversity and co-evolution of technological multiples: Steam turbines and Pelton water wheels." *Social Studies of Science* 8: 183–210.

Constant, E. W., II. 1980. *The Origins of the Turbojet Revolution*. Baltimore: Johns Hopkins University Press.

Constant, E. W., II. 1983. "Scientific theory and technological testability: Science, dynamometers, and water turbines in the 19th century." *Technology and Culture* 24: 183–198.

Constant, E. W., II. 1984. "Communities and hierarchies: Structure in the practice of science and technology," in *The Nature of Technological Knowledge: Are Models of Scientific Change Relevant?* R. Laudan, ed. Dordrecht: Reidel, 27–46.

Copeland, L. D. 1981. "Do we really understand what our customers want?" in *Symposium über Kreiseltechnik 1981*. Stuttgart: Deutsche Gesellschaft für Ortung und Navigation, A1.0–A1.13.

Coulter, J. 1979. *The Social Construction of Mind*. London: Macmillan.

Coulter, J. 1983. *Rethinking Cognitive Theory*. London: Macmillan.

Croon, L. 1939. *Das Fahrrad und seine Entwicklung*. Berlin: VDI-Verlag.

Crosstalk, Proceedings of the Fourteenth Joint Services Data Exchange Group for Inertial Systems. 1980. Clearwater Beach, Fla.

Crozier, M., and Friedberg, E. 1977. *L'Acteur et le système*. Paris: Editions du Seuil. English translation: *Actors and Systems: The Politics of Collective Action*, Chicago: University of Chicago Press, 1980.

Daniels, C. E., and Wertheimer, A. J. 1980. "Therapeutic significance of the drug lag." *Medical Care* 18: 754–765.

Dawkins, R. 1978. *The Selfish Gene*. London: Granada.

De Mey, M. 1982. *The Cognitive Paradigm*. Dordrecht: Reidel.

Denhard, W. G. 1963. "Floated single-degree-of-freedom integrating gyros," in *Air, Space and Instruments: Draper Anniversary Volume*, S. Lees, ed. New York: McGraw-Hill, 464–497.

Denhard, W. G. 1971. "Technology of tomorrow's commercial air traffic control," in *Proceedings of the Institute of Navigation National Air Meeting*. Washington, D.C.: Institute of Navigation, 1–21.

Dennett, D. C. 1978. *Brainstorms: Philosophical Essays on Mind and Psychology*. Cambridge, Mass.: MIT Press.

Denoix, L. 1966. "Characteristiques des Navires de l'Epoque des Grandes Decouvertes," in *Actes du Cinquieme Colloque International d'Histoire Maritime, Les Aspects Internationaux de la Decouverte Oceanique aux XVe et XVIe siecles*, M. Mollat and P. Adam, eds. Paris: SEVPEN, 137–147.

de Solla Price, D. J. 1969. "The structure of publication in science and technology," in *Factors in the Transfer of Technology*, W. H. Gruber and D. G. Marquis, eds. Cambridge, Mass.: MIT Press, 91–104.

de Vries, G. H. 1984. "International workshop on new developments in the social study of technology—A personal report." *EASST Newsletter*, November, 3: 12–18.

Diesel, E. 1953. *Diesel: Der Mensch, Das Werk, Das Schicksal*. Stuttgart: Reclam.

Diffie, B. W., and Winius, G. D. 1977. *Foundations of the Portuguese Empire, 1415–1580*. Minneapolis: University of Minnesota Press.

Dolman, H., and Bodewitz, H. 1985. "The sedimentation of a scientific concept: The use of citation data." *Social Studies of Science* 15: 507–524.

Donald, I. 1969. "On launching a new diagnostic science." *American Journal of Obstetrics and Gynecology* 103: 609–628.

Donald, I. 1974a. "Apologia: How and why medical sonar developed." *Annals of the Royal College of Surgeons of England* 54: 132–140.

Donald, I. 1974b. "Sonar: The story of an experiment." *Ultrasound in Medicine and Biology* 1: 109–117.

Donald, I. 1976. "The ultrasonic boom." *Journal of Clinical Ultrasound* 4 (5): 323–328.

Donald, I. 1980. "Medical sonar: The first 25 years," in *Advances in Ultrasound Diagnosis*, A. Kurjak, ed. Amsterdam: Excerpta Medica, 4–20.

Donald, I., and Brown, T. 1961. "Demonstration of tissue interfaces within the body by ultrasonic echo-sounding." *British Journal of Radiology* 34 (405): 539–546.

Donald, I., MacVicar, J., arid Brown, T. 1958. "Investigation of abdominal masses with pulsed ultrasound." *The Lancet* 1: 1188–1194.

Dosi, G. 1982. "Technological paradigms and technological trajectories: A suggested interpretation of the determinants and directions of technical change." *Research Policy* 11: 147–162.

Dosi, G. 1984. *Technical Change and Industrial Transformation*. London: Macmillan.

Douglas, S. J. 1985. "Technological innovation arid organizational change: The navy's adoption of radio, 1899–1919," in *Military Enterprise and Technological Change: Perspectives on the American Experience*, M. Roe Smith, ed. Cambridge, Mass.: MIT Press, 117–173.

Draper, C. S., 1947. "Fundamental possibilities and limitations of navigation by means of inertial space references." Meteor Report 9. Cambridge, Mass.: MIT Guided Missiles Program, February.

Draper, C. S. 1959. "Submarine inertial navigation: A review arid some predictions." Paper presented to Polaris Steering Task Group, 22 October 1959. Charles Stark Draper Laboratory Library, CSD-107.

Draper, C. S. 1975. "Importance of research directed toward the development of ultimate performance for inertial system components." Cambridge, Mass.: Charles Stark Draper Laboratory, January, P-030.

Draper, C. S. 1977. "The evolution of aerospace guidance technology at the Massachusetts Institute of Technology, 1935–1951: A memoir," in *Essays on the History of Rocketry and Astronautics: Proceedings of the Third through Sixth History Symposia of the International Academy of Astronautics*, R. Cargill-Hall, ed. Washington, D.C.: NASA, vol. 2. 219–252.

Dreyfus, H. 1965. *Alchemy and Artificial Intelligence*. Santa Monica, Calif.: Rand Corporation.

Dreyfus, H. 1979. *What Computers Can't Do: The Limits of Artificial Intelligence*, second edition. New York: Basic Books.

Dreyfus, H., and Dreyfus, S. 1985. "From Socrates to expert systems: The limits of calculative rationality." *Technology in Society* 6: 217–233.

Dubois, J. H. 1972. *Plastic History USA*. Boston: Cahners.

Duffy, R. A. 1973. "Summary of new developments at the Draper Laboratory," in *Inertial Navigation Components and Systems* (AGARD Conference Proceedings 116), NATO Advisory Group for Aerospace Research and Development. Neuilly-sur-Seine: AGARD, 3-1–3-5.

Duisberg, C. 1933. *Meine Lebenserinnerungen*. Leipzig: Verlag Reclarn.

Dukes, M. N. G. 1979. "The strong, the weak and the wicked," in *Drug Assessment, Criteria and Methods*, J. Z. Bowers and G. P. Velo, eds. Amsterdam: Elsevier/North-Holland, 1–17.

Dunlop, D. 1972. "Medicines, governments and doctors." *Drugs* 3: 305–313.

Dunlop, J. B. 1888. "An improvement in tyres of wheels for bicycles, tricycles, or other road cars." British Patent 10607. Filed on July 23, 1888.

Dussik, K. T., Dussik, F., and Wyt, L. 1947. "Auf dem Wege zur Hyperphonographie des Gehirnes." *Wiener Medizinische Wochenschrift* 97: 425–429.

Elias, N. 1970. *Was ist Soziologie?* München: Juventa Verlag.

Elster, J. 1983. *Explaining Technical Change*. Cambridge: Cambridge University Press.

Elster, J. 1984. *Ulysses and the Sirens*. Cambridge and Paris: Cambridge University Press and Editions de la Maison des Sciences de l'Homme.

Elzen, B. 1985. "De ultracentrifuge: op zoek naar patronen in technologische ontwikkeling door een vergelijkin van twee case-studies." *Jaarboek voor de Geschiedenis van Bedrijf en Techniek* 2: 250–278.

Elzen, B. 1986. "Two ultracentrifuges: A comparative study of the social construction of artefacts." *Social Studies of Science* 16: 621–662.

Enos, J. L. 1962. *Petroleum Progress and Profits.* Cambridge, Mass.: MIT Press.

Erickson, J. 1982. "The Soviet view of deterrence: A general survey." *Survival* 24: 242–250.

Feigenbaum, E. A., and McCorduck, P. 1984. *The Fifth Generation.* London and Sydney: Pan Books.

Feldman, E. J. 1985. *Concorde and Dissent.* Cambridge: Cambridge University Press.

Ferguson, E. 1974a. "An historical sketch of central heating." Unpublished.

Ferguson, E. 1974b. "Toward a discipline of the history of technology." *Technology and Culture* 15: 13–30.

Fincke, M. 1977. *Arzneimittelprüfung—Strafbare Versuchsmethoden.* Heidelberg-Karlsruhe: Müller.

Firestone, F. A. 1945. "The supersonic reflectoscope for interior inspection." *Metal Progress* 48: 505–512.

Fleck, J. 1982. "Development and establishment in artificial intelligence," in *Scientific Establishments and Hierarchies. Sociology of the Sciences Yearbook*, N. Elias, H. Martins, and R. Whitley, eds. Dordrecht: Reidel, vol. 6, 169–217.

Fleck, J. 1984. "Artificial intelligence arid industrial robots: An automatic end for utopian thought?" in *Nineteen Eighty Four: Science Between Utopia and Dystopia. Sociology of the Sciences Yearbook*, E. Mendelsohn and H. Nowotny, eds. Dordrecht: Reidel, vol. 8, 189–231.

Fleck, L. 1935. *Entstehung und Entwicklung einer wissenschaftlichen Tatsache: Einführung in die Lehre vom Denkstil und Denkkollektiv.* Basel: Benno Schwabe. Reprinted by Suhrkamp, Frankfurt am Main, 1980. English translation: *The Genesis and Development of a Scientific Fact*, Chicago: The University of Chicago Press, 1979.

Fodor, J. A. 1975. *The Language of Thought.* New York: Crowell.

Franklin, B. 1960 [1740]. "An account of the new invented Pennsylvanian Fireplace," in *The Papers of Benjamin Franklin*, L. W. Labaree. ed. New Haven: Yale University Press, 419–446.

Freeman, C. 1974. *The Economics of Industrial Innovation.* Harmondsworth: Penguin. Reprinted by Frances Pinter, London, 1982.

Freeman, C. 1977. "Economics of research and development," in *Science, Technology and Society. A Cross-Disciplinary Perspective*, I. Spiegel-Rösing and D. de Solla Price, eds. London and Beverly Hills: Sage, 223–275.

Friedel, R. 1979. "Parkesine and Celluloid: The failure and success of the first modern plastic," in *History of Technology*, A. R. Hall, and N. Smith, eds. London: Mansell, vol. 4, 45–62.

Friedel, R. 1983. *Pioneer Plastic. The Making and Selling of Celluloid*. Madison, Wis.: University of Wisconsin Press.

Fuhrman, E., and Oehler, K. 1986. "Discourse analysis and reflexivity." *Social Studies of Science* 16: 293–307.

Galbraith, J. K. 1971. *The New Industrial State*. Boston: Houghton Mifflin.

Garfinkel, H. 1967. *Studies in Ethnomethodology*. Englewood Cliffs, N.J.: Prentice-Hall.

Geertz, C. 1973. *The Interpretation of Cultures*. New York: Basic Books.

Gerber, M. A. 1978. "Gravity gradiometer: Something new in inertial navigation." *Astronautics and Aeronautics*, May, 18–26.

Gibbons, M., and Johnson, C. 1970. "Relationship between science and technology." *Nature* 227: 125–127.

Gibbs-Smith, C. H. 1960. *The Aeroplane: An Historical Survey of Its Origins and Development*. London: HSMO.

Giedion, S. 1948. *Mechanization Takes Command: A Contribution to Anonymous History*. New York: Oxford University Press.

Gieryn, T. F., and Hirsch, R. F. 1983. "Marginality and innovation in science." *Social Studies of Science* 13: 87–106.

Gilbert, G. N., and Heath, C., eds. 1985. *Social Action and Artificial Intelligence*. Aldershot: Gower Press.

Gilfillan, S. G. 1935. *The Sociology of Invention*. Cambridge, Mass.: MIT Press.

Gille, B. 1978. *Histoire des techniques*. Paris: Gallimard.

Gillen, E. 1977. "Die Sachlichkeit der Revolutionäre," in *Wem ğehört die Welt: Kunst und Gesellschaft in der Wiemarer Republik*. Berlin: NGBK.

Gillespie, B., Eva, D., and Johnston, R. 1979. "Carcinogenic risk assessment in the United States and Great Britain: The case of aldrin/dieldrin." *Social Studies of Science* 9: 265–301.

Gohr, H., and Wedekind, T. 1940. "Der Ultraschall in der Medizin." *Klinische Wochenschrift* 19: 25–29.

Goldman, J. 1982. "The introduction of the cast iron stove in the northeast, 1800–1840." SUNY-Stony Brook, unpublished.

Gooding, D. 1986. "How do scientists reach agreement about novel observations?" *Studies in History and Philosophy of Science* 17: 205–230.

Gordon, D. 1959. "Echo-encephalography: Ultrasonic rays in diagnostic radiology." *British Medical Journal* 1: 1500.

Götschl, J., and Rip, A., eds. 1984. "Problems and perspectives of the study of science and technology in Europe." *Zeitschrift für Wissenschaftsforschung*, special issue 2, vol. 3, April.

Gould, S. J. 1980. *Ever Since Darwin: Reflections in Natural History.* Harmondsworth: Penguin.

Gould, S. J. 1984. *Hen's Teeth and Horse's Toes: Further Reflections in Natural History.* Harmondsworth: Penguin.

Gray, C. 1977. *The Future of Land-Based Missile Forces.* Adelphi Paper 140. London: International Institute for Strategic Studies.

Gray, C. 1982. "The idea of strategic. superiority." *Air Force Magazine*, March, 62–63.

Greenwood, T. 1975. *Making the MIRV: A Study of Defense Decision Making.* Cambridge, Mass.: Ballinger.

Grew, W. 1921. *The Cycle Industry, Its Origin, History and Latest Developments.* London: Pitman & Sons.

Griffin, J. P., and Long, J. R. 1982. "International co-operation in drug evaluation and marketing approval in Western Europe." *Pharmacy International* 3: 153–158.

Gutting, G. 1984. "Paradigms, revolutions, and technology," in *The Nature of Technological Knowledge: Are Models of Scientific Change Relevant?* Dordrecht: Reidel, 47–65.

Hackmann, W. 1985. *Seek and Strike. Sonar, Anti-submarine Warfare and the Royal Navy, 1914–1954.* London: HMSO.

Haeussermann, W. 1981. "Developments in the field of automatic guidance and control of rockets." *Journal of Guidance and Control* 4: 225–239.

Halacsy, A. A., and von Fuchs, G. H. 1961. "Transformer invented seventy-five years ago." *Transactions of the American Institute of Electrical Engineers: Power Apparatus and Systems* 80: 121–125.

Halmos, P., ed. 1972. *The Sociology of Science.* Sociological Review Monograph 18. Keele: University of Keele.

Handlin, D. 1980. *The American Home: Architecture and Society, 1815–1915.* Boston, Mass.: Little Brown.

Hanieski, J. F. 1973. "The airplane as an economic variable: Aspects of technological change in aeronautics, 1903–1955." *Technology and Culture* 14: 535–552.

Hayes-Roth, F. 1985. "Knowledge-based systems: The state of the art in the US." *The Knowledge Engineering Review* 1: 18–27.

Haynes, W. 1954. *American Chemical Industry*, Vol. 2. New York: Van Nostrand.

Hayward, S. 1984. "Is a decision tree an expert system?" in *Research and Development in Expert Systems*, M. Brauner, ed. Cambridge: Cambridge University Press, 184–192.

Healey, P. 1982. "The research funding organization as a focus for science studies." Paper presented to the Science Studies Conference, Oxford, September.

Heidegger, M. 1977. *The Question Concerning Technology and Other Essays*, W. Lovitt, trans. New York: Harper & Row.

Hellige, H. D. 1984. "Die Gesellschaftlichen und historischen Grundlagen der Technikgestaltung als Gegenstand der Ingenieurausbildung." *Technikgeschichte* 51 (4): 281–283.

Hesse, M. 1974. *The Structure of Scientific Inference*. London: Macmillan.

Hessen, B. 1931. "The social and economic roots of Newton's *Principia*," in *Science at the Crossroads*, by N. I. Bukharin, A. F. Joffe, M. Rubinstein, B. Zavadovsky, E. Colman, N. I. Vavilov, W. Th. Mitkewich, and B. Hessen. London: Frank Cass, 147–212.

Hill, C. R. 1973. "Medical ultrasonics: An historical review." *British Journal of Radiology* 46: 899–905.

Hoag, D. G. 1971. "Ballistic-missile guidance," in *Impact of New Technologies on the Arms Race*, B. T. Feld et al., eds. Cambridge, Mass.: MIT Press, 19–106.

Hoddeson, L. 1981. "The emergence of basic research in the Bell Telephone Lab." *Technology and Culture* 22: 512–524.

Holloway, D. 1977. "Military technology," in *The Technological Level of Soviet Industry*, R. Amann et al., eds. New Haven, Conn.: Yale University Press, 407–489.

Holloway, D. 1982. "Innovation in the defence sector: Battle tanks and ICBMs," in *Industrial Innovation in the Soviet Union*, R. Amman and D. Cooper, eds. New Haven, Conn.: Yale University Press, 368–414.

Holmes, J. H. 1967. Discussion in the *American Journal of Obstetrics and Gynecology* 99: 676.

Holmes, J. H. 1980. "Diagnostic ultrasound during the early years of the AIUM." *Journal of Clinical Ultrasound* 8: 299–308.

Homburg, E. 1983. "The influence of demand on the emergence of the dye industry. The roles of chemists and colourists." *Journal of the Society of Dyers and Colourists* 99: 325–333.

Hounshell, D. A. 1975. "Elisha Gray and the telephone: On the disadvantages of being an expert." *Technology and Culture* 16: 133–161.

Hounshell, D. A. 1984. *From the American System to Mass Production, 1800–1932.* Baltimore: Johns Hopkins University Press.

Howry, D. H., and Bliss, W. R. 1952. "Ultrasonic visualisation of soft tissue structures of the body." *Journal of Laboratory and Clinical Medicine* 40: 579–592.

Hughes, T. P. 1969. "Technological momentum in history: Hydrogenation in Germany 1898–1933." *Past & Present*, August, 44: 106–132.

Hughes, T. P. 1971. *Elmer Sherry.* Baltimore: Johns Hopkins University Press.

Hughes, T. P. 1976a. "The development phase of technological change." *Technology and Culture* 17: 423–431.

Hughes, T. P. 1976b. "The science-technology interaction: The case of high-voltage power transmission systems." *Technology and Culture* 17: 646–659.

Hughes, T. P. 1979a. "The electrification of America: The system builders." *Technology and Culture* 20: 124–161.

Hughes, T. P. 1979b. "Emerging themes in the history of technology." *Technology and Culture* 20: 697–711.

Hughes, T. P. 1983. *Networks of Power: Electrification in Western Society, 1880–1930.* Baltimore: Johns Hopkins University Press.

Hummon, N. P. 1984. "Organizational aspects of technological change," in *The Nature of Technological Knowledge: Are Models of Scientific Change Relevant?* R. Laudan, ed. Dordrecht: Reidel, 67–82.

Hunter, L. C. 1949. *Steamboats on the Western Rivers: An Economic and Technological History.* Cambridge, Mass.: Harvard University Press.

Hyatt, J. W. 1870. "Improvement in treating and molding pyroxyline." US Patent 105338. Filed on July 12, 1870.

Hyatt, J. W. 1914. "Address of acceptance of the Perkin Medal." *Journal of Industrial and Engineering Chemistry* 6: 158–161.

Hyde, B. 1986. "An interview study of pregnant women's attitudes to ultrasound scanning." *Social Science and Medicine* 22 (5): 587–592.

Illich, I. 1975. *Medical Nemesis: The Exploration of Health.* New York: Wiley.

Illinois Institute of Technology. 1968. *Technology in Retrospect and Critical Events in Science (TRACES)*. Chicago: ITT Research Institute.

Jacob, F. 1970. *La logique du vivant*. Paris: Gallimard.

Jelsma, J., and Smit, W. A. 1986. "Risks of recombinant DNA research: From uncertainty to certainty," in *Impact Assessment Today*, H. A. Becker and A. L. Porter, eds. Van Arkel: Utrecht, 715–741.

Jenkins, R. V. 1975. "Technology and the market: George Eastman and the origins of mass amateur photography." *Technology and Culture* 16: 1–19.

Jenkins, R. V. 1976. *Images and Enterprise: Technology and the American Photographic Industry, 1839–1925*. Baltimore: Johns Hopkins University Press.

Jenkins, R. V. 1985. "Edison and the art of invention." Paper presented at the Seventeenth International Congress of History of Science, Berkeley, July–August.

Jevons, F. R. 1976. "The interaction of science and technology today, or, is science the mother of invention?" *Technology and Culture* 17: 729–742.

Jewkes, J., Sawers, D., and Stillerman, R. 1969. *The Sources of Invention*. London: Macmillan.

Johnston, R. 1972. "The internal structure of technology," in *The Sociology of Science*, P. Halmos, ed. Keele: University of Keele, 117–130.

Johnston, R. 1984. "Controlling technology: An issue for the social studies of science." *Social Studies of Science* 14: 97–112.

Jokisch, R., ed. 1982. *Techniksoziologie*. Frankfurt am Main: Suhrkamp.

Jones, B. 1984. "The division of labour and the distribution of tacit knowledge in the automation of metal machining," in *Design of Work in Automated Manufacturing Systems*. T. Martin, ed. Oxford: Pergamon, 232–251.

Jones, B., and Wood, S. 1985. "Tacit skills, division of labour and new technology." *Sociologie du Travail* 4: 402–421.

Jones, P. 1940. *History of the Consolidated Edison System, 1878–1900*. New York: Consolidated Edison Co.

Kaldor, M. 1981. *The Baroque Arsenal*. New York: Hill and Wang.

Kaplan, F. 1984. *The Wizards of Armageddon*. New York: Simon & Schuster.

Kaufman, M. 1963. *The First Century of Plastics: Celluloid and Its Sequel*. London: The Plastics Institute.

Keep, W. J. 1916. "History of stoves." Baker Library, Harvard University, unpublished.

Keller, W. 1985. "Imperfect science, important conclusions." *New York Times*, July 28, 4E.

Kessler, H. 1969. *Walther Rathenau: His Life and Work*. New York: Howard Fertig.

Kidder, J. T. 1982. *The Soul of a New Machine*. London: Allen Lane.

Kienle, G. 1974. *Arzneimittelsicherheit und Gesellschaft*. Stuttgart: Schattaner.

Kleeberg, W. 1891. "Uber die Einwirkung des Formaldehyds auf Phenole." *Annalen der Chemie* 263: 283–286.

Kline. S. 1985. "Research, invention, innovation, and production: Models and reality." *Research Management* 28 (4): 36–45.

Knoll & Co. 1907. "Improvements relating to the manufacture of resin-like products from phenols and formaldehyde." British Patent 28,009. Filed on December 19, 1907.

Knoll & Co. 1908. "Verfahren zur Beschleunigung der Erhärtung von Kondensationsprodukten, aus Phenolen und Aldehyden." Deutsches Patent 214194. Filed on July 4, 1908. Addition filed on October 21, 1908.

Knorr, K. D., Krohn, R., and Whitley, R., eds. 1980. *The Social Process of Scientific Investigation, Sociology of the Sciences Yearbook*, vol. 4. Dordrecht: Reidel.

Knorr-Cetina, K. D. 1981. *The Manufacture of Knowledge: An Essay on the Constructivist and Contextual Nature of Science*. Oxford: Pergamon.

Knorr-Cetina, K. D., and Mulkay, M. J., eds. 1983. *Science Observed: Perspectives on the Social Study of Science*. London and Beverly Hills: Sage.

Koestler, A. 1964. *The Act of Creation*. New York: Macmillan.

Kranakis, E. F. 1982. "The French connection: Giffard's injector and the nature of heat." *Technology and Culture*, January, 23: 3–38.

Kränzlein, G. 1935. *Werden, Sein und Vergehen der künstlichen organischen Farbstoffe*. Stuttgart: Verlag von Ferdinand Enke.

Kras, R. J., Gras, S., Lokin, D. H. A. C., Rube, M. M. J. T., Zweers van der Elst, F. J., and Tange, J. A. 1981. *Bakeliet—techniek, vormgeving, gebruik*. Rotterdam: Museum Boymans-Van Beuningen, catalogue 288.

Kreilkamp, K. 1971. "Hindsight and the real world of science policy." *Science Studies* 1: 43–66.

Krohn, W., Layton, E. T., and Weingart, P., eds. 1978. *The Dynamics of Science and Technology, Sociology of the Sciences Yearbook*, vol. 2. Dordrecht: Reidel.

Kubler, G. 1962. *The Shape of Time: Remarks upon the History of Things*. New Haven: Yale University Press.

Kuhn, T. 1970. *The Structure of Scientific Revolutions,* second edition. Chicago: Chicago University Press.

de Lamettrie, J. O. 1748. *L'homme machine.* Leiden.

Landstrom, B. 1978. *Sailing Ships in Words and Pictures from Papyrus Boats to Full-Riggers.* London: Allen and Unwin.

Lane, F. C. 1934. *Venetian Ships and Shipbuilders of the Renaissance.* Baltimore: Johns Hopkins University Press.

Lane, F. C. 1973. *Venice, a Maritime Republic.* Baltimore: Johns Hopkins University Press.

Langrish, J., Gibbons, M., Evans, W. G., and Jevons, F. R, 1972. *Wealth from Knowledge.* London: Macmillan.

Latour, B. 1983. "Give me a laboratory and I will raise the world," in *Science Observed: Perspectives on the Social Study of Science,* K. D. Knorr-Cetina and M. J. Mulkay, eds. London and Beverly Hills: Sage, 141–170.

Latour, B. 1984. *Les microbes, querre et paix, suivi de irréductions.* Paris: Métaillé. English translation: *The Pasteurization of French Society, followed by Irréductions. A Politico-Scientific Essay,* Cambridge, Mass.: Harvard University Press, 1987.

Latour, B. 1986. "Visualization and cognition: Thinking with eyes and hands." *Knowledge and Society: Studies in the Sociology of Culture Past and Present* 6: 1–40.

Latour, B. and Woolgar, S. 1979. *Laboratory Life. The Social Construction of Scientific Facts.* London and Beverly Hills: Sage; second edition published as *Laboratory Life. The Construction of Scientific Facts,* Princeton, N.J.: Princeton University Press, 1986.

Laudan, R., ed. 1984a. *The Nature of Technological Knowledge: Are Models of Scientific Change Relevant?* Dordrecht: Reidel.

Laudan, R. 1984b. Introduction to *The Nature of Technological Knowledge: Are Models of Scientific Change Relevant?* R. Laudan, ed. Dordrecht: Reidel, 1–26.

Laudan, R. 1984c. "Cognitive change in technology and science," in *The Nature of Technological Change: Are Models of Scientific Change Relevant?* R. Laudan, ed. Dordrecht: Reidel, 83–104.

Law, J. 1984a. "International workshop on new developments in the social studies of technology," *4S Review,* Winter, 2: 9–13.

Law, J. 1984b. "Sur la tactique du controle sociale: une introduction a la theorie de l'acteur-reseau," in *La Leqitimite Scientifique, Cahiers Science, Technologie, Societe.* Paris: CNRS, vol. 4, 106–126.

Law, J. 1985. "On politicians and planes: A memo on the TSR.2." Keele: Department of Sociology, University of Keele, unpublished.

Law, J. 1986a. "On the methods of long distance control: Vessels, navigation and the Portuguese route to India," in *Power, Action and Belief: A New Sociology of Knowledge?* J. Law, ed. London: Routledge and Kegan Paul, 234–263.

Law, J, ed. 1986b. *Power, Action and Belief: A New Sociology of Knowledge?* Sociological Review Monograph 32. London: Routledge and Kegan Paul.

Law, J., and Lodge, P. 1984. *Science for Social Scientists.* London: Macmillan.

Layton, E. T. 1972. "Mirror image twins: The communities of science and technology in nineteenth century America," in *Nineteenth Century American Science: A Reappraisal,* G. H. Daniels, Jr., ed. Evanston, Ill.: Northwestern University Press.

Layton, E. 1977. "Conditions of technological development," in *Science, Technology and Society: A Cross-Disciplinary Perspective,* I. Spiegel-Rösing and D. de Solla Price, eds. London and Beverly Hills: Sage, 197–222.

Layton, E. 1978. "Millwrights and engineers," in *Dynamics of Science and Technology,* W. Krohn, E. Layton, and P. Weingart, eds. Dordrecht: Reidel, 61–87.

Lazonick, W. 1979. "Industrial relations and technical change: The case of the self-acting mule." *Cambridge Journal of Economics* 3: 231–262.

Lebach, H. 1909. "Uber Resinit." *Zeitschrift für Angewandte Chemie* 22: 1598–1601.

Lederer, L. 1894. "Eine neue Synthese von Phenolalkoholen." *Journal für Praktische Chemie, Neue Folge* 50: 223–226.

Leslie, S. W. 1980. "Thomas Midgley and the politics of industrial research." *Business History Review* 14: 480–503.

Levine, A. S. 1982. *Managing NASA in the Apollo Era.* Washington, D.C.: NASA.

Liljestrand, A. 1977. Discussion in *Trends and Prospects in Drug Research and Development,* Eleventh CIOMS Round Table Conference, Z. Bankowski arid J. F. Dunne, eds. Geneva: Scrip/CIOMS, 110.

Lilley, S. 1973. "Technological progress and the Industrial Revolution 1700–1914," in *The Fontana Economic History of Europe,* C. M. Cipolla, ed. London: Collins/Fontana Books, vol. 3, 187–254.

Lindqvist, S. 1984. *Technology on Trial: The Introduction of Steam Power Technology into Sweden, 1715–1735.* Uppsala: Almqvist & Wiksell International.

Livermore, H. V. 1953. *Portugal and Brazil: An introduction.* Oxford. Clarendon Press.

Louis, M. R. 1981. "A cultural perspective on organizations: The need for and consequences of viewing organizations as culture-bearing milieux." *Action Research Management* 2: 246–258.

Ludwig, G. D. 1950. "The velocity of sound through tissues and the acoustic impedance of tissues." *Journal of the Acoustical Society of America* 22: 862.

Luhmann, N. 1973. *Vertrauen.* Stuttgart: Ferdinand Enke Verlag.

Luhmann, N. 1984. *Soziale Systeme.* Frankfurt: Suhrkamp Verlag.

Lynch, M. 1985a. *Art and Artefact in Laboratory Science: A Study of Shop Work and Shop Talk in a Research Laboratory.* London: Routledge and Kegan Paul.

Lynch, M. 1985b. "Discipline and the material form of images: An analysis of scientific visibility." *Social Studies of Science* 15: 37–66.

MacKenzie, D. 1978. "Statistical theory and social interest: A case study." *Social Studies of Science* 8: 35–83.

MacKenzie, D. 1984. "Marx and the machine." *Technology and Culture* 25: 473–502.

MacKenzie, D. 1985. "The missile accuracy system." Paper presented at the International Congress of the History of Science. Berkeley, August.

MacKenzie, D. 1989. "From Kwajalein to Armageddon? Testing and the social construction of missile accuracy," in *The Uses of Experiment*, D. Gooding, T. Pinch, and S. Schaller, eds. Cambridge: Cambridge University Press.

MacKenzie, D., and Wajcman, J., eds. 1985. *The Social Shaping of Technology.* Milton Keynes: Open University Press.

Magalhaes-Godinho, V. 1969. *L'economie de l'empire portuqais aux XVe et XVI siecles.* Paris: SEVPEN.

Manasse, O. 1894. "Uber eine Synthese aromatischer Oxy-Alkohole." *Berichte der Deutschen Chemischen Gesellschaft* 27: 2409–2413.

Manwaring, T., and Wood, S. 1985. "The ghost in the labour process," in *Job Redesign: Critical Perspectives on the Labour Process*, D. Knights, H. Wilmott, and D. Collinson, eds. London: Gower, 171–196.

March, J. G., and Simon, H. A. 1958. *Organizations.* New York: Wiley.

Matthis, A. R. c. 1920. *Insulating Varnishes in Electrotechnics.* London: John Heywood.

Mayr, O. 1976. "The science-technology relationship as a historiographic problem." *Technology and Culture* 17: 663–673.

McCorduck, P. 1979. *Machines Who Think.* San Francisco: Freeman.

McCrea, F. B., and Markle, G. E. 1984. "The estrogen replacement controversy in the USA and UK: Different answers to the same question?" *Social Studies of Science* 14: 1–26.

McKenna, R. 1963. *The Sand Pebbles*. London: Gollancz.

Mensch, G. 1979. *Stalemate in Technology: Innovations Overcome Depression*. New York: Ballinger.

Merz, C. H. 1908. "Power supply and its effects on the industries of the north-east coast." *Journal of Iron and Steel Institute*, September, 4.

Meyer, S. M. 1983–1984. *Soviet Theatre Nuclear Forces*. Adelphi Papers 187 and 188. London: International Institute for Strategic Studies.

Michael, A. 1883–1884. "Action of aldehydes on phenols." *American Chemical Journal* 5: 338–349.

Moncrief, F. J. 1979. "SALT verification: How we monitor the Soviet arsenal." *Microwaves*, September, 41–51.

Mowery, D. C., and Rosenberg, N. 1979. "The influence of market demand upon innovation: A critical review of some recent empirical studies." *Research Policy* 8: 103–153.

Mueller, F. K. 1960. *A History of Inertial Guidance*. Redstone Arsenal, Ala.: Army Ballistic Missile Agency.

Mulkay, M. J. 1979a. "Knowledge and utility: Implications for the sociology of knowledge." *Social Studies of Science* 9: 63–80.

Mulkay, M. J. 1979b. *Science and the Sociology of Knowledge*. London: Allen and Unwin.

Mulkay, M. J., and Milič, V. 1980. "The sociology of science in East and West." *Current Sociology*, Winter, 28: 1–342.

Murphy, J. J. 1873. "Instinct: A mechanical analogy." *Nature*, April 24, 7: 483.

Needham, J. 1928. *Man a Machine, in Answer to a Romantical and Unscientific Treatise Written by Sig. Eugenio Rignano and Entitled "Man Not A Machine."* New York: Kegan Paul.

Nelson, R. R., and Langlois, R. N. 1983. "Industrial innovation policy: Lessons from American history." *Science* 219: 814–818.

Nelson, R. R., and Winter, S. G. 1977. "In search of a useful theory of innovation." *Research Policy* 6: 36–76.

Nelson, R. R., and Winter, S. G. 1982. *An Evolutionary Theory of Economic Change*. Cambridge, Mass.: The Belknap Press of Harvard University Press.

"The new German Drug Law." *Internist*, Special issue, 21: 301–345 (1980).

Nickles, T. 1981. "ERISS and international sociology of science." Presented to the Sixth Annual Meeting of the Society for Social Studies of Science, Atlanta, November.

Nickles, T. 1982. "How discovery is important to cognitive studies of science." Paper presented to the Philosophy of Science Association Meeting, Philadelphia, October.

Noble, D. F. 1977. *America by Design: Science, Technology and the Rise of Corporate Capitalism.* New York: Knopf.

Noble, D. F. 1979. "Social choice in machine design: The case of automatically controlled machine tools," in *Case Studies on the Labour Process*, A. Zimbalist, ed. New York: Monthly Review Press, 18–50.

Noble, D. F. 1984. *Forces of Production: A Social History of Industrial Automation.* New York: Knopf.

Oakley, A. 1984. *The Captured Womb: A History of the Medical Care of Pregnant Women.* Oxford: Blackwell.

Ogburn, W. F. 1945. *The Social Effects of Aviation.* Boston: Houghton Mifflin.

Ogburn, W. F., and Meyers Nimkoff, F. 1955. *Technology and the Changing Family.* Boston: Houghton, Mifflin.

Oldman, D., and Ducker, C. 1984. "The non-reproduceability of ethno-methods: Can people and computers form a society?" in *Social Action and Artifical Intelligence*, G. N. Gilbert and C. Heath, eds. Aldershot: Gower Press, 128–141.

Palermo, D. S. 1973. "Is a scientific revolution taking place in psychology?" *Science Studies* 3: 211–244.

Parkes, A. 1855. "Manufacture of elastic and adhesive compounds." British Patent 2359. Filed on October 22, 1855. Sealed on January 25, 1856.

Parkes, A. 1865a. "On the properties of Parkesine, and its application to the arts and manufacture." *Journal of the Society of Arts*, December, 14: 81–86.

Parkes, A. 1865b. "Preparing compounds of gun cotton and other substances, &c." British Patent 2675. Filed on October 28, 1864. Sealed on January 31, 1865.

Parry, J. H. 1963. *The Age of Reconnaissance.* London: Weidenfeld and Nicolson.

Parsons, T. 1968. "Social systems." *Encyclopedia of the Social Sciences* 15: 458–472.

Paskoff, P. F. 1983. *Industrial Evolution: Organization, Structure and Growth of the Pennsylvania Iron Industry, 1750–1860.* Baltimore: Johns Hopkins University Press.

Passer, H. C. 1953. *The Electrical Manufacturers, 1875–1900.* Cambridge, Mass.: Harvard University Press.

Pavitt, K. 1984. "Sectoral patterns of technical change: Towards a taxonomy and a theory." *Research Policy* 13: 343–373.

Peirce, J. 1951. *Fire on the Hearth: The Evolution and Romance of the Heating Stove.* Springfield, Mass.: Pond-Ekberg.

Peters, T., and Austin, N. 1985. *A Passion for Excellence.* New York: Random House.

Pickering, A. 1984. *Constructing Quarks—A Sociological History of Particle Physics.* Chicago and Edinburgh: University of Chicago Press and Edinburgh University Press.

Pinch, T. J. 1977. "What does a proof do if it does not prove? A study of the social conditions and metaphysical divisions leading to David Bohm and John von Neumann failing to communicate in quantum physics," in *The Social Production of Scientific Knowledge*, E. Mendelsohn, P. Weingart, and R. Whitley, eds. Dordrecht: Reidel, 171–215.

Pinch, T. J. 1985. "Recent trends in the history of technology." *BSHS Newsletter,* January, 16: 19–21.

Pinch, T. J. 1986. *Confronting Nature: The Sociology of Solar-Neutrino Detection.* Dordrecht: Reidel.

Pinch, T. J., and Bijker, W. E. 1984. "The social construction of facts and artefacts: Or how the sociology of science and the sociology of technology might benefit each other." *Social Studies of Science* 14: 399–441. Also published in Serbo-Croatian as "Društveno Proizvodenje Činjenica I Tvorevina: O Cjelovitom Pristupu Izučavanju Znanosti I Tehnologije," *Gledišta, časopis za društvenu kritiku i teoriju*, March–April, 25: 21–57 (1984).

Polanyi, M. 1967. *The Tacit Dimension.* New York: Anchor.

Polanyi, M. 1968. "Life's irreducible structure." *Science* 160: 1308–1312.

Popper, K. R. 1968. *The Logic of Scientific Discovery.* London: Hutchinson.

Popper, K. 1972. *Objective Knowledge.* Oxford: Oxford University Press.

Putnam, H. 1960. "Minds and machines," in *Dimensions of Mind*, S. Hook, ed. London: Collier-Macmillan, 148–179.

Quarterly Progress Report of the Acoustics Laboratory, 1951. Cambridge, Mass.: Massachusetts Institute of Technology, 24.

Ragan, R. R. 1980. *Innovation at the Draper Laboratory or an Environment for Innovation.* Cambridge, Mass.: Charles Stark Draper Laboratory, October, P-1226.

Ravetz, A. 1968. "The Victorian coal kitchen and its detractors." *Victorian Studies* 11: 428–445.

Ravetz, J. R. 1971. *Scientific Knowledge and Its Social Problems*. Oxford: Oxford University Press.

Baron Rayleigh (J. W. Strutt). 1877–1878. *The Theory of Sound*, 2 vols. London: Macmillan.

Redman, L. V., and Mory, A. V. H. 1931. "The Bakelite Corporation." *Journal of Industrial and Engineering Chemistry* 23: 595–597.

Reich, L. S. 1980. "Industrial research and the pursuit of corporate security: The early years of Bell Labs." *Business History Review* 14: 504–529.

Reich, L. S. 1983. "Irving Langmuir and the pursuit of science and technology in the corporate environment." *Technology and Culture* 24: 199–221.

Reisner, S. J. 1978. *Medicine and the Reign of Technology*. Cambridge: Cambridge University Press.

Rignano, E. 1926. *Man Not a Machine: A Study of the Finalistic Aspects of Life*. London: Paul, Trench, Trubner.

Robbins, D., and Johnston, R. 1976. "The role of cognitive and occupational differentiation in scientific controversies." *Social Studies of Science* 6: 349–368.

Roberts, E. B. 1968. "A basic study of innovators: How to keep and capitalize on their talents." *Research Management* 11: 249–266.

Robinson, C. A. 1984. "Parallel programs advance small ICBM." *Aviation Week and Space Technology*, March 5, 14–17.

Ropohl, G. 1979. *Eine Systemtheorie der Technik: Zur Grundlegung der Allgemeinen Technologie*. München und Vienna: Hanser.

Rosenberg, D. A. 1983. "The origins of overkill: Nuclear weapons and American strategy, 1945–1960." *International Security* 7: 3–71.

Rosenberg, N. 1963. "Technological change in the machine tool industry, 1840–1910." *Journal of Economic History* 23: 414–446.

Rosenberg, N., ed. 1971. *The Economics of Technological Change*. Harmondsworth: Penguin.

Rosenberg, N. 1972. *Technology and American Economic Growth*. Evanston: Harper.

Rosenberg, N. 1976. *Perspectives on Technology*. Cambridge: Cambridge University Press.

Rosenberg, N. 1982. *Inside the Black Box: Technology and Economics*. Cambridge: Cambridge University Press.

Rosenberg, N., and Vincenti, W. G. 1978. *The Britannia Bridge: The Generation and Diffusion of Knowledge*. Cambridge, Mass.: MIT Press.

Rosenfield, L. C. 1941. *From Beast-Machine To Man-Machine: Animal Soul in French Letters from Descartes to La Mettrie.* New York: Oxford University Press.

Rudwick, M. 1976. "The emergence of a visual language for geological science." *History of Science* 14: 149–165.

Ryle, G. 1949. *The Concept of Mind.* London: Hutchinson.

Sahal, D. 1981. *Patterns of Technological Innovation.* Reading, Mass.: Addison-Wesley.

Sapolsky, H. M. 1972. *The Polaris System Development: Bureaucratic and Programmatic Success in Government.* Cambridge, Mass.: Harvard University Press.

Schmookler, J. 1966. *Invention and Economic Growth.* Cambridge, Mass.: Harvard University Press.

Schmookler, J. 1972. *Patents, Invention and Economic Change, Data and Selected Essays.* Z. Griliches and L. Hurwicz, eds. Cambridge, Mass.: Harvard University Press.

Scholz, L. with a contribution by G. von L. Uhlmann, 1977. *Technik-Indikatoren, Ansätze zur Messung des Standes der Technik in der industriellen Produktion.* Berlin and München: Duncker & Humblot.

Schot, J. 1985. "De ontwikkeling van de techniek als een variatieen selectieproces. De meekrapteelt en-bereiding in het licht van een alternatieve techniekopvatting." Master's thesis, Erasmus University of Rotterdam, unpublished.

Schot, J. 1986. "De meekrapnijverheid: de ontwikkeling van de technick als een proces van variatie en selectie," in *Jaarboek voor de Geschiedenis van Bedrijf en Techniek*, E. S. A. Bloemen, W. E. Bijker, W. van den Brocke, et al., eds. Utrecht: Stichting, vol. 3, 43–62.

Schroeer, D. 1985. "Quantifying technological imperatives in the arms race," in *Reassessing Arms Control*, D. Carlton and C. Schaerf, eds. London: Macmillan, 60–75.

Schuler, M. 1923. "Die Störung von Pendel and Kreiselapparaten durch die Beschleunigung des Fahrzeuges." *Physikalische Zeitschrift* 24: 344–350.

Schumpeter, J. 1971 [1928]. "The instability of capitalism," in *The Economics of Technological Change*, N. Rosenberg, ed. Harmondsworth: Penguin, 13–42.

Schumpeter, J. 1942. *Capitalism, Socialism and Democracy.* New York: Harper & Row. Reprinted 1974 by Unwin University Books, London.

Schutz, A. 1962. *The Problem of Social Reality. Collected Papers*, vol. 1. The Hague: Martinus Nijhoff.

Schutz, A. 1964. *Studies in Social Theory. Collected Papers*, vol. 2. The Hague: Martinus Nijhoff.

Schwartz Cowan, R. 1983. *More Work for Mother: The Ironies of Household Technology from the Open Hearth to the Microwave*. New York: Basic Books.

Scientific Advisory Board, Office of the Chief of Staff, US Air Force. 1949. "Seminar on automatic celestial and inertial long range guidance systems." MIT, February 1–3.

Searle, J. 1980. "Minds, brains and programs." *The Behavioral and Brain Sciences* 3 (3): 417–457.

Sestre Levante Workshop. 1977. "Towards a more rational regulation of the development of new medicines." *European Journal of Clinical Pharmacology* 11: 233–238.

Shapin, S. 1979. "The politics of observation: Cerebral anatomy and social interests in the Edinburgh phrenology disputes," in *On the Margins of Science: The Social Construction of Rejected Knowledge*, R. Wallis, ed. Keele: University of Keele, 139–178.

Shapin, S. 1980. "Social uses of science," in *The Ferment of Knowledge*, G. S. Rousseau and R. Porter, eds. Cambridge: Cambridge University Press, 93–139.

Shapin, S. 1982. "History of science and its sociological reconstructions." *History of Science* 20: 157–211.

Shapin, S. 1984. "Pump and circumstance: Robert Boyle's literary technology." *Social Studies of Science* 14: 481–520.

Shapin, S., and Schaffer, S. 1985. *Leviathan and the Air-Pump: Hobbes, Boyle and the Experimental Life*. Princeton: Princeton University Press.

Shapley, D. 1978. "Technology creep and the arms race: ICBM problem a sleeper." *Science* 201: 1102–1105.

Sherwin, C. S., and Isenson, R. S. 1966. *First Interim Report on Project Hindsight: Summary*. Washington, D.C.: Office of the Director of Defense Research and Engineering.

Sherwin, C. S., and Isenson, R. S. 1967. "Project hindsight: A Defense Department study of the utility of research." *Science* 156: 1571–1577.

Sherwood Taylor, E. 1957. *A History of Industrial Chemistry*. London: Heinemann.

Shrum, W. 1984. "Scientific specialities and technical systems." *Social Studies of Science* 14: 63–90.

Simon, H. A. 1969. *Sciences of the Artificial*. Cambridge, Mass.: MIT Press.

Slater, J. M. 1966. *Twenty Years of Inertial Navigation at North American Rockwell*. Anaheim, Calif.: Autonetics Division, North American Rockwell.

Smith, J. Maynard. 1958. *The Theory of Evolution*. Harmondsworth: Penguin.

Smith, M. Roe. 1977. *Harpers Ferry Armory and the New Technology: The Challenge of Change*. Ithaca, N.Y.: Cornell University Press.

Smith, M. Roe, ed. 1985. *Military Enterprise and Technological Change: Perspectives on the American Experience*. Cambridge, Mass.: MIT Press.

Sperry, E. 1930. "Spirit of invention in an industrial civilization," in *Toward Civilization*, Charles A. Beard, ed. New York: Longmans Green, 47–68.

"The Stanley Exhibition of Cycles." 1889. *The Engineer* 67: 157–158.

"The Stanley Exhibition of Cycles." 1890. *The Engineer* 69: 107–108.

Staudenmaier, J. M., SJ. 1983. "What SHOT hath wrought and what SHOT hath not: Reflections on 25 years of the history of technology." Paper presented to the Twenty-fifth Annual Meeting of SHOT.

Staudenmaier, J. M., SJ. 1985. *Technology's Storytellers: Reweaving the Human Fabric*. Cambridge, Mass.: MIT Press.

Stern, N. 1981. *From ENIAC to UNIVAC: An Appraisal of the Eckert-Mauchly Computers*. Bedford, Mass.: Digital Press.

Stowe, Harriet Beecher. 1964. "House and home papers." *Atlantic Monthly* 13: 42.

Strasser, S. 1982. *Never Done: A History of American Housework*. New York: Pantheon. ch. 3.

Suchman, L. 1985. *Plans and Situated Actions: The Problem of Human-Machine Communication*. Palo Alto: Xerox Corporation, ISL-6.

Suppe, F. 1974. *The Structure of Scientific Theories*. Urbana: University of Illinois Press.

Tarr, J. A., and Lamperes, B. C. 1980–1981. "Changing fuel use behavior and energy transitions. The Pittsburgh smoke control movement, 1940–1950: A case study in historical analogy." *Journal of Social History* 14: 78–96.

Taylor, E. R. G. 1956. *The Haven-Finding Art: A History of Navigation from Odysseus to Captain Cook*. London: Hollis and Carter.

Taylor, E. S., Holmes, J. H., Thompson, H. E., and Gottesfeld, K. R. 1964. "Ultrasound diagnostic techniques in obstetrics and gynecology." *American Journal of Obstetrics and Gynecology* 90: 655–669.

Temin, P. 1964. *Iron and Steel in Nineteenth Century America*. Cambridge, Mass.: MIT Press.

Temmin, P. 1980. *Taking Your Medicine: Drug Regulation in the United States*. Cambridge, Mass.: Harvard University Press.

Thinius, K. 1976. "Entwicklung der Reaktion zwischen Formaldehyde und Phenol zu einem Productionszweig der Plasindustrie zwischen 1900 und 1930." *Plaste und Kautschuk* 23: 746–749.

Toulmin, S. 1972. *Human Understanding*, vol. 1. Oxford: Oxford University Press.

Touraine, A. 1968. *Le mouvement de mai ou le communisme utopique*. Paris: Le Seuil.

Touraine, A. 1973. *Production de la société*. Paris: Le Seuil.

Touraine, A. 1979. *La phrophétie antinucléaire*. Paris: Le Seuil.

"The trouble with gravity." 1950. *Time*, January 2, 55: 54.

Turkle, S. 1982. "The subjective computer: A study in the psychology of personal computation." *Social Studies of Science* 12: 173–205.

Turkle, S. 1984. *The Second Self: The Human Spirit in a Computer Culture*. New York: Simon & Schuster.

Uhlmann, L. 1978. *Der Innovationsprozess in westeurpäischen Industrieländern. Band 2: Den Ablauf industriellen Innovationsprozesses*. Berlin and München: Duncker and Humblot.

Unger, R. W. 1980. *The Ship in the Medieval Economy, 600–1600*. London: Croom Helm.

USAEC. 1955. "Studies in methods and instruments to improve the localisation of radioactive materials in the body with special reference to the diagnosis of brain tumours and the use of ultrasonic techniques." USAEC Final Progress Report, July 1, 1952, to June 30, 1953. Report AECU-3012. Minneapolis: University of Minnesota Press.

US Bureau of the Census. 1940. *Census of Housing, 1940. General Characteristics*, Washington, D.C.: Government Printing Office, table 12, 42.

US Bureau of the Census. 1950. *Census of Housing, 1950. General Characteristics*. Washington, D.C.: Government Printing Office, table Q, xxxiii.

US Bureau of the Census. 1970. *Census of Housing, 1970. General Characteristics*. Washington, D.C.: Government Printing Office, table 24, 18.

Usher, A. P. 1971. "Technical change and capital formation," in *The Economics of Technological Change*, N. Rosenberg, ed. Harmondsworth: Penguin, 43–72.

van den Belt, H. 1985. "A. W. Hofman en de Franse Octrooiprocessen rond anilinerood: demarcatie als sociale constructie." *Jaarboek voor de Geschiedenis van Bedrijf en Techniek* 2: 64–86.

van der Meer, F. B. 1983. *Organisatie als Spel*. Enschede: Boerderij-cahier, Twente University of Technology.

van der Meer, F. B. 1986. "Social simulation: A research methodology and learning strategy for social impact assessment," in *Impact Assessment Today*, H. A. Becker and A. L. Porter, eds. Van Arkel: Utrecht, 701–714.

Vartanian, A. 1960. *La Mettrie's* L'Homme Machine: *A Study in the Origins of an Idea*. Princeton, N.J.: Princeton University Press.

Veatch, R. M. 1981. "Federal regulation of medicine and biomedical research: Power, authority and legimimacy," in *The Law-Medicine Relation*, S. F. Spicker, J. M. Healey, and H. T. Engelhardt, eds. Dordrecht: Reidel, 75–91.

Vincenti, W. G. 1979. "The air propeller tests of W. F. Durand and E. P. Lesley: A case study in technological methodology." *Technology and Culture* 20: 712–751.

Vincenti, W. G. 1984. "Technological knowledge without science: The innovation of flush riveting in American airplanes, c. 1930–c. 1950." *Technology and Culture* 25: 540–576.

Vincenti, W. G. 1986. "The Davis wing and the problem of airfoil design: Uncertainty and growth in engineering knowledge." *Technology and Culture* 26.

Wade, O. 1977. Discussion in *Trends and Prospects in Drag Research and Development*, Eleventh CIOMS Round Table Conference, Z. Bankowski and J. F. Dunne, eds. (Geneva: Scrip/CIOMS), 107.

Wallis, R., ed. 1979. *On the Margins of Science: The Social Construction of Rejected Knowledge*. Sociological Review Monograph 27. Keele: University of Keele.

Wardell, M. W. 1981. "Patients' needs in drug development and regulation," in *Papers in Science and Public Policy I*, B. L. Rosenberg, ed. New York: Academy of Sciences, vol. 368, 187–203.

Waters, D. W. 1980. *Science and the Techniques of Navigation in the Renaissance*. Maritime Monographs and Reports 19–1974, second edition. Greenwich: National Maritime Museum.

Weber, M. 1921. *Gesammelte politische Schriften*. Munich: Drei Masken Verlag.

Weber, M. 1949. *The Methodology of the Social Sciences*, E. A. Shils and H. A. Finch, eds. Glencoe: Free Press.

Weingart, P. 1984. "The structure of technological change: Reflections on a sociological analysis of technology," in *The Nature of Technological Knowledge: Are Models of Scientific Change Relevant?* R. Laudan, ed. Dordrecht: Reidel, 115–142.

Weizenbaum, J. 1976. *Computer Power and Human Reason*. San Francisco: Freeman. Reprinted 1984 by Pelican, London.

Welbank, M. 1983. *A Review of Knowledge Acquisition Techniques for Expert Systems*. Martlesham Heath: British Telecom Research Laboratories.

Wells, P. N. T. 1978. "History," in *Handbook of Clinical Ultrasound*, M. de Vlieger, ed. New York: Wiley, 3–13.

Westrum, R. 1983. "What happened to the old sociology of technology?" Paper presented to the Eighth Annual Meeting of the Society for Social Studies of Science, Blacksburg. Virginia, November.

Whitley, R. D. 1972. "Black boxism and the sociology of science: A discussion of the major developments in the field," in *The Sociology of Science*, P. Halmos, ed. Keele: University of Keele, 62–92.

Wild, J. J. 1950. "The use of ultrasonic pulses for the measurement of biologic tissues and the detection of tissue density changes." *Surgery* 27: 183–188.

Wild, J. J., and Neal, D. 1951. "The use of high frequency ultrasound waves for detecting changes in texture of living tissues." *The Lancet* 1: 655–657.

Wildes, K. L., and Lindgren, N. A. 1985. *A Century of Electrical Engineering and Computer Science at MIT, 1882–1982*. Cambridge, Mass.: MIT Press.

Wilhelm, P. W. 1985. "Coupling mechanisms and integration processes." Paper presented at the Conference on the Social Direction of Public Sciences, Amsterdam, November.

Wilhelm, P. W., and Bolk, H. 1986. "Strategy and the art of structure maintenance: On the interplay of strategy and structure in science policy," in Impact Assessment Today, H. A. Becker and A. L. Porter, eds. Van Arkel: Utrecht, 741–758.

Williamson, B. 1983. *The Shopfloor Politics of New Production Technology*. London: Heinemann.

Williamson, O. E. 1979. "Transaction-cost economics: The governance of contractual relations." *The Journal of Law and Economics* 22: 233–261.

Williamson, O. E. 1981. "The economics of organization: The transaction cost approach." *American Journal of Sociology* 87: 548–577.

Winch, P. G. 1958. *The Idea of a Social Science*. London: Routledge.

Winner, L. 1977. *Autonomous Technology: Technics Out of Control as a Theme in Political Thought*. Cambridge, Mass.: MIT Press.

Wise, G. 1980. "A new role for professional scientists in industry." *Technology and Culture* 21: 408–429.

Wittgenstein, L. 1953. *Philosophical Investigations*. Oxford: Blackwell.

Woodforde, J. 1970. *The Story of the Bicycle*. London: Routledge and Kegan Paul.

Woolgar, S. 1981. "Interests and explanation in the social study of science." *Social Studies of Science* 11: 365–394.

Woolgar, S. 1982. "Laboratory studies: A comment on the state of the art." *Social Studies of Science* 12: 481–498.

Woolgar, S. 1985. "Why not a sociology of machines? The case of sociology and artificial intelligence." *Sociology* 19: 557–572.

Worden. E. C. 1911. *Nitro-cellulose Industry*. New York: Van Nostrand.

Wright, L. 1964. *Home Fires Burning: The History of Domestic Heating and Cooking*. London: Routledge and Kegan Paul.

Wynne, B. 1976. "C. G. Barkla and the J phenomenon: A case study of the treatment of deviance in physics." *Social Studies of Science* 6: 307–347.

Wynne, B. 1983. "Reflecting the issues of risk and public acceptance: The social viability of technology." *Futures*, February, 15: 13–32.

Yearley, S. 1982. "The relationship between epistemological and sociological cognitive interests: Some ambiguities underlying the use of interest theory in the study of scientific knowledge." *Studies in the History and Philosophy of Science* 13: 353–388.

Yearley, S. 1984. *Science and Sociological Practice*. Milton Keynes: Open University Press.

Yoxen, E. J. 1986. "Form and strategy in biology: Reflections on the career of C. H. Waddington," in A *Critical History of Embryology*, T. J. Horder, J. A. Witkowski, and C. C. Wylie, eds. Cambridge: Cambridge University Press, 309–330.

Zimbalist, A. ed. 1979. *Case Studies on the Labour Process*. New York: Monthly Review Press.

Contributors

Note: The biographical sketches are the originals from the 1987 edition, with one added sentence describing each contributor's current affiliation.

Wiebe E. Bijker is research fellow at the Twente University of Technology. He is interested in the history and sociology of technology and also in applying recent results of science and technology studies to secondary school education. With K. Kortland, J. van der Rijst, and A. J. de Wever, he prepared the three-volume physics textbook series *Exact Natuurkunde* (Meulenhoff Educatief, 1982, 1983, 1985). As of 2011, Bijker is Professor of Technology and Society, Maastricht University.

Henk J. H. W. Bodewitz teaches science and society courses at the University of Groningen. His research interests include innovation and regulation in biomedicine. His publications include, with Han Dolman, "The sedimentation of a scientific concept: The use of citation data" (1985) and, with N. F. Wieringa, M. ten Ham, and H. Buurma, several articles on the regulation and use of beta blockers in the Netherlands and Scandinavia. Since 2001, he has been active as an independent consultant in the fields of health care and sustainable energy (Groningen, the Netherlands).

Henk Buurma works at the apothecary affiliated with the University of Leiden. He has taught science and society courses at the University of Groningen and is interested particularly in the pharmaceutical industry. As of 2011, Buurma is Director of SIR Institute for Pharmacy Practice and Policy and of Apotheek Stevenshof (both in Leiden) and has an affiliation with the Department of Pharmacoepidemiology en Clinical Pharmacology of the University of Utrecht.

Michel Callon is Professor of Sociology at the Ecole Nationale Superieure des Mines in Paris and Director of the Centre de Sociologie de l'Innovation. His research focuses on the formation of economic markets, and he was involved in the development of a scientific research policy in France in the early 1980s. With John Law and Arie Rip, he has edited *Mapping the Dynamics of Science and Technology* (Macmillan, 1986). As of 2011, Callon has been retired for one year but is still studying sociotechnical controversies and the role of economics in the organization of economic activities.

H. M. Collins is Director of the Science Studies Centre at the University of Bath. His major interest is the sociology of scientific knowledge, and most of his work is based on case studies of physics and parapsychology. He is currently interested in artificial intelligence. His latest book is *Changing Order: Replication and Induction in Scientific Practice* (Sage, 1985). As of 2011, Harry Collins is Distinguished Research Professor and Director of the Centre for the Study of Knowledge Expertise and Science at Cardiff University, and a major research grant from the European Research Council has helped him sidestep compulsory retirement.

Edward W. Constant II is Associate Professor of History at Carnegie Mellon University. His book *The Origins of the Turbojet Revolution* (Johns Hopkins University Press, 1980) was awarded the 1982 Dexter Prize of the Society for the History of Technology. As of 2011, Constant is retired but still occasionally tries to pay attention.

Gerard H. de Vries is Senior Lecturer in the Philosophy of Science at the University of Groningen. His current research is on the sociogenesis of the social sciences. He is the author of *Sociale orde: regels en de sociologie* (Boom, 1977) and *De ontwikkeling van wetenschap* (Wolters-Noordhoff, 1985). As of 2011, De Vries is Professor of Philosophy of Science, University of Amsterdam, and a member of the Scientific Council for Government Policy (WRR), the main think-tank for long-term policy issues of the Dutch government.

Thomas Parke Hughes is Professor of the History of Modern Science and Technology at the University of Pennsylvania and member of the American Academy of Arts and Sciences. He has been an adviser to many government and private agencies, including the Atomic Energy Commission and NASA. His latest book, *Networks of Power: Electrification in Western Society, 1880–1930* (Johns Hopkins University Press, 1983), was awarded the 1985 Dexter Prize. As of 2011, Hughes has retired as a professor and lives in Charlottesville, VA.

John Law is Senior Lecturer in Sociology at the University of Keele. He has published widely in the sociology of science, including, with Peter Lodge,

Science for Social Scientists (Macmillan, 1984). As of 2011, Law is Professor of Sociology at the Open University and a Director of the UK's Economic and Social Research Council Centre for Research on Socio-Cultural Change.

Donald MacKenzie is Lecturer in Sociology at Edinburgh University. He is the author of *Statistics in Britain, 1865–1930: The Social Construction of Scientific Knowledge* (Edinburgh University Press, 1981), and he is co-editor of *The Social Shaping of Technology* (Open University Press, 1985). In 2011, MacKenzie is still at the University of Edinburgh, where he now holds a personal chair in Sociology.

Trevor J. Pinch is Lecturer in Sociology at the University of York. His current research interests include the sociology of scientific instrumentation, the sociology of technology, and the rhetoric of selling, sales promotion, and advertising. His latest book is *Confronting Nature: The Sociology of Solar-Neutrino Detection* (Reidel, 1986). As of 2011, Pinch is Professor of Science and Technology Studies and Sociology at Cornell University.

Arie Rip teaches in the Department of Science Dynamics at the University of Amsterdam. His research focuses on the implications (and applications) of recent sociology of science and knowledge to scientific and technological development, science policy, controversy studies, and technology assessment. He is the author of *Wetenschap als Mensenwerk* (Ambo, 1978). As of 2011, Rip has been retired for five years, but has remained active, e.g., leading a research program on technology assessment of nanotechnology.

Ruth Schwartz Cowan is Professor of History and Director of Women's Studies at the State University of New York at Stony Brook. Her book *More Work for Mother: The Ironies of Household Technology from the Open Hearth to the Microwave* (Basic Books, 1983) was awarded the 1984 Dexter Prize. In 2002, Cowan became Janice and Julian Bers Professor of the History and Sociology of Science at the University of Pennsylvania; she plans to retire from that position in 2012.

Henk van den Belt worked for some years at the University of Nijmegen on the history of the synthetic dye industry. His main interest is in the sociology of science and technology, and his publications include "A. W. Hofman en de Franse Octrooiprocessen rond anilinerood: demarcatie als sociale constructie" (1985). As of 2011, Van den Belt is Assistant Professor of Applied Philosophy, Wageningen University.

Steve Woolgar is Lecturer in Sociology at Brunel University. His current work is on reflexivity and science and on artificial intelligence. With Bruno

Latour, he has written *Laboratory Life: The Construction of Scientific Facts* (Princeton University Press, second edition, 1986). As of 2011, Woolgar is Professor of Marketing and Head of Science and Technology Studies at Saïd Business School, University of Oxford.

Edward Yoxen teaches courses in the history, philosophy, and social relations of biology at the University of Manchester. His latest book is *Unnatural Selection? Coming to Terms with the New Genetics* (Heineman, 1986). As of 2011, Yoxen is President, Gravitas Consulting, based in Stoddard, New Hampshire, USA.

Name Index

Subject Index

Note: Page numbers in italics refer to illustrations.